IRRIGATION PRINCIPLES AND PRACTICES

IRRIGATION PRINCIPLES AND PRACTICES

Fourth Edition

Vaughn E. Hansen
Consulting Engineer
Vaughn Hansen Associates
Salt Lake City, Utah

Formerly Professor of Civil Engineering and
Director of Engineering Experiment Station and
Utah Water Research Laboratory
Utah State University, Logan, Utah

Orson W. Israelsen
Late Emeritus Professor, Civil and Irrigation Engineering
Utah State University, Logan, Utah

Glen E. Stringham
Professor of Agricultural and Irrigation Engineering
Utah State University, Logan, Utah

JOHN WILEY & SONS
New York • Chichester • Brisbane • Toronto

Library of Congress Cataloging in Publication Data

Hansen, Vaughn E.
    Irrigation principles and practices.

    Third ed. by O. W. Israelsen and V. E. Hansen.
Includes bibliographies and index.
1. Irrigation. I. Stringham, G. E., joint
author. II. Israelsen, Orson Winso, 1887–1968,
joint author. III. Israelson, Orson Winso, 1887–
1968. Irrigation principles and practices. IV. Title.
S613.I75 1980   631.7   79-13319
ISBN 0-471-03058-9

The Fourth Edition is dedicated to Orson W. Israelsen and to those leaders throughout the world who are applying their time and talents to produce from the natural resources of the earth the food and fiber needed by their fellow citizens.

PREFACE

In this edition of *Irrigation Principles and Practices* the basic underlying principles that govern irrigation practices are stressed. Generalized concepts are discussed, and newly formulated practices are examined.

We show that the scope of irrigation is worldwide, extending from the sources of the water supplies to the drainage outlets. Sources and the storage and measurement of water are described as are wells, pumping, and the conveyance of water. Basic soil and water relations, including storage, movement, and drainage of waterlogged soils are presented. The significance of saline and alkali problems are stressed as are the principles of consumptive use of water. Questions of when to irrigate and how much water to apply are answered in a manner that applies to any crop in any locality. The discussion also examines concepts of irrigation efficiency, and principles of surface and trickle irrigation. We emphasize the importance of social, administrative, and legal aspects of irrigation and include guiding concepts for their evaluation. The interrelationship of soil, water, climate, and irrigation management is considered to be of fundamental importance.

Metric units are used in the chapter discussions. In the appendixes of English units we include not only the usual conversion tables but also show the key equations, tables, and figures in their English counterparts for readers who prefer the English system.

The problems at the end of the book have been revised to present a broader coverage of the subject matter, and problems have been added to illustrate the new material of the fourth edition.

We are grateful for the privilege of having been closely associated professionally with Dr. Orson W. Israelsen. His inspiration and leadership were deeply appreciated by those of us who were fortunate enough to have been his students.

We thank those who offered suggestions for revision, called attention to errors in the prior edition, and supplied pictures and reference material.

Excellent secretarial assistance was rendered by Dorothy Sharp and Lavon Gardner.

Salt Lake City, Utah   Vaughn E. Hansen
December 1979   Glen E. Stringham

CONTENTS

IRRIGATION PRINCIPLES
AND PRACTICES

CHAPTER 1

IRRIGATION— WORLDWIDE

Irrigation is an age-old art. Historically, civilization has followed the development of irrigation. Civilizations have risen on irrigated lands; they have also decayed and disintegrated in irrigated regions. Most men who are well informed on irrigation are certain of its perpetuity, as long as it is intelligently practiced. Others think that a civilization based on agriculture under irrigation is destined sooner or later to decline, because some ancient civilizations based on irrigation have declined. Most ancient cultures depending upon irrigated agriculture declined because of lack of political and community stability so essential to irrigated agriculture. The duration of civilized peoples is probably dependent on many factors, of which a permanently profitable agriculture is vitally important. Some of the principles and practices essential to permanent and profitable agriculture under irrigation are considered in this volume.

1.1 CENTURIES OF IRRIGATION

The antiquity of irrigation is well documented throughout the written history of mankind. Genesis mentions Amraphel, King of Shinar, a contemporary of Abraham, who is probably identical with Hammurabi, sixth king of the first dynasty of Babylon. He developed laws, bearing the name of Hammurabi, indicating that the people had to depend upon irrigation for existence. One of the laws of Hammurabi states that if a man neglects to strengthen his bank of the canal and waters carry away the meadow, the man in whose bank the breach is opened shall render back the corn which he has caused to be lost.

The letters of Hammurabi about 2000 B.C. reveal a busy, governmental administrator who wastes no words when instructing his officials:

To Sid-Indiannam, Hammurabi speaks as follows: Gather the men that have fields along the Damanum Canal to clear out the Damanum Canal. Within this month, let them complete the digging of the Damanum Canal.

Further mention of irrigation is found in Second Kings 3:16–17:

And he said, Thus saith the Lord, Make this valley full of ditches.
For thus saith the Lord, Ye shall not see wind, neither shall ye see
rain; yet that valley shall be filled with water, that ye may drink, both
ye, and your cattle, and your beasts.

An ancient Assyrian Queen, supposed to have lived before
2000 B.C. is credited with directing her government to divert the
water of the Nile to irrigate the desert lands of Egypt. The in-
scription on her tomb is:

I constrained the mighty water to flow according to my will and led
its waters to fertilize lands that had before been barren and without
inhabitants.

Irrigation canals supposed to have been built under this Queen
of Assyria are still delivering water. Thus, there are records and
evidence of continuous irrigation for thousands of years in the
valleys of the Nile and for comparatively long periods likewise in
Syria, Persia, India, Java, and Italy.

Egypt claims to have had the world's oldest dam, 108 meters
long and 12 meters high, built 5000 years ago to store water for
drinking and irrigation. Basin irrigation introduced on the Nile
about 3300 B.C. still plays an important part in Egyptian agriculture.

In China, where reclamation was begun more than 4000 years
ago, the success of early kings was measured by their wisdom
and progress in water-control activities. King Yu, of Hsia dynasty
(2200 B.C.), was elected king by the people as a reward for his
outstanding work in water control. The famous Tu-Kiang Dam,
still a successful dam today, was built by a man named Mr. Li
and his son in the Chin dynasty (200 B.C.) and provides irrigation
water for about one-half million acres of rice fields. The water
ladder, a widely used pumping device in China and neighboring
countries, is believed to have been invented about the same
time. Its inventor was worshipped as a god by country carpen-
ters. The Grand Canal, 1125 meters long, was built by the Sui
Empire, 589–618 A.D.

The practice of irrigation in India antedates the historical epic
by an indeterminate period. There are reservoirs in Ceylon to the
south of India more than 2000 years old. Writings at 300 B.C. indicate that the whole country was under irrigation and very pros-
perous because of the double harvests which the people were
able to reap each year.

The Spaniards on their first entrance into Mexico and Peru found elaborate provisions for storing and conveying water supplies which had been used for many generations. Their origin was almost lost even to tradition. Extensive irrigation works also existed at that time in the southwestern United States. The early Spanish missionaries brought knowledge of irrigation from their Mediterranean homes. Irrigation was practiced also by trappers, miners, and frontiersmen in many places in the western United States although no effort was made to develop an agricultural economy based on irrigation until the Mormon Pioneers entered Salt Lake Valley in July, 1847. Under the Mormons, irrigation was a cooperative undertaking, with communities being located on the streams issuing from the mountains. Community ditches were constructed to serve both outlying agricultural areas and garden plots in the towns.

The pressure of survival and the need for additional food supplies have necessitated a rapid expansion of irrigation throughout the world. Even though irrigation is more important in the more arid regions of the earth, it has become increasingly important in humid regions.

1.2 ARID REGIONS OF THE WORLD

Areas requiring irrigation are very extensive and encompass portions of every continent of the world. The arid belt is roughly divided into two parts: The northern belt extends across the western part of the United States and Mexico, across Spain, southern France, Italy, and Greece into the Asia Minor area, thence over most of India and into China. The southern arid belt encompasses the portion of South America on the west side of the Andes Mountains, the southern part of that continent and most of South Africa, joins with the upper belt through the Arabian peninsula and India, and then extends south, encompassing essentially all of Australia.

Civilization has existed and now exists in these areas solely because of the art and science of irrigation. The irrigated acreage of the world, however, is not restricted to these areas. Some of the most profitably irrigated agricultures in the world are located in areas normally thought to have sufficient rainfall. These are areas such as central Brazil, Central America, the West Indies, and western and southern parts of Africa which have considerable annual rainfall, but during 6 months of the year have practically no rainfall. Other areas have periods of 2 weeks to 2 months of drought which necessitate irrigation if a profitable and diversified agriculture is to be practiced. Therefore, irrigation is no longer a regional practice of arid countries, but is becoming a basic part of well-developed agriculture throughout the world. Having ample water available for crop production in humid regions as well as arid regions, contributes to profitable and productive agriculture. The basic principles of irrigation are the same, whether practiced in arid or humid climates.

1.3 IRRIGATION DEFINED

Irrigation generally is defined as the application of water to soil for the purpose of supplying the moisture essential for plant growth. However, a broader and more inclusive definition is that irrigation is the application of water to the soil for any number of the following eight purposes

1. To add water to soil to supply the moisture essential for plant growth.

2. To provide crop insurance against short duration droughts.

3. To cool the soil and atmosphere, thereby making more favorable environment for plant growth.

4. To reduce the hazard of frost.

5. To wash out or dilute salts in the soil.

6. To reduce the hazard of soil piping.

7. To soften tillage pans and clods.

8. To delay bud formation by evaporative cooling.

Irrigation may be accomplished in five different ways: (1) by flooding; (2) by means of furrows, large or small; (3) by applying water underneath the land surface through subirrigation, thus causing the water table to rise; (4) by sprinkling, (5) or by trickle systems.

Water to supply moisture essential for plant growth may come from five sources, none of which should be ignored when irrigation water requirements are estimated:

1. Precipitation.

2. Atmospheric water other than precipitation.

3. Flood water.

4. Ground water.

5. Irrigation.

Failure to consider all five sources and the proportion of water that each supplies to total plant needs may result in faulty design of an irrigation system. In some areas one of the five sources may supply the major portion of plant needs; in other areas two or more will contribute appreciable amounts of water for plant growth.

Precipitation

To be of greatest benefit, precipitation should have the following characteristics:

1. Amounts should be sufficient to replace moisture depleted from the root zone.

2. Frequency should be often enough to replenish the soil moisture before plants suffer from lack of moisture.

3. Intensity should be low enough so that the applied water can be absorbed by the soil.

In only a few locations will precipitation fulfill all of the above requirements at all times to produce maximum yields. Failure of precipitation to fill these requirements is resulting in increasing amounts of irrigation in arid and in humid areas.

When planning irrigation projects, special precautions should be taken not to misinterpret "average" precipitations during the month and year. Averages can be very misleading, since every area has its wet and dry months, and wet and dry years, which are either above or below normal. Figure 1.1 shows the variations in precipitation that can occur from year to year. One-third of the time over a 40-year period, rainfall during the critical month of July was equal to or less than 50 percent of the average. Hence, long-time averages have little value in predicting how much moisture to expect in any month or year. Irrigation systems must be designed to provide for anticipated drought periods.

One of the most interesting things in nature is its variation, its changes from time to time and place to place. There is no uniformity in nature or in the rainfall; they are continuously chang-

Figure 1.1. Precipitation varies from year to year. Long-time averages are of little help in predicting the amount of rain to expect in a given month or year. Every area has its "wet" and "dry" years.

ing from day to day, week to week, month to month, and from year to year. These changes are of vital concern to agriculture in both arid and humid regions. Rainless periods of 2 or more weeks during the crop-growing season frequently occur in humid-climate states. During a 10-year period in Michigan, there were, on the average, seven periods each year, from 1 to 2 weeks, in which there was no rainfall. In Iowa there were eight such periods; in Wisconsin, Minnesota, Illinois, and Indiana, six; and in Ohio, five. A rainless period of 2 to 3 weeks' duration occurred, on the average, twice each year in Minnesota and once in each of the other states. Rainless periods of 3 weeks or more are comparatively rare in this "humid" region.

In Missouri during a 77-year period, one-fourth of the years have been very dry and have caused excessive losses of crops. During 40 years there were 15 dry years in which the Missouri corn crop was much below the average. In 3 of these years yield was less than half of the average and during 10 of the 15 years it was less than three-fourths of the average.

The need for irrigation has been brought forcibly to the attention of farmers throughout the world because of severe droughts that have affected much of the area. Although sufficient rainfall may be available for the growing of crops in normal years, it has been found through costly experience that short periods without rainfall have ruined crops which would otherwise have brought ample returns to the farmers.

Since precipitation records are always somewhat limited, care should be exercised not to extrapolate the records too far without adequate consideration of variations which may occur. Wide variations, of course, are readily observable through vegetative changes and from the knowledge of local residents. However significant this extreme, variations may well occur without being noticeably evident.

Atmospheric Water Other Than Precipitation

In some parts of the world the contribution of atmospheric water in forms other than precipitation is significant. In portions of western Australia, sufficient dew forms to produce good pasture. In the Negeb desert southwest of the Dead Sea, dew is largely responsible for summer growth of grapes. The scant rainfall, 10 to 28 centimeters annually, is augmented with 100 to 250 nights of dew a year.

Archaeologists have found rock lattices in the Negeb which were built around vines and trees to "wring" water out of the air in sufficient quantities to irrigate the plants.

On the island of Hawaii grass was observed to be twice as tall under a wire fence where dew formed than in the adjacent field.

Atmospheric conditions which generally prevail to make this source of water significant are: (1) considerable dew formation, (2) fog and clouds, and (3) high humidity. These conditions reduce the plant's water needs by reducing the forces causing water to transpire from the plant. Dew is especially effective in reducing the amount of water moving through the plant, and in some cases, dew is absorbed by the plant. Water which evaporates from the ground and foliage normally reduces by a like amount the water which would have been withdrawn from the soil by the plant. Fog, drip, and dew amounted to 28-centimeter depth of water at Cascade Head, Oregon, during 142 foggy days. Over a 6-year period 23 centimeters were deposited annually at Coshocton, Ohio. Forty centimeters are estimated to be a reasonable upper limit for the amount of atmospheric moisture that may condense during the year. The contribution of atmospheric moisture, in forms other than precipitation, should not be overlooked when considering the need for additional water for agricultural production.

Flood Water

Flood water is similar in some respects to irrigation water, but it is not supplied by man. As floods pass over the surface of the land, water is absorbed by the soil and stored for subsequent use by plants. In some regions agricultural production is wholly dependent upon flood water.

Ground Water

Ground water is water beneath the soil surface where voids in the soil are substantially filled with water. Upward movement of ground water by capillarity from the water table into the root zone can be a major source of water for plant growth.

To be most effective without seriously restricting growth, ground water should be near but below the depth from which the major portion of the plant's water needs are extracted. If ground water is within the normal root zone, plant growth is definitely suppressed. If ground water is too near the surface, the land's ability to economically produce most crops becomes almost nil. However, a water table within the lower portion of the root zone may supply a considerable amount of water and thereby reduce the cost of irrigation more than it offsets the loss of production. The optimum depth of the water table is that depth which gives the maximum economic return.

1.4 IRRIGATION IN HUMID-CLIMATE AREAS

Whereas irrigation is essential in arid regions, justification of humid-climate irrigation depends upon amount and occurrence of precipitation. Rainfall is generally sufficient but seldom coincides completely with crop needs. Rainfall excesses and drought are common occurrences. Several factors must be carefully considered in humid regions.

1. Application of irrigation water makes drainage more critical and sometimes essential.

2. Irrigation may be followed shortly by a heavy storm resulting in excess water on the land and in the soil.

3. Irrigation will generally extend the harvest season increasing the hazard of wet weather during harvest.

4. Good farm management becomes more critical.

5. Farm machinery requirements are increased because of a shorter time when fields are dry.

6. The irrigation system is a supplemental system increasing the cost per unit of benefit.

7. Increased erosion and greater runoff will occur.

8. Water source and quality and amount of labor available are important and must be carefully evaluated.

Advantages of adequate moisture, possible double cropping, a wider variety of crops to grow, and the possibility of better crop quality must be balanced against increased cost and other adverse factors.

Investments of funds for irrigation facilities in humid and arid regions are basically long-term investments. Some irrigation systems have increased profits enough in the first year of operation to pay the entire cost of installation, but most systems require a number of years to recover the original investment. Climatic conditions, marketing conditions, and management are among the factors that determine the rapidity with which the initial cost is repaid. The crops under irrigation, the depth of their rooting system, the readiness with which they are damaged by drought or respond to irrigation, and their values are also important. Some soils show the effect of dry weather much more quickly than others. In wet years, however, a large investment is tied up in irrigation equipment without increased financial return.

When deciding whether or not it is desirable to install an irrigation system, the average probable use should be carefully considered. Irrigation, even in humid areas, should be thought of as a normal farm operation. Seldom is it possible to secure and install equipment in short enough time to save a crop already suffering from lack of water.

1.5 EXTENT OF IRRIGATION

The area of land irrigated in the world is more than 233 million hectares. The results of comprehensive studies of irrigated land areas by the International Commission for Irrigation and Drainage are given in Table 1.1. Total cultivated and irrigated areas and also percentages of cultivated areas that are irrigated are shown. Total area cultivated is estimated to be 233,637,000 hectares which is 16 percent of the cultivated area. Five countries—China, India, Pakistan, the Soviet Union, and the United States—have the largest irrigated areas, the total being about 161,000,000 hectares which is 68 percent of the world's irrigated area.

Irrigation has been steadily increasing over the world as shown in Fig. 1.2. In 1949 the estimated land irrigated was 106 million hectares whereas by 1974 the estimate was 234 million hectares.

Boerma[1] estimates that on 46 million hectares significant improvement in productive capacity would result from irrigation renovation and improvement within the next 10 years, and new

[1] A. H. Boerma, "The World Could Be Fed," *Journal of Soil and Water Conservation*, Vol. 30, No. 1, 1975.

irrigation could be undertaken on 23 million hectares. He states appropriately that along with bringing new land into production, productivity of existing farm land should be increased by extending irrigation facilities and by improving existing irrigation systems that operate at less than 50 percent efficiency. In some regions such as north Africa and the Near East where most river water is already in use, the main emphasis must be on developing ground water. In large parts of Africa south of the Sahara and in Latin America most rivers as yet have been used very little for irrigation. Modest investment could yield substantial returns. This also is true for some counties of the Far East. In other

TABLE 1.1   Irrigated Areas in Countries Having More Than One Million Hectares of Irrigated Land

Country	Cultivated Land 1000 ha	Irrigated Land 1000 ha	Percent
1. Afghanistan	7,980	2,900	31
2. Argentina	26,028	1,555	6
3. Australia	44,610	1,581	4
4. Bulgaria	4,516	1,001	24
5. Chile	4,632	1,500	32
6. China and Formosa	111,167	76,500	69
7. Egypt	2,852	2,852	100
8. India	164,610	38,969	24
9. Indonesia	18,000	6,800	38
10. Iran	16,727	5,251	31
11. Iraq	10,163	4,000	39
12. Italy	14,409	3,500	29
13. Japan	5,446	2,626	48
14. Korea	2,311	1,070	46
15. Mexico	23,817	4,200	18
16. Pakistan	21,700	12,400	57
17. Peru	2,979	1,116	37
18. Philippines	11,145	1,090	10
19. Spain	20,626	2,435	12
20. Sudan	7,000	2,520	25
21. Thailand	11,200	3,170	28
22. Turkey	26,068	1,724	7
23. USA	192,318	21,489	11
24. USSR	232,609	11,500	5
25. Venezuela	5,214	1,000	19
26. Vietnam	5,083	3,040	60
27. Other countries	463,790	17,848	4
World's total	1,457,000	233,637	16

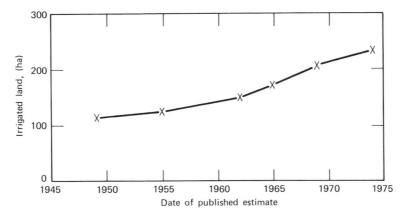

Figure 1.2   Estimates of world irrigated land showing steady and consistent growth.

countries irrigation progress depends on costly public works to control flows of large rivers.

1.6 FUTURE GROWTH OF IRRIGATION

As the population of the world continues to increase, the demand for food and fiber for the people will also increase. Men and women with knowledge of irrigation will be called upon to find the solution of world food and fiber problems. Water must be provided for more of the earth. Arid lands become highly productive with irrigation water. Productivity of the lands now producing food and fiber under natural rainfall can generally be increased considerably by application of irrigation water. However, the capital investment necessary to reclaim remaining lands of the earth will far exceed the investment that has been made to supply water for areas now being irrigated. The largest reservoirs, the longest canals, and the most expensive tunnels and inverted siphons are yet to be built. As long as population is materially increasing each decade, demand for further utilization of water for irrigation will also increase.

1.7 SCOPE OF IRRIGATION SCIENCE

Irrigation science is not restricted to application of water to soil. The scope of irrigation science extends from the watershed to the farm and on to the drainage channel. The watershed yielding the irrigation water, the stream conveying the water, the management and distribution of the water, and the drainage problems arising from irrigation practices are all of concern to the irrigationist. Observing one portion of an irrigation system without considering its other components will lead to faulty design and inadequate preparation.

Watershed characteristics, including the nature of vegetation and retentive ability of soils and subsoils, are all important in their influence upon the yield of irrigation water. Likewise, characteristics of the stream conveying water are of importance. Oftentimes management of this stream, the diversion structures built upon it, and remedial measures to reduce seepage losses and consumptive use along the stream and canals are of importance. Diversion structures, measuring devices, and conveyance channels are extremely important. Layout of the irrigation system on the farm, method of control, and disposition of excess and waste waters also have vital significance.

Proper disposition of excess waters is of equal importance to acquisition of irrigation water. Too often irrigation projects are designed without adequate preparation to utilize waste waters. Water is often needlessly lost in conveyance and by excess application. Both surface and subsurface waste water should be utilized on lower lands of the project or pumped to higher lands for reuse as long as these waters are suitable for agricultural production. The design of surface and subsurface drainage systems is of vital importance in maintaining high productivity of irrigated lands.

The natural hydrologic balance of a valley always is upset by the introduction of irrigation water to the land of the valley. Consequently, drainage problems will arise, usually both surface and subsurface problems, which have not heretofore been present. These problems should be anticipated and their consequences included in original economic considerations.

1.8 ECONOMICS OF IRRIGATION

Economics is important in evaluating irrigation practices, for irrigation is largely for the purpose of increasing profit. Higher profits resulting from more efficient production will ultimately result in lower prices for consumers, and lower prices result in more consumption of food and fiber; greater availability of food and fiber results in higher standards of living for the peoples of the earth. These factors should be kept constantly in mind. Irrigation projects make the world a better place in which to live.

CHAPTER
2
SOURCES AND STORAGE OF IRRIGATION WATER

Basically, rain and snow are the sources of all water. That portion which is not used at the point where it falls, flows over the surface of the land or seeps into the ground and augments the ground-water supply. Therefore, the rain or snow which is not used becomes a potential source of either surface or underground water for irrigation. Irrigation water may also come from waste water, which is water that is not used consumptively by agriculture, industry, and municipalities.

The success of every irrigation project rests largely on the adequacy and dependability of its water supply. In irrigated regions public agencies should make continuous long-term records of precipitation, stream flow, and ground-water storage as a basis for intelligent and complete utilization of all water resources.

2.1 VALLEY AND MOUNTAIN PRECIPITATION

Rain and snow which fall in irrigated valleys are valuable as sources of water to be stored directly in the soil. In some valleys winter precipitation provides enough water to germinate seeds and maintain growth of young plants for several weeks. In these valleys, perennial crops also make substantial growth early in the season by using winter precipitation which has been stored in the soil. Other arid region valleys receive so little winter precipitation that farmers must irrigate the soil before seeding their crops in order to ensure a sufficient amount of moisture to germinate seeds and start satisfactory growth. Such valleys must depend almost wholly on rain and snow that falls in adjacent mountain areas as a source of their water supply for irrigation. In nearly all irrigated regions valley precipitation is relatively unimportant as a direct source of irrigation water. Precipitation in mountain areas, as illustrated in Fig. 2.1, constitutes a major source of water supply. Conveyance of water from mountainous sources to valley lands presents to the people of arid regions

Figure 2.1. Snow stored on the watershed represents a potential irrigation water supply. (Photograph by Bert V. Allen.)

very interesting and yet perplexing problems. Painstaking study is needed of seasonal and annual water yields of each mountainous area on which rain and snow fall. Complete and economical utilization of nearly all natural resources are inseparably connected with the water yield of the watershed, its conveyance to places of use, and its economical use, whether for power, irrigation, recreation, or domestic and industrial purposes. There is need for a wider recognition of the fact that intelligent solution of watershed-yield problems and of economical conveyance of water will appreciably advance the general welfare.

2.2 WATER-SUPPLY STUDIES

Accumulation of dependable information concerning water supply demands intelligent and painstaking endeavor and continuous effort. Inadequate water supply has contributed significantly to financial stress and failure of irrigation projects. Overoptimism and conclusions based on insufficient knowledge of watershed yield have been common and expensive follies.

Overestimates of water supply for various projects are frequently reflected in the small areas of land actually irrigated when compared to the area of land intended to be irrigated in the original project. Occurrence of climatic cycles, which cannot as yet be predicted with precision, together with wide variations of precipitation and stream flow from one time to another, complicates the problem of economically using all available water every year. Yet arid-region communities may intelligently adjust their irrigation practices to some extent on the basis of reliable information concerning water supplies. Investigations on a watershed should be conducted to determine the yield and supply of water available for further irrigation expansion.

2.3 SNOW SURVEYS AND THEIR BENEFITS

Snow remaining on the surface of the land provides excellent storage of water. For example, 10 centimeters of snow normally holds approximately 1 centimeter of water and may hold as much as 3 centimeters. Hence, a snowfall of 10 centimeters with 1 centimeter of water on a 10-square kilometer area, will provide for the storage of 10,000 hectare-centimeters of water. This amount of water would fill a surface reservoir covering 100 hectares to a depth of 1 meter. Not infrequently does the snowfall lie 3 meters deep on a water shed. Also in the spring of the year the water held in 10 centimeters of snow usually is in excess of 1 centimeter. When and how fast this vast quantity of water is released by the melting of the snow is of utmost importance to water users. As the stream becomes more fully developed, it becomes increasingly important to know how much water can be expected.

It is important to know whether or not the water supply will be adequate or inadequate and in what manner runoff will occur. In years of limited water supply, cropping and irrigation plans may be modified, less land may be irrigated, crops that use less water may be planted, and early-maturing crops may be substituted for those requiring a longer season.

When water supplies are above normal, additional lands may be brought under irrigation or more intensive farming may be practiced. These and other benefits from snow surveys have resulted in a great expansion of snow survey network to obtain more reliable data more quickly and safely.

Since many remote locations must be sampled once a month, over-snow vehicles (see Fig. 2.2) have been developed that will

Figure 2.2. The trackmaster developed by the Utah Scientific Research Foundation for off-highway travel to snow courses.

travel over muddy roads, rocks, snow, and ice, carrying the surveyors into and out of the area safely. Helicopters are also being used to reach remote locations (see Fig. 2.3). Also considerable effort is now being devoted to developing sensing elements that will modulate an electric circuit and thereby permit transmission of snow depth and water content by radio to central locations.

Since forecasts must be made quickly after surveys are made, techniques of analysis have been developed utilizing high-speed computers.

Representative snow courses are usually established in small mountain meadows where the snow cover is least influenced by shifting winds. The course consists of 10 to 15 observation points spaced about 15 to 30 meters apart. Depth and water content of the snow are determined at each observation point.

Snow is sampled with a lightweight seamless aluminum tube. The bottom of the tube is tipped with a circular saw-edge cutter for penetrating hard snow or icy crusts. The tube is forced into the snow and then removed, holding within the tube a core of snow, and weighed as shown in Fig. 2.4. Data collected are correlated with one or more of the following variables to predict the peak and total runoff: present and antecedent stream flow, temperature, water content of the soil mantle, and rainfall.

Figure 2.3. Helicopter being used to survey snow conditions in remote locations. (Courtesy Soil Conservation Service, U.S.D.A.)

2.4 SURFACE RESERVOIRS

Surface reservoirs are built to store irrigation water for use when the natural flow of a stream is not sufficient to meet irrigation demands. Winter and spring runoff can be impounded until needed for crop growth. In large reservoirs excess runoff in wet years can be stored until needed in dry years.

All storage dams must be built with spillways large enough to convey the maximum anticipated flood flows. The larger the structure and the reservoir, the greater the danger to life and property, and the safer must be the spillway.

The capacity of each reservoir is fixed largely by the natural conditions of the canyon or valley in which water is to be stored, together with the height a dam must be to store the quantity of water needed and economically available. Capacities vary from a few hundred hectare-meters for reservoirs on small streams (as shown in Fig. 2.5), to millions of hectare-meters. Likewise, dams constructed for storage of irrigation water vary from a few meters in height, built at a low cost, to massive masonry structures over 200 meters high. Increasing demands for water will necessitate construction of higher, longer, and more expensive dams than have thus far been built.

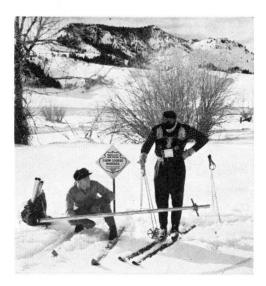

Figure 2.4. Driving a sampling tube into the snow, measuring the depth of snow, and weighing the tube and snow core to determine water content. (Courtesy Soil Conservation Service, U.S.D.A.)

2.5 SMALL EARTH DAMS

Small dams are valuable for impounding water from springs and small streams so that it can be more efficiently utilized. Small reservoirs built for irrigation frequently have multiple advantages. Often they supply water for fish, provide stock water, and supply nesting cover and food for ducks and geese. Ponds can also add general beauty and interest as well as utility to the farming units.

When selecting a reservoir site, several factors should be considered. The dam should be as short as possible and located on dry firm soil void of roots and brush. The height of the dam should be consistent with water supply, economy of the project, and diversion and spillway requirements. A good supply of medium-textured soil is advisable for the dam. When heavy clay is used the dam may crack when dry, and sand cannot be used because it will not hold water. The site should be readily accessible and should permit construction of a good spillway and reservoir. Consideration should be given to the hazard to life and property which would occur in case an excess flood washes out the dam and releases the water in the reservoir.

Figure 2.5. Government reservoirs like this one contribute to irrigation advancement. (Courtesy Bureau of Reclamation.)

The reservoir should be located on material that will not allow excessive seepage. Shale and rock outcroppings frequently have cracks that will develop solution channels and result in serious loss of water. Trees and shrubs should be removed from the area to be inundated. The size of the reservoir should be consistent with the water supply and the amount of water needed.

The spillway is a very critical segment of the design and is the feature which frequently causes failure of dams. Design considerations for spillways are complex and must fit site conditions. Therefore, competent professional engineering help should be secured to design and construct the spillway and dam to ensure that the most economical design is used consistent with materials at hand, site, and flood-water flows.

2.6 SEDIMENTATION OF RESERVOIRS

The useful life of irrigation reservoirs may be shortened by accumulation of sediment. Once sediment has accumulated, the site is essentially of no further value for storage of water. Since good damsites are very limited, it is not only a serious financial loss when a site is no longer of value, but also a grave, often irreplaceable loss of a natural resource. In irrigation-project planning, allowance should be made for the effect of sediment accumulation upon the usefulness of the reservoir, and all practical steps should be undertaken to minimize the rate of sedimentation.

Devices and methods used to control silting of reservoirs are settling basins, by-pass canals, off-channel locations, vegetated streams, venting density currents, flood sluicing, dredging, draining, and flushing. Most of these methods are useful only under special site conditions. Watershed protection and special reservoir design which will permit utilizing one or more of the foregoing means of prevention are the two most useful approaches to controlling sediment.

2.7 REDUCING EVAPORATION LOSSES

Loss of water by evaporation is a very serious problem in arid regions. Critical shortage of water in desert areas has stimulated interest in ways to reduce evaporation. Extensive research is being conducted to discover practical methods. In the seventeen western states of the United States estimates are that 45 percent as much water evaporates as is used for irrigation. Hence, the incentive is great to reduce evaporation in arid regions, particularly with the increasing competition for water supplies.

Several technical problems must be solved before using monomolecular film becomes a practical method for general use to reduce evaporation. Materials are costly, microorganisms consume the material, and wind drags the film across the water surface often depositing the chemical on the shore of the reservoir. Suitable chemicals will undoubtedly be found and better techniques of application be developed so that this method can be used more widely to conserve water supplies. It is now being used in a number of areas to conserve water supplies for livestock.

The use of underground reservoirs for storage of water is increasing and will continue to increase as the water supplies become more valuable and surface reservoir storage sites become more limited and more costly. As in the case of the monomolecular films, technical advances are needed to facilitate full use of underground reservoirs. Nevertheless, the absence of an exposed water surface from which evaporation can occur is a major asset. Development of ground-water basins, and techniques used for their exploration, are discussed in a later section.

2.8 PHREATOPHYTE PROBLEMS

Phreatophytes are water-loving plants growing along stream courses and on wet soils having high water tables where an abundant supply of water is available. They have little or no economic value, and they consume water which normally could be used beneficially. The problem is particularly acute in arid regions where phreatophytes abound and where water is critically needed for economic development. Estimates have been made that for every 10 cubic meters of water used for agricultural crops in the southwestern United States, 8 cubic meters are used by native vegetation such as cattails, tules, willows, salt cedar, cottonwoods, salt grass, greasewood, baccharis, and mesquite.

Water used by phreatophytes is high and increases as the depth to the water table decreases. Occurrence and growth of most species are reduced by increasing the depth to the water table. The thicker the capillary fringe above the water table, the more favorable is the growing condition. Therefore, even though the water table may be 3 meters below the surface, a normally shallow-rooted plant like salt grass may be able to obtain ample water from a thick capillary fringe which brings water near enough to the ground surface. Cattails and tules prefer water on or near the surface. Salt grass does best when the water table does not exceed 2 to 3 meters. Willows, salt cedar, and cottonwood prefer a water table not more than 3 meters below the surface. Greasewood does best where the water table is less than 4 meters, whereas mesquite has been known to send its roots to a depth of 10 meters and even 30 meters to water.

Salt cedar, mesquite, and baccharis grow best in warm climates. Willows, cottonwood, and salt grass grow rather generally, and greasewood is ususally found in cold or dry climates.

Two general approaches are used to control phreatophytes. One is to remove the water supply by lowering the water table, channeling the water, or piping the water around or through critical areas. The other approach is to mechanically or chemically prevent plant growth. Both methods have been useful and successful in varying degrees, but better techniques are needed and more knowledge is required for effective control. Methods of eradication must be adapted to the local physical and economic conditions.

2.9 RAINMAKING OR CLOUD-SEEDING

One of the most interesting developments in man's search for additional water supplies is cloud-seeding, or rainmaking. The components parts are:

1. Reaching a decision to undertake cloud-seeding.

2. Recognizing a suitable opportunity.

3. Delivering the required treatment to the clouds.

4. Evaluating what has been accomplished.

Most cloud-seeding has dealt with clouds containing a significant amount of water in the form of super-cooled liquid droplets (colder than zero degrees Celsius) and low ambient counts of natural freezing nuclei. Silver iodide particles have been found to be very effective in inducing precipitation when introduced into a super-cold cloud. Ice crystals are formed upon particles in the air such as dust, salt, and silver iodide. These ice crystals attract other crystals or droplets and grow in size until they fall toward the earth as snow, or melt and fall as rain.

Silver iodide has the advantage that it can be vaporized on the ground surface and carried aloft into the clouds by naturally rising air streams existing in storm situations suitable for cloud-seeding. Better results, but at higher cost, are secured by using an airplane to deliver silver iodide to the desired location in or near the super-cooled cloud where thermal currents will carry particles into contact with super-cooled droplets.

Warm clouds can also be induced to produce more precipitation. Precipitation occurs through a process of coessence of droplets into a mass large enought to fall toward the ground. Seeding with hydroscopic materials such as finely divided common salt or droplets of concentrated ammonium nitrate-urea solution has been found to be effective.

Cloud-seeding is now being done over extensive areas with an average increase in rainfall of from 10 to 15 percent. When terrain and cloud cover are favorable, rainfall can be economically increased by cloud-seeding.

Considerable effort is being devoted toward improving techniques of control and measurement. A better understanding of storm activity is being obtained. Surely rainmaking efforts and success will increase with the need for weather control and with increased technical knowledge of the process. Mankind will ever strive to control the weather, and by so doing will make available the greatest source of water yet undeveloped—an essentially untouched resource, the clouds.

2.10 RIVER DEVELOPMENT BY PUMPING

Extensive development of a river can be accomplished by pumping water to higher lands for irrigation purposes. One of the principal advantages of pumping is that the desired development can usually be undertaken without the heavy capital costs of dams and gravity diversions.

Increasing cost of energy required for pumping and growing shortages impact the feasibility of river development by pumping. Adjacent landowners can develop the needed water as rapidly as finances and equipment permit. Many developing areas have made excellent progress by staged development financed largely by private funds. Large complex government projects are not required.

2.11 SALINE-WATER CONVERSION

Another source of usable water is found in the saline waters of the world. Until recently these waters were reclaimed only by the natural processes of evaporation and subsequent condensation as precipitation. Here, as in rainmaking and reduction of evaporation, the pressure for conservation and development of our water resources has led to significant technological advances. Great emphasis now is being placed on conversion of saline water to water fit for human, animal, plant, and industrial use.

Inorganic salts dissolved in water have definite physical and chemical properties which are well understood. These known properties are utilized by a variety of techniques to develop practical and useful ways of reclaiming saline waters. The challenge is to develop practical and efficient techniques of removing the salt.

Theoretical calculations showing the cost of sea-water conversion under optimum conditions with 100 percent efficiency are very informative. To remove the salt from 1 cubic meter of sea water requires at least 0.7 kilowatthours of electrical energy. At 3 cents per kilowatthour the cost is 2.1 cents for 1 cubic meter of water or $210 per hectare-meter.

When efficiency, capital investment, and operating expenses are also considered, economical conversion of saline water presents a major challenge to man's ingenuity. New and cheaper sources of energy and the increasing value of water will certainly promote practical development. Major support has been given to this development by many governments of the world, and results should yield techniques of increasing value.

Water need not be pure for all uses. With proper methods of management irrigation water containing considerable salt can be used for permanently irrigated agriculture. Water with a satisfactory degree of purity may be reclaimed from saline water at a fraction of the cost that is needed to obtain pure water. Table 2.1 shows the general relationship of energy consumption to purity of feedwater reclaimed with a commercial membrane demineralizer. It may not be long before practical, usable methods are available to reclaim the brackish waters of inland areas at a cost that is compatible with the economy of agriculture in areas of short water supply.

2.12 IMPORTANCE OF GROUND WATER

Silting of reservoirs, availability of good ground water storage sites and growing lack of surface storage sites, prevention of evaporation, and development of additional water will all contribute to more extensive and intensive development of underground water. Numerous natural advantages are inherent in utilizing ground water and underground storage reservoirs.

Many countries are blessed with copious water supplies. However, almost every country and section of the earth is plagued with the problem that water supplies are not evenly distributed, neither geographically nor seasonally. Full utilization of the total water supply will require complete water control. Surface water

TABLE 2.1 Increase of Energy Consumption with Increase of Salinity for a Commercial Membrane Demineralizer, Based on a Product Purity of 500 mg/l. (Courtesy of Ionics, Inc.)

Feedwater Salinity (mg/l)	Approximate Production Rate (m³/day)	Approximate Energy Consumption (kwh/m³)
900	106	0.8
1,500	53	2
2,000	38	3
3,000	26	4
4,000	21	5
5,000	17	7
6,000	15	10
8,000	11	11
10,000	9	13
15,000	7	21
20,000	5	26
35,000	3	48

control is well underway. Numerous dams, diversions, and conveyances have been built. Nevertheless, the control of subsurface water is still in its infancy. For economic reasons subsurface water supplies generally are not developed until surface supplies have been reasonably well utilized.

Pumping water from underground sources is widely practiced and is a well-established method of obtaining irrigation water. Lowering the ground-water table following extensive pumping for irrigation in some places has proved valuable as a means of drainage. In other localities the ground-water surface has been lowered so much by pumping that deepening of wells has become necessary. Lowering ground-water levels increases pumping lift and makes water increasingly expensive. In some areas shortages have become sufficiently acute to justify curtailed use.

2.13 RECHARGING GROUND-WATER RESERVOIRS

Critical shortages of underground water due to limited natural recharge, small storage capacity, and overuse have stimulated efforts to recharge ground-water reservoirs with surface waters. Flood flows which would otherwise have been lost are diverted and applied to the land, thus providing water to seep into under-

ground reservoirs. Winter flow of streams, sewage, and industrial water may also be utilized to excellent advantage for recharging these reservoirs. Full conservation and use of available
water supplies requires an integrated use of surface and subsurface waters and storage facilities.

Systematic flooding of land surfaces overlying or draining into
underground reservoirs is a recognized method of water storage.
Usually the water is spread over the land to expedite infiltration
into the soil and percolation downward to the ground water. Figure 2.6 shows a well-designed detention facility to retard flood-
flow and permit maximum recharge of the ground-water basin.
Water percolates into the ground-water reservoir to be stored
until needed for irrigation and municipal purposes.

Infiltration has been increased by using organic residues and
grasses to condition ground surfaces to prolonged periods of
submergence. Grasses appear to maintain a more pliable and
open condition on soil surfaces than do presently used soil con-
ditioners, particularly where the sediment content of the water is
considerable. High infiltration rates during the recharging operation are essential.

A less common method of recharging utilizes shafts, pits,
trenches, or wells to convey water to the gravel and rock materials underground. The Peoria recharge pit[1] has been very successful in obtaining an infiltration sufficient to absorb 2.5 to 2.9
cubic meters per hectare (20 to 25 meters depth per day) by using
pea gravel as a filter. When sand was used the rate was only 0.7
cubic meters per hectare.

South of Los Angeles, a permanent water supply of 37 million
cubic meters per year of acceptable water quality has been developed from the reclamation of waste water and desalinization
of sea water. Reclaimed water is then injected into basin aquifers through a series of injection wells along the coast line. Positive protection from sea water intrusion and recharge of ground-
water aquifers are benefits resulting from this project.

2.14 SAFE YIELDS FROM GROUND-WATER RESERVOIRS

Early withdrawals of ground water were based upon water
needs and costs. Serious lowering of the water level (or pressure)
brought alarm and resulted in attempts to define "safe yield."

---

[1]Suter, Max., "The Peoria Recharge Pit: Its Development and Results,"
   Proc. Am. Soc. Civil Eng., No. 1102, November 1956

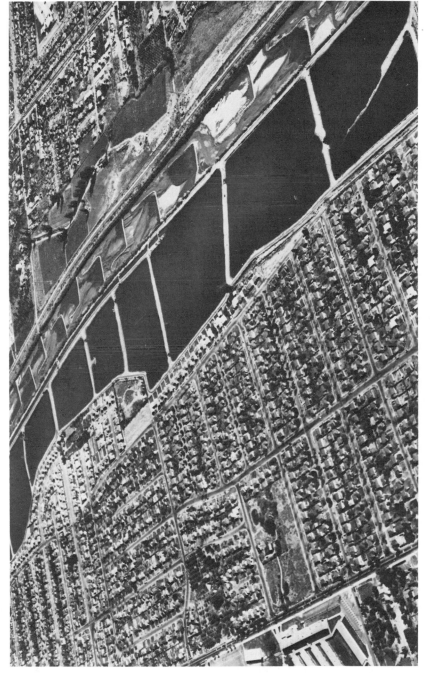

Figure 2.6. Water spreading for ground-water recharge along the San Gabriel River in California. "Hook" levels in river allow maximum detention time. (Courtesy Los Angeles County Flood Control District.)

Demands from adjacent water users resulted in many courts restricting water use to the extent that the level (or pressure) of the
water in the aquifer is maintained. In essence, these court decisions guarantee not only the right of the user to the water, but
also the energy represented by the elevation of water in the well.
Ground-water resourses cannot be effectively utilized if natural
water levels and pressures must be maintained. Usually only a
small fraction of the available water can be withdrawn under
these restrictions. For full utilization of surface and subsurface
water resources, ground-water levels and pressures must be
lowered during periods of drought or periods of limited surface
supplies and elevated by recharging the reservoirs during
periods of water surplus.

In some areas where recharging possibilities are limited, the
concept of "mining" the water is being accepted as a necessary
and desirable procedure. Water is being considered, therefore, in
the same category as other natural resources—a resource which
may be expended in the interest of current development.

An outstanding example of this philosophy exists in the Central Valley development in California. Depletion of ground water
was not only permitted but encouraged by some of the more
progressive irrigation leaders. The plan was to utilize the
"mined" ground water to build a strong taxable economy which
could support later development and transportation of distant
waters to areas of the valley which would develop a water shortage. This plan of ultimate development and integration of water
resources of the state has been implemented, and supplemental
waters have been diverted to replace ground-water reserves
which were seriously depleted.

With development of new techniques of exploration for ground
water, with increasing control and understanding of weather,
and with a relative decrease of cost of energy, water supplies
can be made available to areas of shortage which at the present
time appear to be impractical. Therefore, the definition of "safe-
yield" is changing and will continue to change as economic and
technological conditions change. More surplus and reclaimed
surface waters can be stored in subsurface reservoirs when additional capacities have been created in these reservoirs by prior
withdrawal of existing waters.

Safe-yield also implies a wise balance where excessive and
uneconomical pumping lifts are not imposed, nor serious deterioration in water quality permitted. It is important to consider
"safe-yield" in the light of total water resources: surface and subsurface, developed and potential.

2.15 FEASIBILITY OF GROUND-WATER DEVELOPMENT

Undue lowering of ground water results in higher pumping lifts and sometimes prohibitive pumping costs. Wells may need to be deepened and pumps lowered in order to obtain sufficient quantities of water. The extent of irrigation pumping from ground-water supplies should therefore be determined on the basis of thorough, long-time investigations of the quantity of annual inflow or recharge to ground-water streams, basins, or reservoirs.

Essential decisions concerning development of ground-water supplies for irrigation can be made after checking "yes" or "no" on twelve important points, listed in the following questionnaire. The answers shown are generally considered as indicating that ground-water supplies can be developed satisfactorily.

	Yes	No
A. Availability, quality, and depth of water:		
1. Is a plentiful supply of water available?	yes	____
2. Is water of the right quality available to permit production of the desired crops?	yes	____
3. Is water available at a depth that will permit economical pumping?	yes	____
B. Trend of the water table:		
4. Is the water table stable?	yes	____
5. Is it rising?	____	no
6. Is it declining?	____	no
7. Is development in the area likely to bring about withdrawals of water seriously in excess of the natural recharge?	____	no
C. Legal or natural protection:		
8. Do the statutes or court decisions provide legal or administrative protection in your area against excessive depletion of water supplies, or	yes	____
9. Is the area protected against overexpansion and depletion by "natural" controls?	yes	____
D. Cost of operation:		
10. Will the prospective production under irrigation bring enough returns to pay the increased costs of irrigation farming?	yes	____

E. Land requirements:
    11. Is the land physically suitable for irriga-
        tion considering topography, productivity,
        water-holding ability?   <u> yes </u> <u>   </u>
    12. Is the land suitable for the types of crops to
        be produced?   <u> yes </u> <u>   </u>

2.16 CHANGES IN GROUND-WATER STORAGE

The change that can be expected to occur in ground-water
storage can be shown in terms of the components to that change.
The general ground-water equation is helpful in relating and
focusing attention upon the various sources of supply and disposition of water to and from the ground-water reservoir:

$$\Delta S = (Q_i + S_i + P) - (Q_o + S_o + W + C) \qquad (2.1)$$

in which ΔS = change in storage
              Q_i = surface inflow
              Q_o = surface outflow
              S_i = subsurface inflow
              S_o = subsurface outflow
              P = precipitation
              W = withdrawal
              C = consumptive use (evapotranspiration)

Usually all of these elements can be changed and moderated
by man. Therefore, all must be considered when a plan of operation of a ground-water reservoir is being developed. This is true
not only because these factors can be varied by management,
but because each one either contributes water to or withdraws it
from the reservoir. Certainly all are involved when safe yields
are being considered.

2.17 GROUND-WATER INVESTIGATIONS

The greater the detailed knowledge of the ground-water reservoir, the more effectively and efficiently ground waters can be
developed, removed, and utilized. Physical features of ground-
water basins are hard to ascertain under ideal conditions. Therefore, all available methods for gleaning factual data should be
used. Four apporaches can normally be made: (1) geological observations and investigations; (2) physical aspects of the existing
ground water, such as pressures, elevations, movements, temperatures, quality; (3) geophysical methods of exploration; and
(4) logs of existing wells. All of these are complementary, and all

are needed to obtain the complete picture of an underground reservoir.

Geological investigations and observations can outline physical features of the surface and are a good index to subsurface conditions. Physical surface features and geological patterns are often indicative of what can or cannot be reasonably expected beneath the surface, particularly with regard to the occurrence of a suitable aquifer from which water can be withdrawn. Coarse-textured, highly permeable material or fractured sedimentary deposits are essential in the aquifer if available water is to be economically withdrawn for irrigation.

Aerial maps are very helpful to outline physical features. When the aerial map is viewed through a stereoscope, surface relief is portrayed clearly, and physical characteristics of the terrain become very evident.

Knowledge of local well-drillers should not be overlooked. Well-drillers accumulate a wealth of information through their well-drilling activities.

Frequently, existing wells are available wherein water levels, water temperatures, water quality, and ground-water gradients can be ascertained. Well-logs are also extremely informative. These data, when plotted in plan and in profile, often permit extrapolation with reasonable confidence to new locations.

Fractures in rock are often important in ground-water development. Fractures not only contain water but also serve as conduits for collection and movement of ground-water. Surface observations and use of stereoscopes and geophysical tools are useful in determining the extent of subsurface fractures.

Geophysical methods are based upon detecting differences in the physical properties of surface and subsurface materials. Electrical resistivity and seismic refraction are the two methods most commonly used. Both require special equipment. Experience is essential to interpret the usually complex results properly. More compact and reliable equipment has extended the use of both methods to studies of water supplies, whereas formerly these concepts were employed only in oil-well exploration.

Electrical resistivity methods involve measurement of changes in resistivity. The resistance of rock formations to electrical current varies widely. In the more porous formations water content is the major variable affecting resistivity. The apparent resistivity is compared to the electrode spacing as an indication of the change in properties with depth. Wells drilled on the basis of the results of resistivity surveys, conducted by competent and trained personnel, consistently yield from 75 to 90 percent success. This method has also been used for detecting the advance of salt-water intrusion and locating perched water tables. Resistivity has been used also as an index to the seepage from canals.

Seismic refraction involves creating waves in the earth's surface by impact or explosion, and measuring the time required for the wave to travel a known distance. Waves are reflected and refracted by changes in the properties of conducting media in a manner similar to light waves. The time interval for the first shock to reach varying distances from the point of origin is analyzed to determine subsurface properties. Because the seismic method requires special equipment and a greater degree of training of operators and is more complex to interpret, it has been applied to ground-water explorations only on a relatively limited basis.

CHAPTER 3

BASIC SOIL-WATER RELATIONS

Knowledge of soil and water relationships is valuable to all who have the opportunity to improve irrigation practices, including the irrigators who desire to obtain the best use of water available for their farms.

Chapters 3, 4, and 5 are devoted to a consideration of the relationships of soils and water, with special reference to their influence on irrigation and drainage practices. Soil and water relationships of special importance in irrigated regions include the capacities of well-drained soils in the field to retain water available for plants, and the flow, or movement, of water in soils. The salinity and alkali conditions, together with translocation and concentration of soluble salts due to the movement and evaporation of soil water, also constitute important soil and water relationships.

The functions of soil moisture in plant growth are very important. Excessive volumes of water in soils retard or inhibit plant growth and make drainage essential. Sterility of arid-region soils is usually caused by deficient amounts of water. Irrigation is an artificial means of preventing deficiencies in the moisture content of soils. Intelligent irrigation practices, based on knowledge of soil and moisture, is also a means of preventing, or at least retarding, the occurrence of water-logging of soils. Students of irrigation are interested in the basic physical laws that influence the distribution, storage, flow, and consumptive use of water in unsaturated soils.

Intermolecular forces and tensions in unsaturated soils are influenced by soil texture and structure. These forces give rise to "capillary phenomena" in unsaturated soils which are of special interest in the study of soil and water relations. The close contact of soil particles and particles having thin water films results in attractive forces of large magnitude.

3.1 SOIL TEXTURE

The sizes of particles making up a soil determine its texture. These particles range in size from fine gravel to clay. Particles larger than 1.00 millimeter in diameter are gravel, particles from

0.05 to 1.00 millimeter are sand, particles from 0.002 to 0.05 millimeter are silt, and smaller than 0.002 millimeter are clay. Most soils contain a mixture of sand, silt, and clay. If sand particles dominate, the soil is called sand. If clay particles dominate, it is called clay. Silts fall between clays and sands. Loams are medium-textured soils having about equal amounts of clay, silt, and sand particles.

Sand grains feel gritty to the fingers and can be distinguished without difficulty by the unaided eye. Silt, barely visible to the naked eye, has the appearance and feel of flour. The individual particles of the clay fraction, many of which are inorganic colloids, are not distinguishable by the eye, and a large portion of them are too small to be seen under a microscope. It is this fraction that makes soils swell and become sticky when wet, and shrink and become brittle on drying.

Where the main soil-forming process is due to mechanical disintegration, relatively coarse-grained particles are found. The size of particles is designated by the word texture. Sandy soils are classed as coarse-textured, loam soils as medium-textured, and clay soils as fine-textured. The texture of a soil has a very important influence on the flow of soil water, the circulation of air, and the rate of chemical transformations which are of importance to plant life. The farmer is unable to modify the texture of his soil by any practical means. The size of soil particles has a great influence on crop production the world over, but to the irrigation farmer it is particularly important, because it determines in a large measure the depth of water he can store in a given depth of soil.

3.2 SOIL STRUCTURE

Soils in which the particles are relatively uniform in size have comparatively large spaces between the particles, whereas those soils in which the size of the grains varies greatly may become more closely packed, and thus the volume of spaces between the grains is restricted. Fine-grained irrigated soils, if properly managed, function as granules, each consisting of many soil particles, whereas in coarse-textured soils each particle functions separately. The existence of granules assures a desirable soil structure. Excessive irrigation, plowing, or otherwise working fine-textured soils, when either too wet or too dry, tends to break down these granules. A soil worked too wet is said to be puddled and has a poor structure. Favorable structure in fine-textured soils is essential to the satisfactory movement of water and air.

Favorable soil structure is recognized by soil scientists as the key to soil fertility. Adequate amounts of chemical nutrients in soils, though essential to crop production, do not assure satisfactory plant growth and crop yields. The permeability of soils to water, air, and roots, provided by favorable soil structure, is equally important to crop growth as are adequate supplies of nutrients.

Soil structure is fundamentally built by wetting and drying, freezing and thawing, or combinations of these conditions. Penetrating roots build soil structure by removing water from the soil, which causes drying and permits subsequent wetting. The primary functions of organic matter and of humus in soil is to add stability to soil aggregates, serving as a cushion against the shock of tillage.

Some of the more important practices to be followed to maintain and improve the structure of irrigated soils are as follows: (1) plow below compacted layers, not at the same depth each year; (2) allow as much time as practical for soil and air to interact after plowing before giving preplanting irrigation or before preparing seed bed; (3) return all possible organic matter to the soil; (4) follow a good crop rotation on legumes, cash crops, and fibrous-rooted crops; (5) reduce cultivation and tillage operations to a minimum.

Remember that modern implements such as rubber-tired wheels and disks are used to compact soils for roads, airports, and race tracks. Likewise, compaction will occur on farm soils when these implements are used. Hence, they should be used to a minimum, consistent with the needs for seed-bed preparation and weed control.

3.3 REAL SPECIFIC GRAVITY

The real specific gravity of a soil is a dimensionless quantity and is defined as the ratio of the density of a single soil particle to the density of a volume of water equal to the volume of the particle of soil. Particle density is the same (in metric units) as real specific gravity and is the weight per unit volume of a soil particle with dimensions of grams per cubic centimeter. The specific gravity of the common soil-forming minerals varies from 2.5 to more than 5. A few minerals, such as quartz and feldspar, usually make up the bulk of a soil. The real specific gravity of soils which have a low percentage of organic matter varies but little, and approaches closely an average of 2.65. Some irrigated soils, which are formed largely of organic matter, have a real specific gravity of 1.5 to 2.0, depending on the amount of mineral matter present.

3.4 APPARENT SPECIFIC GRAVITY

The apparent specific gravity of a soil is defined as the ratio of the weight of a given volume of dry soil, air space included, to the weight of an equal volume of water. This ratio is known also as the "volume weight" or "bulk density." Whereas apparent specific gravity is a dimensionless quantity, being weight of soil per weight of water, bulk density is grams per cubic centimeter or mass per unit volume. Therefore, the dimensions are not equal. However, since 1 gram of water fills a volume of 1 cubic centimeter at normal temperatures, the two terms do have equal numerical values.

The apparent specific gravity is influenced by structure; i.e., the arrangement of soil particles, and by its texture and compactness. Apparent specific gravity is a soil property of great importance to the irrigation farmer, as will be shown more fully in connection with the capacity of soils to retain irrigation water.

Compacting a soil of fixed real specific gravity will increase the apparent specific gravity, because it reduces the space between soil particles and thereby reduces the volume of pore space. When working with irrigated soils, it is necessary to know their apparent specific gravity in order to account for the water applied in irrigation, since it is impractical to measure, by direct means, the volume of water which exists in the form of soil moisture in a given volume of soil. It is necessary to measure the weight of water in a given weight of soil by observing the loss of weight in drying, and then convert the weight percentage so obtained to a volume percentage by use of the apparent specific gravity; thus, the volume of water in a given volume of soil may be determined.

The usual method for determining apparent specific gravity is to obtain a soil sample of known volume. This is usually done by driving a tube with a cutting edge into the soil and obtaining an uncompacted core within the tube. Sometimes pits are dug and blocks of soil are obtained directly. Also, a hole can be bored with an auger and all the soil that has been removed can be dried and weighed. The volume is determined by measuring the average size of the hole or by placing a flexible rubber or plastic tube in the hole and determining the volume of water required to fill it.

Another method of measuring apparent specific gravity is by the gamma-ray absorption technique. Absorption of gamma rays has been found to be independent of the chemical composition of soils for the energy used and to depend almost entirely upon the density of soil. Cobalt 60, a relatively inexpensive and stable radio isotope, is an ideal source of gamma radiation. A portable survey meter is satisfactory as a counting device, and small, stable detectors are readily available. These have been combined into commercially available units. The source and detector are lowered into a prepared hole to the depth to be tested; the counts per minute are calibrated to yield apparent specific gravity readings.

3.5 PORE SPACE

If a solid sphere having a diameter of 1 cubic centimeter is placed in a cubical box having a capacity of 1 cubic centimeters, there will remain an air space of 0.476 cubic centimeters, or 47.6 percent of the total volume. The same air space will be left by any number of solid spheres of any diameter if arranged in vertical columns. If, however, the spheres are arranged in oblique order, there will remain only 25.9 percent air space. These facts show that considerable variation of pore space between spherical particles may result from change in arrangement, but they do not show the maximum range or variation that may occur in soil. There is great variation in the size and shape of soil particles, and these differences in particles influence the closeness of contact and the interpacking of small particles between large ones, thus determining the total percentage of pore space, which for convenience is represented by the symbol n. In general, coarse-textured, gravelly, and sandy soils have a smaller percentage of total pore space, and fine-textured clay loams and clays have a greater percentage. It is not unusual in irrigated soils for the pore space to vary from 35 percent to 55 percent. To compute the percentage pore space n, it is necessary to know the real and the apparent specific gravity of the soil. The ratio of the apparent to the real specific gravity gives the proportionate space occupied by the soil, and this ratio subtracted from unity gives the pore space. The pore space on a percentage basis is given by the equation:

$$ n = 100 \left(1 - \frac{A_s}{R_s} \right) \qquad (3.1) $$

where n = the percentage pore space
A_s = the apparent specific gravity
R_s = the real specific gravity, approximately 2.65 for most agricultural soils

The term porosity, equivalent to pore space, is much used in soil mechanics. It is defined as the ratio of the volume of voids (air-

and water-filled space) to the total volume of soil plus water and air.

Pore space has a direct bearing upon productive value of soils because of its influence upon water-holding capacity and upon the movement of air, water, and roots through the soil. When the pore space of a productive soil is reduced 10 percent, movement of air, water, and roots is greatly restricted and growth is very seriously impeded.

3.6 INFILTRATION

A property of soils, of great importance to irrigators, is the time rate at which water will percolate into soil, or rate of infiltration. Usually, the infiltration rate is much higher at the beginning of a rain or irrigation than it is several hours later. It is influenced by soil properties and also by moisture gradient. Moisture tension, explained in the following chapters, may be zero near the surface of soil shortly after wetting and may be very high a few centimeters below, thus causing a large downward force (in addition to gravity) pulling the water into the unsaturated soil. Several hours after wetting, these differences in tension may be very small, and gravity then becomes the dominant force causing infiltration. The decrease of infiltration with time after wetting a soil is of importance in rainfall-runoff studies and in irrigation. Water standing on gravelly or coarse sandy soils percolates into the soil so rapidly that the water surface may be lowered several inches an hour. On fine-textured clay soils, water may collect and stand on soil, seemingly with very little infiltration, for many days. Desirable infiltration rates are between these two extremes. A convenient means of expressing infiltration is in terms of centimeters lowering of water surface per hour. For example, if a hectare of level land at 9 o'clock is covered with water to a depth of 5 centimeters and at 10 o'clock the water is but 2 centimeters deep, the infiltration rate is 3 centimeters per hour, neglecting evaporation losses.

3.7 INTAKE

The rate of infiltration from a furrow into the soil is referred to as the intake rate. This term indicates that infiltration occurs under a particular soil surface configuration. Intake rate is therefore influenced by furrow size and shape, whereas infiltration rate applies to a level surface covered with water. Hence, whenever the configuration of a soil surface influences the rate of water entry, the term intake rate should be used rather than the term infiltration rate to refer to the rate of entry of water into soil.

3.8 PERMEABILITY

One of the important properties of soils is the ability to convey water flow through the pore spaces caused by a given force. The permeability[1] of soil is defined as the velocity of flow caused by a unit gradient. Permeability is not influenced by the gradient, and this is an important point of difference between permeability and infiltration. Also the term permeability is used for designating flow through soils in any direction. Permeability is influenced most by the physical properties of the soil. Changes in water temperature influence permeability slightly. In saturated field soils permeability varies between wide limits: from less than 25 centimeters per year in compact clay soils, up to several thousand meters per year in gravel formations. For unsaturated soils the moisture content is one of the dominant factors influencing permeability. Permeability is a velocity having the physical dimensions of length divided by time.

Infiltration, intake, and permeability are discussed in greater detail in Chapter 5.

3.9 DEPTH OF SOIL

The importance of having an adequate depth of soil in which to store satisfactory amounts of irrigation water at each irrigation should be emphasized. Soils of arid regions are relatively deep compared to humid-region soils. There are many areas of productive shallow soils in irrigated regions underlain at depths of 30 to 100 centimeters by coarse gravel, hardpan, or other formations in which plants obtain little or no sustenance. Shallow soils require frequent irrigations to keep crops growing. Excessive deep percolation losses usually occur from irrigation when shallow soils overlie coarse-textured, highly permeable sands and gravels. Deep soils of medium texture and loose structure permit plants to root deeply, provide for storage of large volumes of irrigation water in the soil, and consequently sustain satisfactory plant growth during relatively long periods between irrigations. The volume of water actually absorbed by some plant roots and consumed to produce a crop may be practically the same for shallow and deep soils. Yet more water is required during the

[1]Permeability and conductivity are frequently used interchangeably.

crop-growing season to irrigate a given crop on a shallow soil than is required for the same crop on a deep soil. The larger number of irrigations required for shallow soils and greater unavoidable water losses at each irrigation on shallow soils account for differences in practical water requirements for different soils during the season. The relationship of depth of soils to their water capacities is considered in more detail in Chapter 4.

3.10 PLANT FOOD COMPOUNDS

In order to produce satisfactory yields of crops, soils must have adequate supplies of available plant nutrient elements. Many chemical elements are essential to plant growth. Calcium, carbon, hydrogen, iron, magnesium, nitrogen, oxygen, potassium, phosphorus, and sulfur are of major importance because they are required in considerable amounts. Because of the sparse or scanty growth of native vegetation on the virgin soils of arid regions, nitrogen is relatively deficient. Plants absorb nitrogen in the form of soluble nitrates which are dissolved in the soil water. In order to assure an adequate supply of available nitrogen, irrigated soils must contain sufficient amounts of nitrogenous matter which may be supplied from barnyard manure, from the growing of legume crops as green manures, or from commercial fertilizers. Soil moisture content, soil structure, and soil aeration should be made favorable to the bacterial activity which is essential to the formation of nitrates.

The low rainfall of arid regions results in a comparatively small amount of leaching, and hence, arid soils are usually high in the more important mineral plant-food elements, particularly calcium and potassium. The open structure of most arid soils permits very favorable aeration to great depths, and consequently favorable bacterial activity occurs at much greater depths in arid-than in humid-climate soils.

3.11 EXCESS SOLUBLE SALTS

The arid-region conditions of low precipitation, high evaporation, and relatively small amounts of soil leaching, although favorable to the occurrence of satisfactory quantities of calcium, phosphorus, and potassium in soils, result in the accumulation of excessive quantities of soluble salts that retard or inhibit plant growth. Soils having excess soluble salts are designated saline soils, and those having excessive exchangeable sodium are designated alkali or sodic soils. Even such compounds as sodium nitrate and potassium nitrate which normally function as plant nutrients, if accumulated in the soil in excessive amounts, become toxic to plants. There are many areas of highly productive soils in arid regions which, because of good drainage, are entirely free from salinity and alkali troubles. Moreover, some of these areas probably never will be adversely influenced by the occurrence of the accumulation of excess soluble salts or of exchangeable sodium. In other areas these accumulations are the cause of wide-spread sterility and barrenness of arid-region soils. A large amount of research has been directed toward the solution of the salinity and alkali problem in arid regions. The waters of some arid-region streams and rivers contain appreciable amounts of soluble salts, the actual amounts being influenced by the soil from which the waters flow into streams and rivers. Because of the importance of salinity and alkali in arid-region soils and their relations to irrigation practices, Chapter 6 is devoted to a consideration of these topics.

3.12 SURFACE TENSION

Surface tension is due to unbalanced molecular forces. In any body of water the particles in the interior of the liquid are attracted equally in all directions by the other particles of the liquid, as illustrated by the particle at point A in Fig. 3.1. A particle on the water surface, on the contrary, is not attracted equally on all sides, since the molecules of the air surrounding the particle exert less attraction upon the water-surface particle than is exerted by the interior particles of the liquid. There is consequently a resultant inward attraction along a line perpendicular to the surface of the liquid as illustrated in points B and C of Fig. 3.1.

Since the surface tension of water causes a suction on water within the soil, the term "suction head" is being used by many researchers instead of "tension head." Suction head is generally expressed in terms of an equivalent length of vertical water column.

Figure 3.1. Illustrating the fact that surface tension is due to unbalanced molecular forces.

3.13 TENSION HEADS

The water in the capillary tube illustrated in Fig. 3.2 is held in the position shown at a height h_t above the water surface by an upward force due to surface tension in the water.
Let

F_u = the total upward force
F_d = the total downward force

Then $F_u = F_d$ because the water is at rest. For the purpose of this analysis, the angle of contact between the water and the glass tube α in Fig. 3.2 is considered zero. Then the upward force is equal to the circumference of the tube times the surface tension of the water, and the downward force is equal to the volume of water times its specific weight w. Using symbols and equations,

$$F_d = \pi r^2 h_t w = F_u = 2\pi r T$$

It follows that

$$h_t = \frac{2T}{rw} \tag{3.2}$$

The length h_t in Equation 3.2 is the height to which water with weight per unit volume of w will rise in a capillary tube of radius r as the result of the surface tension of the water T. This is here designated the tension head. Unsaturated soil pores, like the

Figure 3.2. Showing that water rises in a capillary tube, owing to surface tension, to a height well above the surface of the water in the container.

capillary tube of Fig. 3.2 develop concave water surfaces and suction forces or tensions that tend to pull water in soils from points of high moisture percentages to points of lower moisture percentages.

Multiply each side of Equation 3.2 by w and there results a negative pressure or suction p

$$p = wh_t = \frac{2T}{r} \qquad (3.3)$$

Each side of Equation 3.3 is a force intensity (F/L^2) due to tension.

Using metric units for T and w of Equation 3.2, since the surface tension of water is 75.6 dynes per centimeter and its specific weight is 980 dynes per cubic centimeter, it follows that

$$h_t = \frac{2 \times 75.6}{980r} = \frac{0.15}{r} \qquad (3.4)$$

Equation 3.4 gives the height in centimeters to which water at 4°C will rise in a capillary tube of radius r centimeters.

Applying essentially the same reasoning as underlies Equation 3.4 to ideal soils in which the capillary tubes are triangular in cross section, the maximum height of rise of water in centimeters is given approximately by the equation:

$$h_t = 0.75/r, \quad \text{and} \quad h = 1.5/d \qquad (3.5)$$

in which r is the radius and d the diameter in millimeters of the soil particles.

In soils the pores are irregular in size. Furthermore, the liquid-vapor interface, where the surface tension forces act, is not only irregular in shape but usually inclined to the horizontal. Thus, the actual heights that water can rise in a soil by capillary action is usually less than the theoretical heights computed from Equation 3.5. Under average conditions capillarity usually acts freely to 150 centimeters, frequently to 300 centimeters, and rarely to 10 meters.

Equation 3.5 shows that film tension of soil moisture increases as the radius of a capillary water film decreases. Soils of a given texture and structure having low percentages of water have, therefore, high moisture tensions, whereas the same soils having high percentages of water have low moisture tensions. These facts explain observed soil and water relationships such as, for example, the decrease in capillary water with increase in distance above a water table when at equilibrium.

3.14 SOIL-MOISTURE TENSION

The tension force of water in unsaturated soils has been described by several different expressions, such as soil-pull, the force of suction, and capillary tension.

The moisture content of the soil corresponding to a particular tension force is influenced by soil texture, soil solution, and temperature. Soil texture influences particularly the curvatures of capillary water films, and the other two factors of surface tension of the liquid; thus, all influence tensions.

As shown in the previous section, the tension head $h_t = 2T/wr$. The surface tension T and specific weight w can be measured, but it is not feasible to measure the radius r in soils; hence, h_t is measured by means of the tensiometer.

The relationships of soil-moisture content to tension for typical sandy, loam, and clay soils are presented Fig. 3.3.

3.15 SOIL-MOISTURE CONTENT

Direct methods of measuring moisture content of soils, although tedious and costly, have substantial value. The practice is to bore or drive to desired depths with a soil auger or a soil tube, to remove samples of the moist soil and place it in cans provided with covers, and to take the soil samples to a laboratory for weighing and drying. Samples of 100 or more grams of moist soils are kept in an oven having a temperature of 105° to 110°C, until the soil is free from moisture. The loss of weight in drying,

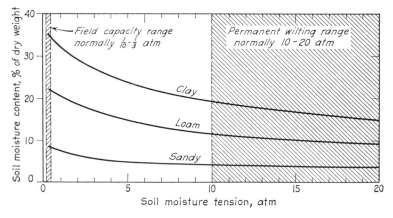

Figure 3.3. Typical curves of soil moisture variation with tension.

divided by the weight of the water-free soil, yields the moisture percentage on the dry-weight basis, represented by the symbol P_w. For example:

$$\text{Weight of moist soil} \quad = 100 \text{ g}$$
$$\text{Weight of water-free soil} = \;\; 80 \text{ g}$$
$$\text{Loss of weight in drying} = \;\; 20 \text{ g}$$

Then P_w (i.e., the moisture content on the dry-weight basis) = 100 (20/80) = 25%.

Measurements of soil-moisture content are sometimes reported on the wet-weight basis; that is, in the above example, the moisture percentage on the wet-weight basis is 20. The apparent advantage of the wet-weight basis is the simplicity of computation if 100 grams of moist soil is used. In reality, however, the wet-weight basis is irrational because the reference base for the percentage computation, i.e., the weight of a wet sample of soil, varies according to the moisture content of the soil.

Interpretations of the significance and influence of different quantities of water in soil, both in relation to water storage and to plant growth, are facilitated by converting the moisture percentage on the dry-weight basis to the volume basis. The percentage on a volume basis is defined as the volume of water per unit volume of space within the body of soil. For example, if a cube of space within the soil contains 25 percent air, 25 percent water, and 50 percent solid soil particles, the percentage of moisture on the volume basis, represented by the symbol P_v, is 25. Drying the soil in an oven, as a means of extracting the water from it, results in a loss of water in the form of vapor. It is therefore desirable to convert dry-weight basis moisture percentages P_w to volume percentages P_v. The apparent specific gravity of the soil, represented by a symbol A_s, is defined as the ratio of the weight of a given volume of soil to the weight of an equal volume of water, and:

$$P_v = P_w A_s \tag{3.6}$$

Moisture-content of soil can also be represented as a depth d obtained by multiplying the percentage volume P_v by the depth of soil D:

$$d = \frac{P_v}{100} D \tag{3.7}$$

or combining Equations 3.6 and 3.7.

$$d = \frac{P_w}{100} A_s D \qquad (3.8)$$

Since A_s and P_w are dimensionless quantities, the dimension of D will also be the dimension of d. Hence, if the depth of the soil D is in centimeters, then d also will be in centimeters.

3.16 CLASSES AND AVAILABILITY OF SOIL WATER

Soil water has been classified as hygroscopic, capillary, and gravitational. Hygroscopic water is on the surface of the soil grains and is not capable of significant movement by the action of gravity or capillary forces. Capillary water is that part in excess of the hygroscopic water which exists in the pore space of the soil and is retained against the force of gravity in a soil that permits unobstructed drainage. Gravitational water is that part in excess of hygroscopic and capillary water which will readily move out of the soil if favorable drainage is provided. There is no precise boundary or line of demarcation between these three classes of soil water. The proportion of each class depends on soil texture, structure, organic matter content, temperature, and depth of soil column considered.

Water may also be classified as unavailable, available, and gravitational or superfluous. Such a grouping refers to the availablility of soil water to plants. Gravitational water drains quickly from the root zone under normal drainage conditions. Unavailable water is held too tightly by capillary forces and is generally not accessible to plant roots. Available water is the difference between gravitational and unavailable water.

Water drains from the soil under the constant pull of gravity. Sandy soils drain readily, while clay soils drain very slowly. Hence, one day after irrigating a sandy soil most of the gravitational water has drained out of the soil, whereas clay may require 4 or more days for gravitational water to drain. The rate of drainage is most rapid immediately after irrigating and decreases constantly; nevertheless it continues to drain at a relatively slow rate, even after gravitational water has been removed. On the average, 2 days are required before the rate of drainage decreases rather sharply and gravitational water has been removed from the root zone. This implies, of course, that no restricting layers are in the root zone to impede the downward flow of water.

Relationships between the various terms used in identifying classes and availability of soil water are shown in Fig. 3.4.

Figure 3.4. Classes of soil-water availability to plants and drainage characteristics.

Field Capacity

When gravitational water has been removed, the moisture content of soil is called field capacity. Field capacity cannot be determined accurately because there is no discontinuity in the curve of moisture content versus time. Nevertheless, the concept of field capacity is extremely useful in arriving at the amount of water available in the soil for plant use. Most of the gravitational water drains through the soil before it can be used consumptively by plants.

In practice, field capacity is usually determined 2 days after an irrigation. Therefore, field capacity defines a specific point on the moisture-content time curve. Specifying the time also makes it possible to calculate the water used consumptively by plants while gravitational water is draining from the soil. A soil will come to field capacity more quickly when an active crop is growing than when there are no roots removing water from the soil. Special precaution should be taken not to overlook the amount of water consumptively used by the crop between the time of irrigation and the time at which field capacity is determined. The amount of water thus used when light irrigations are given is very important, and the principle should never be overlooked in irrigation practice.

Field capacity can be measured by determining moisture content of soil after an irrigation which is sufficiently heavy to ensure thorough wetting of the soil to be tested. Observing the decrease in moisture by making moisture determinations at different times after irrigation is valuable in understanding and properly interpreting the field capacity characteristics of soil. However, soils must be well drained before reliable field determinations can be made in this manner. Restricting layers of silt and clay as well as a high water table will impede drainage and give erroneous indications of field capacity.

Since field capacity by definition is a direct function of time, the importance of time in the determination and use of this concept cannot be overemphasized. For example, field capacity measured 2 days after an irrigation cannot be used without modification to estimate the water remaining in the soil in the spring following fall irrigation. Furthermore, the value of field capacity determined when a crop is growing on the land will be somewhat different from the value when the land is bare. The effect of excessive surface evaporation versus little or no surface evaporation will also have its influence. Despite these limitations the concept of field capacity is extremely useful for estimating the volume of water retained in the soil. Precise determinations of field capacity are generally not necessary for field applications.

Soil moisture tension is normally between $1/10$ and $1/3$ atm when the soil is at field capacity. The correct value depends upon the drainage characteristics of the soil and the time after irrigation at which the soil is assumed to reach field capacity. Sandy soils tend to be near $1/10$ atm at field capacity, while clays tend toward $1/3$ atm. Some soils have gone as high as 0.6 atm. For most agricultural soils, a tension of $1/10$ corresponds more closely than does $1/3$ atm to the generally accepted values of field capacity determined by moisture content. The large difference in moisture content and volume of water in the soil between $1/10$ and $1/3$ atm makes it important to define the tension correctly at which field capacity occurs. Considering field capacity to occur at $1/3$ atm tension when it is $1/10$ of an atm leads to gross errors in estimating the volume of water available to plants following irrigation.

Permanent Wilting Point

The soil-moisture content when plants permanently wilt is called the permanent wilting point or the wilting coefficient. The permanent wilting point is at the lower end of the available moisture range. A plant will wilt when it is no longer able to extract sufficient moisture from the soil to meet its water needs. Temporary wilting will occur in many crops on a hot windy day, but the plants recover in the cooler portion of the day. Permanent wilting, as well as temporary wilting, depends upon the rate of water used by the plant, the depth of the root zone, and the water holding capacity of the soil. Permanent wilting will occur at a higher moisture content in a hot climate than in a cool climate. A plant is considered to be permanently wilted when it will not recover after being placed in a saturated atmosphere where little or no consumptive water use occurs.

Field estimates of wilting point can often be made by determining the moisture content of soils in which plants have permanently wilted. This method is subject to more error and requires more judgment than field determinations of field capacity. Allowance must be made for depth and nature of rooting. Furthermore, it is often difficult to find plants in the condition of permanent wilt. For a plant to reach permanent wilting following an irrigation, provided that the plant is using considerable water, will require 1 week in sands to perhaps 4 weeks in clays, and even longer if the plant is deeply rooted.

The tension at which permanent wilting occurs can vary from 7 to as high as 40 atm, depending upon rate of consumptive use, crop, salt content of soil, and soil texture. As the temperature and rates of consumptive use increase, permanent wilting will occur at significantly lower tensions and higher moisture contents. The tension in the soil moisture when the soil is at permanent wilting is generally considered to be 15 atm. Whether in reality it is 10 or 20 atm makes very little difference, since the change of moisture is slight with rather large changes of moisture tension.

As an approximation, the permanent wilting percentage can be estimated by dividing the field capacity by a factor varying from 2.0 to 2.4, depending upon the amount of silt in the soil. For soils of high silt content 2.4 should be used.

Available Moisture

The difference in moisture content of the soil between field capacity and permanent wilting is termed the available moisture. This represents the moisture which can be stored in the soil for subsequent use by plants. Available moisture can be expressed as percentage moisture P_w, as percentage volume P_v, or as depth d, whichever is most convenient.

Often it becomes desirable to refer to the amount of available moisture remaining in or extracted from the soil. This can also be expressed by P_w, P_v, or d, when it is desired to know the quantity of water in the soil. However, when correlating the moisture condition of the soil to crop response, the amount of moisture remaining, or extracted, can be represented as a percentage of the available moisture to obtain a more significant expression of moisture condition in the soil.

Readily Available Moisture

Soil moisture content near the wilting point is not readily available to the plant. Hence, the term readily available moisture has been used to refer to that portion of the available moisture that is most easily extracted by plants, approximately 75 percent of the available moisture.

3.17 FILLING THE AVAILABLE SOIL WATER RESERVOIR

Having determined the amount of water to be added to the soil, the irrigator needs to know how long it will be necessary to apply a given stream of water. The relationship between size of stream, time of application, area to be irrigated, and depth of water to be applied is as follows:

$$qt = 28ad \qquad (3.9)$$

where q = size of stream in l/s
  t = time in hours required to irrigate the area
  a = area irrigated, in ha
  d = depth in cm that the volume of water used would cover the land irrigated, if quickly spread uniformly over its surface

also

$$qt = ad \qquad (3.10)$$

where q is in cubic meters, t in hours, a in ha and d in meters.

When Equation 3.9 is combined with Equation 3.8 it is apparent that:

$$t = \frac{28P_wA_sDa}{100q} \qquad (3.11)$$

where q is in liters per second, t in hours, a in hectares, and d in centimeters.

Also

$$t = \frac{P_wA_sDa}{100q} \qquad (3.12)$$

where q is in cubic meters, t in hours, a in hectares, and d in meters.

Provided the apparent specific gravity A_s is known, it is possible to compute the hours required to add a given moisture percentage P_w to a field of given area a and a soil of certain depth D when using a stream of water of q.

Equation 3.12 has considerable practical utility. For example, an irrigator has at his disposal a stream of 300 cubic meters per hour and he wants to apply enough water to increase the 120-centimeters depth root-zone soil moisture from 10 percent to 15 percent. How many hours will be required if the water is spread uniformly and losses are neglected? Using Equation 3.12 a time of 2.34 hours is required to supply enough water with a 300 cubic meters per hour stream to add 5 percent moisture to 1 hectare-meter of soil. Therefore, in 2.34 × 1.2 = 2.8 hr, enough water is applied to add 5 percent moisture to 120 centimeters of soil, and longer application is likely to result in deep percolation loss provided 5 percent satisfies the field capillary capacity for usuable or available water or fills the capillary reservoir.

3.18 REPRESENTATIVE PHYSICAL PROPERTIES

Normal ranges in average values of several physical properties of irrigated soils described in the preceding section are summarized in Table 3.1. Extremes can and will occur. However, Table 3.1 will be applicable to most areas. It will be particularly helpful to estimate physical properties, and to get a clear understanding of the variation that occurs in the physical properties of normal irrigated soils.

TABLE 3.1 Representative Physical Properties of Soils

Soil Texture	Infiltration and Permeability, I_f (cm/hr)	Total Pore Space (%) N	Apparent Specific Gravity A_s	Field Capacity (%) FC	Permanent Wilting (%) PW	Total Available Moisture[b]		
						Dry Weight (%) $P_w = FC - PW$	Volume (%) $P_v = P_w A_s$	cm/m $d = \dfrac{P_w}{100} A_s D$
Sandy	5 (2.5–25)	38 (32–42)	1.65 (1.55–1.80)	6 (6–12)	4 (2–6)	5 (4–6)	8 (6–10)	8 (7–10)
Sandy loam	2.5 (1.3–7.6)	43 (40–47)	1.50 (1.40–1.60)	14 (10–18)	6 (4–8)	8 (6–10)	12 (9–15)	12 (9–15)
Loam	1.3 (0.8–2.0)	47 (43–49)	1.40 (1.35–1.50)	22 (18–26)	10 (8–12)	12 (10–14)	17 (14–20)	17 (14–19)
Clay loam	0.8 (0.25–1.5)	49 (47–51)	1.35 (1.30–1.40)	27 (23–31)	13 (11–15)	14 (12–16)	19 (16–22)	19 (17–22)
Silty clay	.25 (.03–0.5)	51 (49–53)	1.30 (1.30–1.40)	31 (27–35)	15 (13–17)	16 (14–18)	21 (18–23)	21 (18–23)
Clay	0.5 (.01–0.1)	53 (51–55)	1.25 (1.20–1.30)	35 (31–39)	17 (15–19)	18 (16–20)	23 (20–25)	23 (20–25)

Note: Normal ranges are shown in parentheses.

[a] Intake rates vary greatly with soil structure and structural stability, even beyond the normal ranges shown above.

[b] Readily available moisture is approximately 75% of the total available moisture.

CHAPTER
4
MEASUREMENT OF SOIL MOISTURE

Measurement of water stored in soils and capacity of soils to store water are important in humid and arid regions. That some humid-climate soils produce crops despite the lapse of many days, and sometimes weeks, between periods of rainfall is evidence of their capacity to store available water, since all growing plants require water continuously. In irrigated regions the capacity of soils to store available water for the use of growing crops is of special importance and interest, because the depth of water to apply in each irrigation and the interval between irrigations are both influenced by storage capacity of the soil. Irrigated soils having large water-storage capacity may produce profitable crops in places where, and at times when, the shortage of irrigation water makes it impossible to irrigate as frequently as would be desirable.

Knowledge of the capacity of soils to retain available irrigation water is also essential for efficient irrigation. If the irrigator applies more water than the root-zone soil reservoir can retain at a single irrigation, the excess is wasted. If he applies less than the soil will retain, the plants may wilt from lack of water before the next irrigation unless water is applied more frequently than otherwise would be necessary. Water losses which result from deep percolation below the root zone of crops cannot be seen. They can be measured approximately by subtracting the storage capacity of the various soils from the amount of water applied in single irrigations, less the runoff.

Determining the moisture in unsaturated soils that is available to plants is of major importance and is therefore given special attention here. In localities where the late-season water supply is always low and surface-storage reservoirs are impracticable or economically prohibitive, storage of water in saturated soils in gravitational form is sometimes practical and helpful. Excessive application of water for the purpose of water storage may cause the ground water to rise and become injurious to crops during the early part of the season.

It is important to find the available water capacity for different soils; i.e., the field capacity less the moisture content at the permanent wilting point. Some soils having high field capacities also have high wilting points, thus making the available water capacity rather low and requiring frequent irrigation. The amount of available water that can be stored increases with root-zone depth. Methods of estimating and measuring depths and volumes of water that may be stored and made available to crops are considered in this chapter.

4.1 DRILLING HOLES AND OBTAINING SOIL SAMPLES

Soil Auger

It is highly desirable that irrigation farmers observe by inspection, and sometimes by measurement, the quantities of moisture in their soils. Boring or drilling deeply into soils to determine soil moisture conditions is essential to profitable farm operation.

Two types of soil augers are used: a spiral-shaped bit made from carpenter's standard wood-auger bit, usually about 3 centimeters in diameter, and a post-hole auger, usually 5 to 10 centimeters in diameter, made especially for boring into the soil.

The spiral-bit auger is shown entering the soil surface in Fig. 4.1(a), and soil being removed in Fig. 4.1(b). The screw-point and the side-cutting edges of a carpenter's auger are removed in preparing the bit for soil boring. The spiral auger is light in weight and convenient to carry around. When using it in compact soils, care is necessary to avoid boring too deeply at one time, thus causing the auger to lodge in the soil. The post-hole auger, illustrated in Fig. 4.2, does not involve this danger, because when it

Figure 4.1. A soil auger helps to show when water is needed, and the depth of water penetration from irrigation.

Figure 4.2. Post-hole auger. Augers of this type are often used for studying the distribution of irrigation water in soil. (Courtesy Iwan Brothers.)

is filled with soil it does not easily advance further. However, because of the open sides of both augers shown in Figs. 4.1 and 4.2, soil obtained at a particular depth is contaminated somewhat with other soil as the auger is removed from the hole. For normal farm work this is not serious, but for research activity wherein soil moisture is to be determined, better results are obtained by use of a post-hole auger having the sides enclosed.

Short tubes, about 30 centimeters in length, have been designed so that (1) the tube can be pushed into the soil with a minimum of effort, (2) the soil will enter the tube readily, and (3) the tube can be extracted easily from the soil. Part of the tube is cut out to inspect and remove the soil better. A soil probe of this type together with a screw type auger are shown in Fig. 4.3. Under ideal conditions the probe can be used to obtain cores of undisturbed soil from which apparent specific gravity of the soil can be determined.

Powered augers, tubes, and probes have been used to secure soil samples faster and to conserve manpower. (See Fig. 4.4) Usually these units can sample soil to a depth of at least 3 meters.

4.2 RESISTANCE OF SOIL TO PENETRATION

Irrigators have long been aware that the wetter the soil, the deeper one sinks into the mud. This observation has been used to indicate how well the soil is irrigated. True, it is only an index, but the man who knows his land soon learns to interpret the tendency to penetrate wet soil in terms of the adequacy of the irrigation.

Inserting a shovel into the soil gives a better indication of soil moisture than the tendency to sink into the wet ground. A still better tool is to push a steel rod, about 1 centimeter in diameter into the soil. In many soils the depth that the rod can be pushed into the soil will be a good indication of the depth of wetting. Welding a crossbar on the top of the rod will facilitate insertion and removal.

Figure 4.3. Details of a soil probe with part of the tube cut away for easier insertion, inspection, and removal. A screw-type auger is also shown. (Courtesy Oakfield Apparatus Co.)

4.3 APPEARANCE AND FEEL OF SOIL INDICATE MOISTURE CONTENT

One of the oldest and most widely used methods of estimating soil moisture content is to look at the soil and to feel it. Using the soil auger, samples of soil usually can be obtained readily throughout the root zone of the soil. Table 4.1 can be used as a guide for judging how much available moisture has been removed from the soil and consequently, how much must be added during the irrigation period. Reasonably good estimates of irrigation needs can be obtained by using Table 4.1. In many applications, greater accuracy is not needed, nor is it justified economically. However, as technology improves, more precise and more usable tools become available. As the prices paid for land and its products increase, the feasibility of using more accurate methods of determining moisture content of soil also increases.

Figure 4.4. Powered soil sampler capable of obtaining 10-cm undisturbed core to a depth of at least 3 m. (Courtesy Giddings Machine Co.)

TABLE 4.1 Guide for Judging How Much of the Available Moisture Has Been Removed from the Soil

Soil Moisture Deficiency (%)	Feel or Appearance of Soil and Moisture Deficiency in cm of Water per m of Soil			
	Coarse Texture	Moderately Coarse Texture	Medium Texture	Fine and Very Fine Texture
0 (Field capacity)	Upon squeezing, no free water appears on soil but wet outline of ball is left on hand. 0.0	Upon squeezing, no free water appears on soil but wet outline of ball is left on hand. 0.0	Upon squeezing, no free water appears on soil but wet outline of ball is left on hand. 0.0	Upon squeezing, no free water appears on soil but wet outline of ball is left on hand. 0.0
0–25	Tends to stick together slightly, sometimes forms a very weak ball under pressure. 0.0 to 1.7	Forms weak ball, breaks easily, will not slick. 0.0 to 3.4	Forms a ball, is very pliable, slicks readily if relatively high in clay. 0.0 to 4.2	Easily ribbons out between fingers, has slick feeling. 0.0 to 5.0
25–50	Appears to be dry, will not form a ball with pressure. 1.7 to 4.2	Tends to ball under pressure but seldom holds together. 3.4 to 6.7	Forms a ball somewhat plastic, will sometimes slick slightly with pressure. 4.2 to 8.3	Forms a ball, ribbons out between thumb and forefinger. 5.0 to 10.0
50–75	Appears to be dry, will not form a ball with pressure.[a] 4.2 to 6.7	Appears to be dry, will not form a ball.[a] 6.7 to 10.0	Somewhat crumbly but holds together from pressure. 8.3 to 12.5	Somewhat pliable, will ball under pressure.[a] 10.0 to 15.8
75–100 (100% is permanent wilting)	Dry, loose, single-grained, flows through fingers. 6.7 to 8.3	Dry, loose, flows through fingers. 10.0 to 12.5	Powdery, dry, sometimes slightly crusted but easily broken down into powdery condition. 12.5 to 16.7	Hard, baked, cracked, sometimes has loose crumbs on surface. 15.8 to 20.8

[a] Ball is formed by squeezing a handful of soil very firmly.

4.4 GRAVIMETRIC DETERMINATION OF MOISTURE

Determining soil moisture by weighing the soil before and after drying in an oven has been described in the previous chapter under the heading "Soil Moisture Content." The method is for primary measurement, chiefly limited in its usefulness by the time required to collect samples and dry them in an oven. Usually 24 hours are required for drying. Samples of about 200 grams are most often used, so that they will dry in a reasonable time and with reasonable uniformity. Such small samples naturally may not be, and usually are not, representative of large areas. Hence, several samples must be used to obtain a satisfactorily representative indication of moisture content. To save time and money, infrared heat has been used to dry the soil at a much more rapid rate than is done in a conventional soil oven. However, care must be exercised not to overdry the soil and burn out the organic matter.

Gravimetric absorption blocks have been used to a limited extent. A porous block is placed in a prepared hole in contact with the soil at a desired depth. Moisture moves into or out of the block until the soil and block have essentially equal amounts of moisture. The block is removed periodically from the hole and weighed to determine moisture content.

The principal disadvantages of the absorption-block method are the lag in approaching equilibrium, the high sensitivity required in weighing, and the accompanying fragile nature of the weighing device. Preparing the holes so that blocks can be placed in close contact with the soil and yet be readily removable for weighing is time-consuming and costly.

4.5 USING ELECTRICAL PROPERTIES OF A POROUS BLOCK

Electrical properties of resistance (or conductance), capacitance, and dielectric strength have been used to indicate moisture content. Changes of moisture affect all of these electrical properties. Porous blocks containing desired electrical elements are placed into the soil. As the moisture content of the blocks change, the electrical properties change. Only limited success has been achieved using capacitance or dielectric properties; but practical units have been developed using the resistance or conductance principle.

Many designs have been used successfully and others are now being developed. Not only have the shapes of the blocks been altered, but the size, shape, and configuration of the electrodes have also been varied. In general, nylon and fiber-glass units are more sensitive than gypsum blocks at high moisture content and low tension conditions. Nylon units are most sensitive at tensions less than 2 atm, and gypsum blocks operate best at tensions between 1 to 15 atm. The gypsum blocks are soluble and deteriorate in one to three seasons of use. However, gypsum is less sensitive to soil salts than nylon and fiber glass because of the concentration of soluble gypsum within the water of a gypsum block. Normally there is considerable variation between blocks and considerable change occurs in the calibration during the season, especially in gypsum blocks.

Resistance blocks and small transistorized meters for reading electrical properties are available commercially and are being used in many areas for moisture control. Some of the outstanding examples of their use are found on some of the large sugarcane plantations in Hawaii, where field crews spend their full time reading the blocks and scheduling irrigations. Figure 4.5 shows some typical units with a portable meter. These units are generally installed by digging a hole to the desired depth, placing a unit on its side, backfilling, and tamping the soil in place. In this manner several units can be placed in a single hole. To increase ease of placement and to facilitate removal at the end of the season, individual units have been cast into single rods and the rods inserted into the soil as a unit.

4.6 TENSIOMETERS

A tensiometer consists of a porous cup filled with water and attached to a vacuum gauge or mercury manometer. A hole is bored or dug in the soil to a desired depth, a handful of loose soil is placed into the hole, and the cup pushed firmly into the soil. Additional soil is packed around the cup and around the tube wherever necessary to ensure firm contact with the soil. A temporary connection is soon established between the water inside the cup and the water in the soil outside. As water moves out of the cup because of the suction or tension existing in the soil water, the vacuum created in the cup is registered on the gauge. Conversely, an increase of water in the soil will lower the tension, water will move into the cup, and the gauge will read less tension. Fluctuations in soil moisture are registered by the tensiometer, as long as the tension does not exceed about 0.8 atm. At greater degrees of tension, air enters the closed system through the pores of the cup and the instrument is no longer accurate. The soil moisture tension must be lowered by irrigation or

Figure 4.5 Portable meter and resistance blocks used to measure soil moisture. (Courtesy Industrial Instrument, Inc.)

rainfall and the system filled with water before it will again operate properly.

Considerable progress has been made in the technology of tensiometer construction since the first units were made by Drs. L. A. Richards and Willard Gardner at Utah State University. Units that perform very well, such as those shown in Fig. 4.6, are now available commercially.

Since these units operate satisfactorily only up to tensions of 0.8 atm, they are most useful in sandy soils, where this represents a major portion of the available water, or for crops that must be irrigated frequently. How tensiometers can be used to predict the time of less frequent irrigations is illustrated in the following example: If it takes 3 days following an irrigation for the tensiometer to become inoperative, it may be desirable when growing a particular crop to irrigate at an interval two times this period, or on a 6-day interval for sandy soil or perhaps three times this period, or on a 9-day interval for a silty loam soil. In this case the tensiometer is used to indicate the rate of depletion of soil moisture, but cannot be used for the full moisture range.

Figure 4.6. Tensiometer in a recently harvested field of sugarcane measuring moisture at five depths. (Courtesy Hawaiian Commercial and Sugar Co.)

Because of its narrow range the tensiometer is used for moist and resistance blocks for dryer soil conditions. Sometimes the two instruments are used together, since the blocks become sensitive at about the same degree of moisture content that the tensiometer ceases to function. With both tensiometers and blocks it is desirable to place units within the zone of major root activity so that they will be most sensitive to changes in soil moisture. Yet the moisture available at deeper depths in the soil should not be overlooked when the instrument is placed at shallower depths.

4.7 NEUTRON METHOD OF MEASURING SOIL MOISTURE

Considerable potentialities exist in the neutron method of measuring soil moisture shown in Fig. 4.7. In this method fast neutrons are emitted from a source into the surrounding soil. The fast neutrons are slowed down by the water. The resulting slow neutrons which reach the counting tube are recorded. Fast neutrons are not registered by the counter. The greater the water content of the soil, the greater the number of slow neutrons reaching the counting tube. Hydrogen is the principal element which absorbs fast neutrons. Since water is the principal source of hydrogen in the soil, there exists a very good correlation be-

Figure 4.7. Soil moisture probe. (Courtesy of Kern Stuttler and Tom Fullerton.)

tween the amount of water in the soil and the number of slow neutrons reaching the counter. In practice, a hole is dug with an auger, and a metal tube is driven into the hole to retain the soil. The neutron source and counting device are placed within the hole and lowered to the desired depth. The reading obtained in a unit of time is proportional to the moisture content of the soil surrounding the source and counter. Sufficient holes should be dug to obtain a representative indication of soil moisture.

Some minerals, as well as boron and chlorine, also absorb fast neutrons and thus influence the reading. Boundary geometry also has an influence on the readings. The number of slow neutrons recorded in a given time while the probe is near the surface will be different from a reading obtained at the same moisture content well beneath the surface, because of the geometrical configuration of the soil surrounding the probe. The zone of influence decreases as moisture content increases. Improvements are being made in reducing weight, making the units more dependable, and reducing the radiation exposure hazard. Since readings can be taken as soon as the instrument is placed in a hole, lag time is negligible, resulting in a fast method of measurement. Also repeated nondestructive sampling in the same location is feasible. The neutron method holds considerable promise for an effective practical method of measuring soil moisture.

4.8 USING THERMAL PROPERTIES

Heat conductivity of the soil can be used as an index of soil moisture, since the rate of conductivity depends upon moisture present in the soil. The thermistor, a thermally sensitive resistor, has given the best control, because of the influence of temperature on heat conductivity. A permanent porous material with proper pore size distribution must be developed, into which a thermally sensitive device can be embedded without damaging the heating and sensing element.

4.9 SAMPLING ERROR

Error of sampling has long plagued the investigator as he seeks to determine the amount of moisture in the soil. Obtaining representative samples is a major problem. Uneven growing of plants and nonuniform root penetration must be considered, since they cause variations of moisture content. Texture and structure variations of soils alter the intake, transmission, and retention of moisture. Variations in land-surface configuration change the opportunity for intake of rainfall, and furrowed shape and size alter the rate of intake of irrigation water. All of these factors cause the moisture content to vary from point to point in a field. To obtain a representative soil moisture sample requires that several samples be taken, unless the method of determining soil moisture inherently integrates a large volume of soil. The number of samples required to obtain a representative sample increases as the moisture variation increases.

Another complexity is that essentially all methods of moisture determination are based upon small samples. Individual samples can be expected to vary at least plus or minus 20 percent from the mean of a large number of samples. Concentrated research is needed to develop practical and faster methods of moisture determination which will give more representative values of moisture content of the soil.

FLOW OF WATER INTO AND THROUGH SOILS

Water characteristics considered in Chapters 3 and 4 dealt mainly with the capacity of the soil to hold water. Practical estimates must be made of the amount of water available in the soil for plant use following an irrigation. Out of this necessity for a volumetric expression of available water have grown the useful concepts of field capacity, permanent wilting, and resulting water-holding capacities discussed in the previous chapters. In reality, as was pointed out, these are not static quantities but are only points on a continuous dynamic curve of moisture content versus time.

The dynamic aspects are now considered in more detail as they affect the movement of water into, through, and out of the soil. The phenomenon of water entry is first considered and then the movement through the soil follows. Finally, consideration is given to unsaturated flow and vapor movement within the soil. However, as a basis for the subsequent discussions, the basic laws and equations governing flow are presented first.

5.1 ENERGY IN FLOWING WATER

The principles of mechanical work and energy are applied in the derivation of fundamental formulas for the flow of fluids, including the flow of water in soils. Energy is defined as capacity to do work.

In fluids, energy may be in two forms:

1. Kinetic energy.

2. Potential energy.
   a. Energy resulting from pressure differences.
   b. Energy resulting from elevation differences.

A gram of water flowing at a velocity of v centimeters per second has a kinetic energy of $v^2/2g$ grams-centimeters per gram where g is the acceleration of gravity.

A gram of water having potential energy has the energy only because of the existence of a difference of pressure or a difference of elevation. The potential energy of a gram of water resulting from a pressure difference can be represented by the ratio p/w where p is the pressure per unit area and w is the weight per unit volume. A gram of fluid having a potential energy due to elevation can be represented by y, representing elevation above some datum.

Actually, a gram of water may have both kinetic and potential energy, where the potential energy is composed of both pressure and elevation. The combined energy H can be represented by the following form of the widely used Bernoulli equation showing energy per unit mass:

$$H = \frac{V^2}{2g} + \frac{p}{w} + y \qquad (5.1)$$

Energy per unit mass, having the dimension of length, is used extensively in engineering practice because of its simplicity, especially when applied to open-channel flow and flow through soils. When applying this equation to flow through soils, a simplification can readily be made. Since the velocity through soils is very small, the kinetic energy $(V^2/2g)$ is even smaller and can usually be ignored. Hence, Bernoulli's equation applied to flow through soils reduces to:

$$h = \frac{p}{w} + y \qquad (5.2)$$

which will be used here as the piezometric head. Piezometric head is also frequently referred to as hydraulic head, the two terms being synonymous. This quantity is measured in saturated soil by the simple piezometer and in unsaturated soil by a tensiometer.

Figure 5.1 illustrates flow of water in three directions and the use of piezometers to measure the direction of flow and the magnitude of the energy. Note that flow always occurs in the direction of decreasing piezometric head.

The Darcy equation relates velocity to head loss as:

$$V = k \frac{h_L}{L} \qquad (5.3)$$

As long as the flow is not turbulent, the velocity V and flow through saturated soils is proportional to the soil permeability k and the slope of the piezometric headline h_L/L. The soil permeability k has the dimension of length per unit time L/t, and is a

Figure 5.1. Illustrating horizontal, upward, and downward flow through soil and the use of piezometers to indicate direction of flow and the magnitude of the piezometric gradient.

combined measure of soil and fluid properties. The fluid properties can be made explicit using the following equation:

$$V = \left(\frac{w}{\mu} \, k'\right) \frac{h_L}{L} \tag{5.4}$$

The permeability k' has the physical dimensions of area L^2. When defined, as in Equation 5.4, the permeability k' is influenced only by the size and shape of the soil particles and pores (soil texture and structure) and is independent of the fluid properties: (specific weight w, and viscosity μ). For most studies of the flow of ground water in irrigation and drainage, the influence of specific weight and viscosity is relatively small; hence, explicit inclusion of w and μ as in Equation 5.4 is not essential for general irrigation studies. The k in Equation 5.3, used hereafter, is equal to $(w/\mu) \, k'$ of Equation 5.4.

Using the value of V from Equation 5.3 in the basic rational equation for quantity of flow, $Q = AV$ where Q is volume per unit of time, A is cross-sectional area, and V is velocity perpendicular to A, it follows that

$$Q = Ak \frac{h_L}{L} \qquad (5.5)$$

in which A is the gross area at right angles to the flow direction. The pressure intensity at any point is

$$p = wh$$

and the pressure head

$$h = p/w$$

Let the hydraulic head at point 1 be h_1 and at point 2 be h_2; then:

$$h_1 = \frac{p_1}{w} + y_1$$

$$h_2 = \frac{p_2}{w} + y_2$$

Assume that h_1 is greater than h_2 and that the two points are a distance L apart. Then the hydraulic slope becomes:

$$\frac{H_L}{L} = \frac{h_1 - h_2}{L} = \frac{[(p_1/w) + y_1] - [p_2/w + y_2]}{L} \qquad (5.6)$$

In Equation 5.6, w is the specific weight of the water and p_1/w and p_2/w are the pressure heads; y_1 and y_2 are the elevation heads with respect to a selected datum plane.

Applications of Equations 5.3, 5.5, and 5.6 for ground-water flow in saturated soils are illustrated by two examples of field conditions: the first for flow of unconfined or free ground water in sand under a small piezometric slope, Fig. 5.2, and the second for upward flow of water through a 16-meter stratum of clay over an artesian aquifer of gravel in which the water is under pressure or confined, Fig. 5.3.

Figure 5.2 illustrates unconfined ground water flowing through sandy soil overlying a compact clay. The piezometer at A shows a piezometric head, $h_1 = (p_1/w) + y = 50$ m, and at B, $h_2 = 40$ m. Therefore, as the flow distance is 100 meters, the slope of the piezometric line, by Equation 5.2, is 10/100 and the velocity is $(1/10)k$. Selecting an average k of 350 meters per year, the approximate velocity of flow through the sand is $V = 120$ m/yr, and the quantity of flow through a section 1000 meters wide and 20 meters deep, by Equation 5.5, is

$$Q = Ak \frac{h_L}{L} = (1000 \times 20)350 \frac{10}{100} = 700,000 \text{ m}^3/\text{yr} = 22 \text{ l/s}$$

Figure 5.2. Flow of unconfined ground water in saturated sand overlying a compact clay.

For a second example, the results of piezometer measurements at different elevations in a clay soil overlying an artesian gravel aquifer are presented in Fig. 5.3. The permeability of this clay soil has been measured and found to average 2 meters per year. Then, to find the average velocity of upward flow from the 16-meter soil depth up to the 2-meter depth, in meters per year, use Equations 5.2 and 5.5:

$$\frac{h_L}{L} = \frac{h_1 - h_2}{L} = \frac{7 - 1}{16 - 2} = \frac{6}{14} = 0.43$$

$$V = k\,\frac{h_L}{L} = 2 \times 0.43 = 0.86 \text{ m/yr}$$

Note that the velocity of 0.86 meters per year is the gross apparent velocity since the cross-sectional area A was a gross area rather than the net cross-sectional area of the pores.

If the piezometric slope and the soil permeability, as measured at the place represented by the data of Fig. 5.3, were the same for 1 hectare of land then the annual movement of water from the artesian aquifer, due to upward flow through the clay soil and surface consumptive use, would be $(1 \times 0.86) = 0.86$ ha-m. The values of pressure head and elevation head in this example are presented to show the application of the Bernoulli and Darcy equations.

Figure 5.3. Showing the average pressure head, p/w, in a clay stratum overlying an artesian aquifer.

5.2 PIEZOMETERS TO MEASURE PRESSURE HEADS IN SATURATED SOILS

The pressure heads p/w illustrated in Figs. 5.2 and 5.3 are measured by driving a small-diameter pipe, designated piezometer, to the desired depth. A much used driving hammer and driving head for small diameter pipe is shown in Fig. 5.4.

When the piezometers are driven, a plug of soil forms in the lower end of the pipe. This plug is usually removed by flushing, Fig. 5.5, which consists of pumping water down to the bottom of the piezometer through plastic tubing. This water flows up through the annular space between the tubing and wall of the

Figure 5.4. Detail of driving hammer and piezometer.

pipe and carries with it soil in suspension. Flushing is continued until a small cavity has been formed below the end of the pipe and the water becomes clear. The length of the piezometer is marked on the plastic tubing, and care is taken not to push the tubing more than 5 to 10 centimeters below the end of the piezometer.

In soils of average permeability, water in the piezometer reaches an equilibrium level in a few minutes, but in soils of low permeability several hours may be required for water to reach an equilibrium level.

In highly compacted soils, and in those containing fine gravel that cannot pass upward through the annular space during the flushing operation, a rivet is inserted in the lower end of the piezometer before it is driven. The pipe is then driven to a point about 5 centimeters below the desired level and pulled up about 5 centimeters so that the rivet can easily be punched out with a rod before flushing. This eliminates difficulty sometimes encountered in removing the soil plug formed in the end of the pipe. In most soils, however, this plug can be removed by flushing in less time than is required to punch out the rivet.

The larger the pipe, the larger and more expensive the installation equipment and the longer the reaction time needed for the water level in the piezometer to come to equilibrium. Jetting equipment is also used to install piezometers, especially for the larger sizes. A cavity is formed by the jet ahead of the pipe and the pipe is forced into the soil. Such a technique permits installation of piezometers to deeper depths in heavy clay soils, but this method does have the liability that a good seal is not always obtained around the outside of the piezometer. Hence, in stratified soils having considerable vertical flow under sizable pressures, driven piezometers are generally more reliable than are jetted piezometers.

Reading piezometers with an electric sounder is shown in Fig. 5.6. Electric sounders generally consist of a weighted insulated probe, the tip of which is a conductor connected to the pipe. With a battery in the circuit, a microampmeter will register a current when the tip of the probe touches the surface of the water. The depth at which this occurs is noted on the marked conductor leading from the probe to the surface. Piezometers can also be read by blowing on a plastic-tubing air line. The depth of the end of the tube can be adjusted until the end just touches the water surface and the depth recorded.

Figure 5.5. Flushing the piezometer. (Courtesy U.S. Regional Salinity Lab.)

Figure 5.6. Reading the piezometer with an electrical sounder gauge. (Courtesy U.S. Regional Salinity Lab.)

5.3 SOIL-PERMEABILITY MEASUREMENT

The permeability of saturated soils varies greatly. In irrigation and drainage studies, permeability is the dominant variable, some soil being as much as 100,000 times as permeable as others.

Knowledge of soil permeabilities is essential to progress in studies of water conveyance and water application efficiencies, and in the design of drainage systems for the reclamation of saline and alkali soils.

Permeabilities are influenced by the size and shape of pore spaces through which water flows and by the specific weight and viscosity of the soil water, as shown in Equation 5.4. For ordinary irrigation applications, it is impractical to measure all of the factors that influence permeability, but it is practical and very essential to measure permeability of soils in the laboratory and in the field.

Two of the many types of equipment for measuring permeability are the constant-head and variable-head permeameters.

Constant-Head Permeameter

With a constant head maintained by either continuous inflow or frequent additions of water, steady flow through the soil is obtained. Figure 5.7 illustrates two constant-head permeameters, one for laboratory tests and the other for field studies. Darcy's law for flow of water in soils is applied for computing permeability after measuring volume of flow in unit time t, gross soil cross-sectional area A at right angles to flow, loss of hydraulic head h_L, and flow length L.

In field studies on undisturbed soil, loss of head and flow length sometimes cannot be measured accurately at reasonable cost. If the surface soil consists of a thin layer of low-permeability soil overlying a layer of highly permeable soil, then the loss of hydraulic head may be considered as the distance from water surface to the highly permeable soil and the flow length, as equal to the thickness of the top layer of soil, as indicated in Fig. 5.7.

Solving the Darcy flow Equation 5.5 for the permeability, it follows that

$$k = \frac{QL}{Ah_L} \qquad (5.7)$$

For example, where a permeameter having an area of 1070 square centimeters was used, the flow of water was 9.5 liters in 0.4 hour and the loss in piezometric head flowing through 30 centimeters of soil was 73 centimeters; therefore, the permeability is

(a)–Laboratory   (b)–Field

Figure 5.7. Constant-head permeameters.

$$k = \frac{QL}{Ah_L} = \frac{9.5 \times 1000}{0.4 \times 1070 \times 73} = 9.1 \text{ cm/hr} = 800 \text{ m/yr}$$

Variable-Head Permeameter

The variable-head permeameter is adapted to the measurement of permeability of fine-textured, compact soils of low permeability. It consists of a cylinder with a conical top to which is attached a vertical glass or plastic tube of small diameter. The cylinder is pressed into the soil to a known depth, and then the whole apparatus is filled with water. As the water percolates through the disk of soil in the cylinder, the water in the tube drops. Since the cylinder is usually made with an area 100 or more times that of the tube, a small volume of water percolation registers as a large drop in the tube.

One difficulty encountered in field tests with a permeameter was the tendency of the cylinder to rise because of the pressure exerted on the inside of the cylinder by the column of water in the tube. Temperature changes causing a change of volume within the cylinder will also produce significant changes in readings.

A variable-head permeameter measures permeability of a canal bed or lining, but not the seepage rate from a canal when in use. When comparative results from different locations are needed the depth should be essentially the same in all cylinders. Care should be taken not to puddle the soil when water is added.

Since geometry, flow conditions, and technique influence the readings obtained, the permeameter when used in canal studies is primarily a tool to be used to locate sections where excessive seepage occurs. Data obtained can seldom be used to obtain a quantitative measure of actual canal losses.

5.4 WATER FLOW IN UNSATURATED SOILS

Darcy's law for velocity of flow in saturated soils applies also to unsaturated soils. However, as moisture content decreases, cross-sectional area through which flow occurs also decreases causing permeability to decrease. In soils, whether saturated or unsaturated, water flow is caused by a difference in potential.

Thus, flow occurs in unsaturated soil much as it does in saturated soil. That is, flow always occurs in the direction of decreasing energy, Equation 5.3, and the total flow is the product of the velocity of flow and the cross-sectional area, Equation 5.5. Complications in unsaturated flow come from the decreasing cross-sectional area as the moisture content decreases, and from the increasing difficulty of measuring negative pressure.

For saturated flow through soils, a simple opening which will allow water to move into a pipe is sufficient to measure the energy at the end of the pipe. This method has been referred to as the piezometer method. However, for a suction within the soil, it is necessary to provide a barrier to movement of air from the soil to the measuring system. This barrier is usually a porous cup and has been referred to in Chapter 4 as a tensiometer, since it measures tension.

A further complication of unsaturated flow rises from the movement of water through the soil as vapor, which is even more difficult to measure. As moisture content decreases, the percentage of water moving through the soil as vapor increases. Movement in the vapor phase occurs in the direction of decreasing vapor pressure. Vapor pressure differences can occur from variations in both moisture content and soil temperature. Near permanent wilting percentages, temperature differences are the principal cause of vapor pressure differences, and hence, of moisture movement.

5.5 INTAKE CHARACTERISTICS OF SOILS

In all irrigation methods, except subirrigation, water is applied to the surface of land where it subsequently enters the soil and is stored for later use by plants. The object of irrigation is to get water into soil where it can be stored. Thus, rate of entry of water into soil under field conditions, called intake rate, is of fundamental importance. One of the major problems on irrigated farms is the low intake rate of fine-textured soils which is often reduced further by excessive working of soil and by salinity. Unless sufficient water enters the soil and is stored therein, crop growth is significantly reduced.

Intake rate varies with many factors, including depth of water on the surface, temperature of water and soil, soil structure and texture, and moisture and salinity content of the soil. Retarding layers of soil will greatly influence the rate. Configuration of the surface such as furrow shape and size, as well as the method of application, will be influencing factors also. Hence, intake rate varies from place to place on a field and it also varies with time. Sandy soils may have rates in excess of 25 centimeters per hour, whereas clay soils may have rates approaching zero when soil structure has been practically destroyed by faulty management. Usually the intake rate plotted against time on a logarithmic scale will show as a straight line, and therefore can be represented by the following equation:

$$I = aT^n \tag{5.8}$$

where I is the intake rate and a is a constant and is the value on the ordinate when time T on the abscissa has a value of 1. The constant n is the slope of the line. The algebraic principles are shown in the logarithmic plot of Fig. 5.8. The same data plotted on a linear scale in Fig. 5.8 shows how intake rates decrease with time. Table 5.1 is a tabulation of the intake data plotted in Fig. 5.8.

When the observation of intake extends over long periods, a better representation of the data can usually be obtained by using the equation:

$$I = aT^n + b \tag{5.9}$$

Since n is negative, I decreases with an increase in T. The intake rate I will approach a constant value b as time increases. The constant rate is referred to as final intake rate. How intake rate varies with soil texture was shown in Table 3.1.

Caution should be observed in using final intake rates for irri-

TABLE 5.1.   Tabulation of Intake Data Obtained from a Test Cylinder

Cylinder No. 3					
Time (hr)			Intake (mm)		
Watch	Difference	Cumulative	Depth	Difference	Cumulative
4:15			260		
	1	1		11	11
4:16			249		
	2	3		7	18
4:18			242		
	4	7		8	26
4:22			234		
refill			271		
	8	15		11	37
4:30			260		
	18	33		14	51
4:48			246		
refill			270		
	14	47		8	59
5:02			262		
	27	74		14	73
5:29			248		
	31	105		10	83
6:00			238		
refill			280		
	29	134		11	94
6:29			269		
	28	162		9	103
6:57			260		
	26	188		7	110
7:23			253		
	20	208		5	115
7:43			248		

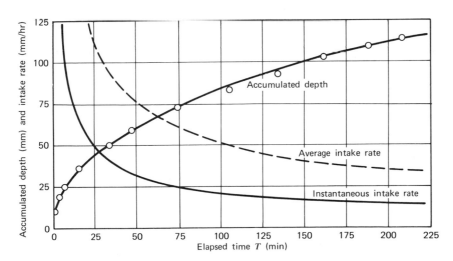

Figure 5.8. Typical intake curves showing different ways of presenting the same data. The method to be used depends upon the use to be made of the results.

gation design. When rather light, frequent irrigations are applied, the irrigation may be completed before the final intake rate is reached. Since initial rates are considerably in excess of final rates, the amount of water entering the soil can best be represented by the accumulated depth of water that has entered the soil. This quantity is represented by the integral of Equation 5.8 as

$$D = \frac{\alpha}{n + 1} \, T^{n + 1} = CT^m \qquad (5.10)$$

or, integrating Equation 5.9 when this equation represents more accurately the intake function, the accumulated depth of water applied becomes

$$D = \frac{\alpha}{n + 1} \, T^{n + 1} + bT = CT^m + bT \qquad (5.11)$$

Dividing the depth Equations 5.10 by time yields the average intake as

$$I_{ave} = \frac{\alpha}{n + 1} \, T^n \qquad (5.12)$$

and Equation 5.11 becomes

$$I_{ave} = \frac{\alpha}{n + 1} \, T^n + b \qquad (5.13)$$

5.6 MEASURING INTAKE RATE

The best method of measuring intake is to obtain direct measurements by recording the water applied less the water flowing from the field. When direct measurements are not feasible, intake cylinders can be used with reasonable success. Figure 5.9 shows a cylinder being used to measure the intake rate.

The cylinders should be at least 25 centimeters in diameter, made of smooth tough steel, strong enough to drive, but thin enough to enter the soil with a minimum of disturbance. Cold rolled steel about 2 millimeters thick (14 gage) works well. Cylinders should be about 30 centimeters long. Since approximately five tests should be made in a given location to obtain a representative sample, it is well to make five cylinders of somewhat different diameters so that all five can be nested together and thereby be made less bulky for handling.

Care should be taken to place the cylinders in an area representative of the field to be evaluated. Cylinders should be carefully driven into the soil to a depth of about 15 centimeters. The

Figure 5.9. A cylinder being used to measure intake rate of the soil.

soil profile should be examined, the moisture estimated or measured, and notes made of soil cover and surface condition.

After water penetrates to the bottom of the cylinder, it will begin to spread radially and the rate of intake will change accordingly. When a principal restricting layer does not lie within the depth of penetration of the cylinder, this radial flow will cause considerable change in the intake rate. Buffer ponds surrounding the cylinder are used to minimize this effect. Buffer ponds can be constructed by forming an earth dike around the cylinder or by driving into the soil a larger diameter cylinder concentric with the intake cylinder. The water levels in both cylinders should be equal and approximately the depth to be expected during the irrigation. When comparative results from different locations are needed, the depth should be essentially the same in all cylinders. Care should be taken not to puddle the soil when water is added to the cylinder or the buffer pond.

The results obtained with cylinders are indicative of the rates to be expected during irrigation provided the surface condition is the same. Closer correspondence is obtained when the soil surface is covered by the irrigation water. Considerable departure usually occurs when the irrigation water is applied by furrows or sprinklers. Hence, the cylinders are generally used to obtain an index from which design values can be obtained on the basis of local experience.

Other methods simulating ponding, furrows, rainfall, or sprinkling have been developed to measure intake rate. Good results have been obtained using these methods, but their practicality depends upon the accuracy desired, and upon the time and money available.

Equation 3.9, $qt = 28ad$, can be used to obtain the following equation:

$$I_{ave} = \frac{d}{t} = \frac{q}{28a} \tag{5.14}$$

where I_{ave} is the average intake in centimeters per hour that would result if the actual intake occurring in the furrow, measured in liters per second, was spread over the area between furrows measured in hectares. Thus, if 0.2 liters per second entered the soil in 30 meters of furrow and the furrows were 50 centimeters apart, the average intake, considering the entire soil surface, would be 4.8 centimeters per hour.

5.7 INFILTRATION AND SOIL WATER MOVEMENT DURING IRRIGATION

Having considered the intake phenomenon and the empirical equations describing its behavior, an analysis of water movements during irrigation will shed further light on the irrigation process.

Experiments point conclusively to the existence of three distinct zones. Since these zones are produced by a dynamic continuum, they are inseparably connected, but nevertheless show distinct characteristics pertaining to the nature of flow through soil. These three zones are the transmission zone, wetting zone, and wetting front.

Within the so-called transmission zone, moisture content is essentially constant regardless of the direction of flow. Degree of saturation (approximately 80 percent) is considerably less than the 100 percent commonly assumed. Since an approximately linear relationship exists between moisture content and permeability over this range of saturation, it is quite possible that a simple expression for permeability in terms of moisture content can be obtained for any particular homogeneous soil to represent effective transmissibility of the transmission zone.

Associated with the constant moisture content throughout the transmission zone is the experimentally verified principle that the gradient of the driving force is also constant throughout this zone. Furthermore, a constant potential gradient implied by a constant permeability throughout the transmission zone aids materially in the development which follows.

When the surface layer of soil is disturbed, as is the case in most irrigations, the degree of saturation and resistance to flow is not the same as in the soil beneath the surface layer. Usually the resistance to flow increases considerably due to structural breakdown and segregation of particles as well as packing of the surface soil.

Moisture content within the wetting zone reduces rapidly as the wetting front is approached. This in turn causes a corresponding decrease in permeability. In a moist soil the wetting zone is deeper or occupies more soil than in a dry soil. This is so because the hydraulic conductivity is greater; also there is a tendency for the displacement of water ahead of the actual incoming flow.

The energy consumed within the wetting zone is approximately constant when the moisture ahead of the wetting zone is constant. Hence, for a given available energy, the gradient of the driving force will be much less in a moist soil than in a dry soil. In other words, dry soil with a low permeability will result in a greater loss of energy per unit length, causing a shallower wetting zone.

Visual observation of the wetting phenomena which take place within the wetting zone indicates that the flow occurs in a very erratic manner. The movement which might be described as a "jumping" movement takes place essentially in increments. This "jumping" movement is more pronounced in drier soils. The advancing or wetting front is in effect a capillary fringe, the foremost part of which has an average moisture content equal to the original soil, while the moisture content of each succeeding cross section behind the wetting front increases until approximately 80 percent of the conducting pores are filled with moisture. The lateral extent of this transition is a measure of the capillary fringe.

When the position of the wetting front is recorded, whether from visual observation or for tensiometer or moisture data, it is usually the forepart of the capillary fringe that is interpreted as the wetting front.

Using the basic principles of flow through soils, it can be shown that the rate of downward movement V_y can be defined by the following equation:

$$V_y = \frac{I_f}{ns} \qquad (5.15)$$

where I_f is the final intake rate, n is the total pore space of the soil, and s is the change in degree of saturation during the irrigation. Hence, for a sandy soil having a final intake rate I_f of 5 centimeters per hour, a porosity of 38 percent, and a degree of saturation of 22 percent, the rate of movement of the water in the soil will be 60 centimeters per hour. Likewise, in a loam soil with $I_f = 1.5$ cm/hr, $n = 47$ percent, and $s = 24\%$, the rate of downward movement is 13 centimeters per hour.

For downward movement the final intake rate I_f is equal to the permeability of the soil at the degree of saturation (about 80 percent) which occurs during irrigation.

When water enters the soil and encounters a restricting soil layer, causing it to move laterally, both the intake rate and the rate of movement throughout the soil will decrease with the square root of time. Thus, when observations of intake are being made, it is well to observe the trend with time as an indication of the nature of movement below the surface. When the intake rate approaches a constant, it can be assumed that water is moving downward in an unrestricted manner. However, when the rate of intake does not approach a constant value, then it can be assumed that the flow is moving laterally, due to restricting layers beneath the surface. Observation of the soil profile should readily verify the existing position and nature of restricting layers.

5.8 NONHOMOGENEOUS ANISOTROPIC SOIL

Soils are seldom homogeneous. Usually they are nonhomogeneous and also anisotropic, meaning that the permeability not only varies from point to point, but that the value of the permeability in one direction is different from the value in another direction, even at a given point within the soil. This lack of uniformity becomes most extreme in alluvial soils. Frequently the horizontal permeability of an alluvial soil is 10 times greater than the vertical permeability because of the natural horizontal orientation of the major axis of particles deposited under water. The presence of tight clay layers further decreases the relative vertical permeability. Thus, it is very important to recognize the nonhomogeneous as well as the anisotropic nature of the soil when permeability measurements are being taken. For example, an undisturbed vertical core of soil may be tested in the laboratory and found to have essentially no vertical permeability. However, in the field this same soil may drain readily due to the presence of interconnected layers of more permeable soil. Flow of water toward wells and drains is predominantly horizontal, and the permeability for horizontal flow is more important than for vertical flow.

Also, because of the extensive variability which can, and does, frequently occur from point to point in a soil, the results of intake or permeability measurements which are essentially point observations should be used cautiously. A sample of soil represents only a very, very small portion of the soil to be evaluated.

Because of these problems of measurement caused by variability in soil and permeability, it is generally not feasible to use precise techniques of permeability measurement for irrigation purposes. More measurements made more quickly at less cost are generally more valuable than a few precise determinations on a few samples which may not be, and probably are not, representative of the field to be evaluated.

5.9 LIMITATIONS OF EQUATIONS

Because of variability which exists in soils, judgment and discretion are essential in applying the equations of this chapter to the solution of practical irrigation and drainage problems. The equations are valid, but the values applied to those equations are not precise. A major challenge exists for the researcher to develop more practical methods of measuring soil properties over extensive areas rather than more precise determinations at a point in the soil.

CHAPTER
6
SALT PROBLEMS IN SOIL AND WATER

Excess soluble salts and alkali and the occurrence of drought cause sterility and barrenness of arid-region soils. Although saline and alkali lands are characteristic of arid regions, they are neither of general occurrence nor of uniform distribution. Reclamation of saline and alkali soils, and prevention of excess salts in fertile areas now irrigated, are of paramount importance in arid regions.

Saline soils are soils having excess soluble salts that make the soil solution sufficiently concentrated to injure plants and impair soil productivity. The term "alkali soil" is applied to soils which have an excess of exchangeable sodium either with or without excessive total soluble salts.

6.1 CLIMATE AND SALINITY

Arid-region soils contain relatively large amounts of soluble salts. The heavy annual rainfalls of humid regions cause water to percolate through the soil and carry to the streams, rivers, and oceans large amounts of soluble mineral substances. The scanty rains of arid regions do not penetrate the virgin arid soils deeply enough to cause appreciable percolation. The greatest depth of penetration of water from natural precipitation in arid regions, either melted winter snow or wet-season rains, is generally found to vary from 25 to 125 centimeters, depending on the depth and time of the precipitation and the nature of the soil. Lack of percolation through arid-region soils, together with excessive evaporation of water in arid regions, gives rise to accumulation on and in the soils of soluble salts that are injurious to plant life. Hence, the basic cause for saline and alkaline soil is insufficient application of water.

6.2 SOURCES AND ACCUMULATION OF SOLUBLE SALTS

Mineral soils are derived largely from the weathering of rocks. There is an intimate relation between salt accumulation and the chemical composition of rocks from which soils are formed. Soils formed directly from salt-bearing rocks usually contain excessive salts.

Some arid-climate soils that were free from excess salts before cultivation have been rendered nonproductive by the use of irrigation water containing excessive quantities of salt. Water that percolates through salt-bearing rocks usually contains an appreciable amount of salt.

Salts continue to accumulate in soils of irrigated areas where greater amounts are brought in than are removed. Irrigation waters contain from 0.5 to 30 kilograms of salt per cubic meter of water. Some irrigation farmers apply only a depth of 50 centimeters of water per season; others, where the summers are long and hot, apply up to 200 centimeters of water or more. Where drainage is not provided, the irrigation of water may add as much as 3 to 25 or more tons of salt each year to a hectare of land.

The most effective method for the removal of salt from soil is by means of water which passes through the root zone of the soil; but, if the amount carried away is less than the amount brought in by irrigation water, salt will accumulate. To prevent salt accumulation, and consequent decrease in crop yields, irrigators must remove as much salt as is brought in. In some areas, an effort is made to spread limited irrigation water supplies over too many acres, with the result that the soil is not wetted below a depth of a few centimeters. In other areas, the ground-water table is so close to the surface as to retard or prevent the leaching of salt from the root-zone soils. In soils having shallow water tables, upward flow of saline ground water results in a continuing accumulation of salt in the surface soil.

The barrenness that results from excessive accumulation of soluble salts is illustrated in Figs. 6.1 and 6.2.

6.3 RELATION BETWEEN CONCENTRATION AND CONDUCTIVITY

The salt content of soil solutions or irrigation waters is usually represented in one of three ways; parts per million (ppm), milliequivalents per liter *(MEQ/L)*, or as electrical conductivity expressed in micromhos per centimeter *(EC × 10⁶)*. Conductivity has the advantage of simplicity since it can be measured readily in the field or laboratory by using a portable electrical conductance meter. The relationship between concentration and conductivity is shown in Fig. 6.3. Cation and anion content of the water determines the actual value.

Figure 6.1. Surface accumulation of salts on lands prevents growth of crops without leaching. (Courtesy Utah Agr. Exp. Sta.)

6.4 SOME BASIC TERMS

In a contribution entitled "Diagnosis and Improvement of Saline and Alkali Soils," the United States Regional Salinity Laboratory presents a glossary of 60 terms together with definitions. The 12 terms most used in describing soils having excessive soluble salts and exchangeable sodium are presented here:

Alkali Soil. Soil which has an excessive degree of saturation with exchangeable sodium, either with or without appreciable amounts of soluble salts.

Figure 6.2. Punjab flax fails to grow on saline spots. (Courtesy U.S.D.A.)

Figure 6.3. Concentration of irrigation waters related to conductivity.

Alkaline. A chemical term referring to "basic" soil reaction where the pH is above 7. The pH is the logarithm of the reciprocal of the hydrogen ion concentration, pH 7 is neutral.

Electrical Conductivity. The reciprocal of electrical resistivity. The resistivity is the resistance in ohms of a conductor which is 1 square centimeter. Therefore, electrical conductivity is expressed in reciprocal ohms per centimeter or mhos per centimeter. The terms "electrical conductivity" and "specific electrical conductance" have identical meaning.

Equivalent Weight (or combining weight with hydrogen). Atomic or formula weight divided by its valence. For example:

$$Na^+ = \frac{23}{1} = 23 \qquad Ca^{2+} = \frac{40}{2} = 20$$

$$Cl^- = \frac{35.5}{1} = 35.5 \qquad SO_4^{2-} = \frac{96}{2} = 48$$

Exchangeable-Sodium Percentage. Degree of saturation of the soil-exchange complex with sodium, defined as follows:

$$ESP = \frac{\text{Exchangeable sodium (milliequivalents per 100-g soil)}}{\text{Cation-exchange capacity (milliequivalents per 100-g soil)}} \times 100$$

Nonsaline-Alkali Soil.[1] Soil for which the exchangeable-sodium percentage is greater than 15, and the conductivity of the saturation extract is less than 4 millimhos per centimeter (at 25°C). The pH values for these soils generally range between 8.5 and 10.

Saline-Alkali Soil. A soil for which the conductivity of the saturation extract is greater than 4 millimhos per centimeter (at 25°C), and the exchangeable-sodium percentage is greater than 15. The pH of the saturated-soil paste may somewhat exceed 8.5.

Saline Soil. A soil for which the conductivity of the saturation extract is greater than 4 millimhos per centimeter (at 25°C), and the exchangeable-sodium percentage is less than 15. The pH of the saturated-soil paste is usually less than 8.5.

Soil Reaction, pH. The pH scale is a measure of the effective concentration of the hydrogen ion. It has long been used as an index of the acidity or alkalinity of soils. A pH value of 7.5 to 8.0 usually indicates the presence of carbonates of calcium and magnesium, and a pH of 8.5 or above usually indicates appreciable exchangeable sodium.

Reference to these terms and definitions will be helpful in a study of this chapter. A summary of conditions for saline, saline-alkali, and alkali soils is shown in Table 6.1.

Saline Soils. For saline soils the electrical conductivity of the saturation extract is more than 4 millimhos per centimeter, and the exchangeable sodium percentage is less than 15. These soils correspond to "white alkali" soils. When adequate drainage is established and the soluble salts are removed by leaching, the soils become normal, that is, nonsaline.

Saline soils may develop from normal soils through the accumulation of salts from applied irrigation water, or by upward moving ground water, or by a combination of both processes.

The chemical characteristics of saline soils are determined by

---

[1]The term commonly used is simply "alkali" soil. Recently the term "sodic" is being used in lieu of "alkali" for clarity of nomenclature.

TABLE 6.1 Conditions for Saline, Saline-Alkali, and Alkali Soils

Salt Condition	Common Term	Salt Index — Conductivity of Saturation Extract (millimhos/cm)	Sodium Index — Exchangeable Sodium Percentage	Hydrogen Ion Index[a] — pH	Reclamation
Saline	White alkali	> 4	< 15	⩽ 8.5	Leaching
Saline-alkali		> 4	> 15	generally about 8.5	Leaching necessary and possible, but as salts are removed the sodium must be replaced to prevent dispersion of soil particles and reduction of permeability.
Alkali (sodic)	Black alkali[b]	< 4	> 15	generally between 8.5–10.0	Low permeability due to dispersion of soil by the sodium requires replacing the sodium to improve the permeability so that leaching can proceed.

[a] Since a pH of 7.0 is neutral, pH values less than 7.0 indicate an acid soil which is common in the nonarid regions. Note that the larger the pH, the less the concentration of hydrogen ions, since pH is the logarithm of the reciprocal of the hydrogen ion concentration.
[b] A black crust forms on the surface of an alkali soil only if organic matter is present. Hence, the term may be misleading in soils of low organic matter.

the kinds and amounts of salts present which largely control the
concentration and osmotic pressure of the soil solution. Sodium
seldom comprises more than half of the soluble cations, and thus
is not absorbed to any appreciable extent. The relative amounts
of calcium and magnesium in the soil solution may vary considerably. Soluble and exchangeable potassium are ordinarily only
minor constituents but may occasionally occur in excess. The
chief anions are chloride, sulfate, and sometimes nitrate. Small
amounts of bicarbonate may occur, but because the pH is 8.5 or
less, soluble carbonates are usually absent. In addition to the
readily soluble salts, saline soils may contain relatively insolu-
ble salts such as calcium sulfate (gypsum), and calcium and
magnesium carbonates.

Owing to the presence of excess salts, and the absence of appreciable exchangeable sodium, the colloids in some saline soils
are highly flocculated, thus providing a favorable structure and
improved permeability to water and air. However, some saline
soils have very low permeabilities.

Saline-Alkali Soils.   Soils for which the conductivity of the saturation extract is greater than 4 millimhos per centimeter, and
the exchangeable-sodium percentage is greater than 15, are designated saline-alkali soils. As long as excess salts are present,
the appearance and properties of these soils are generally similar to those of saline soils. Under conditions of excess salt, the
pH value is seldom higher than 8.5, and the colloids remain flocculated. If the excess soluble salts are temporarily leached downward, the properties of these soils may change markedly (unless
gypsum is present) and become similar to those of nonsaline-
alkali soils. As the concentration of the salts in the soil solution
is lowered, some of the exchangeable sodium hydrolyzes and
forms sodium hydroxide. The sodium hydroxide may change to
sodium carbonate upon reaction with carbon dioxide. Then the
soil usually becomes strongly alkaline (pH above 8.5), the colloids
disperse, and the soil develops a structure unfavorable for water
infiltration and percolation, and for tillage. Although the return
of the soluble salts may lower the pH value and restore the colloids to a flocculated condition, the management of saline-alkali
soils continues to be a problem until the excess salts and exchangeable sodium are removed from the root zone.

Alkali Soils.   When the exchangeable-sodium percentage of
soils is greater than 15, and the conductivity of the saturation
extract is less than 4 millimhos per centimeter, the pH values
generally range between 8.5 and 10 and the soils are designated

alkali soils. These soils correspond to "black alkali" soils. They frequently occur in semiarid and arid regions in small irregular areas, which are referred to as "slick spots." Except when gypsum or another source of soluble calcium is present, the drainage and leaching of saline-alkali soils develop alkali soils. The removal of excess salts in such soils permits hydrolysis of the exchangeable sodium, and may lead to the formation of small amounts of sodium carbonate. The soil organic matter is highly dispersed and distributed over the soil particles, thereby darkening the color. When the soil contains appreciable organic matter, its surface may be quite dark; hence the term "black alkali."

Clay which is partially sodium-saturated is highly dispersible in the absence of flocculating salts and has a tendency to migrate downward through the soil and accumulate at lower levels. As a result, the surface few centimeters of soil may be relatively coarse in texture and friable; but, below where the clay accumulates, the soil develops a dense layer of low permeability. The physical and chemical properties of alkali soils are largely determined by the exchangeable sodium present. As the proportion of exchangeable sodium increases, the soil tends to disperse more and the pH value increases, becoming as high as 10. The soil solution of alkali soils, although relatively low in soluble salts, has a composition which differs greatly from that of normal and saline soils. The anions present consist mostly of chloride, sulfate, and bicarbonate, but small amounts of the normal carbonate usually occur. At high pH values, and in the presence of normal carbonate, calcium and magnesium are precipitated so that the soil solutions of alkali soils usually contain only traces of these cations, sodium being the predominant one. Considerable amounts of exchangeable and soluble potassium occur in some alkali soils.

6.5 CHEMISTRY OF SALTY SOILS

The following clarifies the above statements regarding the sodium (Na) and hydrogen (H) ions:

$$\text{Ca} \; \boxed{\begin{array}{c} \text{Na} \\ \text{Clay particle} \\ \text{Na} \end{array}} \; \text{Na} + H_2O \rightleftharpoons \text{NaOH} + \text{H} \; \boxed{\begin{array}{c} \text{Na} \\ \text{Clay} \\ \text{Na} \end{array}} \; \text{Ca} \qquad (6.1)$$

A clay particle with sodium and calcium ions attached tends to hydrolyze. When a sodium ion is exchanged for a hydrogen ion, and the sodium ion combines with a molecule of water, sodium hydroxide (NaOH) is formed. When carbon dioxide (CO_2) is present in the soil air, it readily reacts with the water to form hydrogen carbonate (H_2CO_3). However, the sodium hydroxide (NaOH) reacts readily with the hydrogen carbonate (H_2CO_3) to form sodium carbonate (Na_2CO_3).

$$H_2CO_3 + 2NaOH \rightleftharpoons Na_2CO_3 + 2H_2O \qquad (6.2)$$

The Na_2CO_3 is gradually removed with extensive leaching, and the soil is left with hydrogen ions having replaced sodium ions. This increase of hydrogen ions is reflected in a lower pH (pH being a reciprocal index of hydrogen ion concentration).

Where soils contain calcium carbonate ($CaCO_3$) or gypsum, calcium is dissolved into the soil solution. This available calcium is exchanged for sodium during the leaching process to obtain a normal soil.

$$Ca \boxed{\text{Clay particle}}\begin{matrix}Na\\Na\\Na\end{matrix} + CaCO_3 \rightleftharpoons Na_2CO_3 + Ca \boxed{\text{Clay particle}}\begin{matrix}Na\\Ca\\Na\end{matrix} \qquad (6.3)$$

When chemicals must be added to the soil to accomplish the exchange of sodium for other ions, the following products can be used:

gypsum ($CaSO_4 \cdot 2H_2O$)
soil sulfur (S)
sulfuric acid (H_2SO_4)
concentrated lime sulfur solution ($CaS_x + H_2O$)
dry lime sulfur (CaS_x)

Availability, cost, and solubility should be considered when determining the most feasible product to use.

6.6 MOVEMENT OF SALTS IN SOILS

If it were possible to maintain a moisture distribution in irrigated soils such that water flow would be continuously downward, there would be relatively little trouble from salinity on irrigated farms. A continuous downward flow of water with adequate drainage would gradually decrease the soluble salts in the upper few feet of soil where plants obtain most of their moisture and food. However, without adequate drainage downward, percolating waters fill the lower soil spaces and cause the water table to rise. During periods between irrigations, a high water table favors the upward capillary flow of water to the land surface, where the water evaporates. The soluble salts carried by the upward-moving water cannot be evaporated, hence they are deposited on or near the soil surface. Salts so deposited may come from salty soil horizons well below the surface. The mere concentration on the surface of salts that are normally distributed through the upper few feet of soil may cause serious salinity.

6.7 INFLUENCE OF WATER TABLE

Irrigation farmers sometimes point out the advantages of keeping the water table within a few feet of the soil surface because of the high crop yields obtained during the early years after it has risen from great depths. The favorable moisture supply from a water table near the soil surface may cause high crop yields; but as a rule, in areas where alkali salts occur, the temporarily favorable condition of the high water table is followed by a serious decrease in yield, if not by complete nonproductivity due to salt concentration.

Water tables can be lowered by preventing excessive seepage losses from canals, by careful and efficient application of water on the farms, and by providing artificial drainage on areas for which natural drainage is inadequate. Keeping the water well below the soil surface is very urgent and not likely to be overemphasized.

6.8 MANAGEMENT OF A HIGH WATER TABLE

Despite hazards of a high water table, good management can obtain a permanently good production with a high water table. The major need is for lighter and more frequent irrigations during the growing season. Sprinkler irrigation is especially well adapted to this condition because of the ability of the irrigator to apply a limited amount of water. Special care must be taken not to apply an excessive irrigation during the growing season, which would result in a rise of water table within the root zone of the crop.

During the dormant season, one or more heavy leaching irrigations will be necessary to remove the excess salts from the soil. During dormancy, a temporary rise of the water table generally will not seriously damage even the perennial crops.

Since a crop will naturally obtain a considerable portion of its needed water from a high water table, the amount of water that must be applied by irrigation is reduced, resulting in a reduction in cost of irrigation. Hence, there are distinct economic benefits arising from a reasonably high water table, but good management is essential if a profitable agriculture is to be maintained.

Not to be overlooked is the reduced root zone available to the plants which usually causes a reduced yield. Therefore, the economic depth for a water table depends upon management, the crops to be grown, and the cost of irrigation. The hazard to productive farming is definitely greater with a higher water table.

6.9 STEPS ESSENTIAL TO RECLAMATION

Temporary control of salts on irrigated land is sometimes practiced by one or more of the following methods:

(a) Plowing salt-surface crusts deeply into the soil.

(b) Removing surface accumulation from the soil.

(c) Neutralizing the effects of certain salts by use of other salts or acids.

Permanent reclamation of saline and alkali land requires four essential steps and the attainment of four basic conditions, namely:

(a) Adequate lowering of water table.

(b) Satisfactory water infiltration.

(c) Leaching excess salts out of the soil.

(d) Intelligent management of soil.

Adequate Lowering of Water Table

All waterlogged lands, whether or not impregnated with alkali, are improved for ordinary crops by lowering the water table. This means a permanent lowering under the farmer's control so that a rise of water above a given elevation in the soil for any length of time may be wholly prevented. The first step in lowering the water table is to learn the source of the water that caused it to rise. In isolated cases on small tracts, it is sometimes possible for one farmer alone, or a small group of farmers, to find the water source and cut it off by construction of one or more intercepting ditches or drains. Usually, in irrigated regions, small waterlogged areas are caused by surface or underground water flowing to them from higher irrigated lands, or from canals, ponds, or reservoirs. The farmer whose holdings are located within large areas of comparatively level waterlogged land cannot, as a rule, lower the water table by his own efforts. For such areas community action is essential.

Satisfactory Intake Rates

The rate of water infiltration into soils depends on soil texture, structure, degree of soil dispersion, and also on the depth to the water table. When adequate drainage has been provided, the structure of the soil and other soil properties are dominant in influencing intake rates. The time required for an adequate depth of water to percolate through the soil may well limit the feasibility of reclamation.

Alkali soils disperse during leaching and often become impermeable as the soluble salts are removed. Chemical amendments are then required to bring about improvement of the soil by replacing the exchangeable sodium with calcium. Gypsum, and under certain conditions, sulfur may be used for this purpose. Considerable gypsum occurs naturally in the soils of some areas, but the amounts and distribution within the soil profile are variable.

Usually the cheapest method of obtaining satisfactory intake rates is to minimize surface tillage. Leaving the soil surface rough and cloddy will materially increase the rate of entry of water through the soil. Avoiding tillage when soil is either too wet or too dry is essential. Working a dry powdery soil is especially harmful to soil structure.

A saline water, not containing appreciable quantities of sodium, will also improve intake rates. For leaching purposes, a saline water may be very useful for increased intake rates; also the saline water often contains the salts that would otherwise be applied as chemical soil amendments. Hence, quality of water may greatly influence intake rates.

Adding chemical amendments to reduce dispersion of soil particles is not the only method of increasing intake rate. Good tillage practice will invariably increase the rate significantly. Also, the value of deep-rooted crops like clover and alfalfa should not be neglected. Getting plant roots to penetrate soil will also materially improve soil permeability to air and water. Not only do these roots leave cavities when they decay, but they also remove water from the soil, causing alternate wetting and drying which are essential for building soil structure.

Information on the rates at which water enters and moves through soil is useful in connection with irrigation methods as well as with improvement of saline and alkali soils. Many factors discussed in Chapter 5 influence these rates, but it is not difficult to make measurements that have practical significance. Intake rates are best measured in the field by flooding or impounding water on the soil surface and measuring the rate of entry into the soil.

Water of the same quality as the irrigation or leaching water should be utilized for infiltration tests in the field; otherwise, the measurements may be misleading. The length of time the test should be conducted, or the depth of water to be applied, will depend on the purpose of the test and the kind of information sought. If it is a matter of appraising an irrigation problem, then the depth corresponding to one irrigation may be sufficient. If getting advance information on infiltration for planning a heavy leaching program is involved, then the full leaching depth is recommended. It often happens that subsurface drainage in the soil is sufficiently slow to reduce infiltration rates considerably. Although small-area tests will give useful information on infiltration during leaching, infiltration values thus obtained will apply to large areas only if underdrainage is adequate.

Permeability as measured in the laboratory is influenced by many factors. Some of these, such as dispersion of the soil, base status, air saturation, and microbial sealing, have been separately studied and partially appraised. The structure and packing of the sample enter directly into the measurement. It appears that air-dried and screened samples may be used if interest is centered primarily on the cultivated layer of soil. Natural structure may be required for significant measurements on the subsurface layers.

Leaching Excess Salts

It is usually essential that large depths of water be applied to saline and alkali lands and be made to percolate through the soil in order to leach out excess salts. Coarse-textured soils of open structure as a rule have sufficiently high permeability to make leaching of alkali salts an easy task after the water table has been sufficiently lowered. Fine-textured compact soils of low permeability predominate in the low-lying waterlogged areas. Consequently, soil permeability is a factor of paramount importance in the leaching of excess soluble salts from most waterlogged soils. Permeability is influenced not only by the texture and compactness of the soil, but also by flocculation or dispersion of the soil particles. The dispersion, and consequently the permeability, is greatly influenced by certain chemical compounds. A very low permeability sometimes follows the leaching of alkali salts, and this decreases the productivity of the soil because of the difficulty of getting air and water to penetrate the soil.

Maintaining a favorable salt balance in the soil requires proper and efficient irrigation methods. Irrigation must provide water for growth of crops and at the same time allow enough water to pass through the soil to leach out the excess salts. Excessive leaching, however, may be detrimental in that plant nutrients, especially nitrates, may be removed from the soil. Overuse of water also adds to the drainage problem. Water applications should be sufficient to maintain a favorable salt balance in the soil, but without excessive leaching of the plant nutrients and without materially adding to the drainage problems.

6.10 LEACHING REQUIREMENTS

The amount of water which must be leached through the root zone to keep a favorable salt balance has been formulated by the staff of the U.S. Salinity Laboratory. The leaching requirement is defined as the fraction of the irrigation water that must be leached through the root zone to control soil salinity at any specified level. The concept is most useful when applied to steady-state water-flow rates, or to total depths of water used for irrigation and leaching over a long period of time. The amount of water required to leach a saline soil will be several times greater than the leaching requirement determined from the following formula to maintain a healthy soil.

$$LR = \frac{D_d}{D_i} = \frac{EC_i}{EC_d} \qquad (6.4)$$

where LR is the leaching requirement, D_d and D_i are the depths of drainage water and irrigation water, respectively, and EC_i and EC_d are the electrical conductivities of the irrigation and drainage waters. When rainfall during the period of observation is appreciable, or when water moves into the root zone from the water table, a weighted average of conductivities for all sources of water should be used.

The leaching requirements for different conductivities and concentrations of irrigation water are shown in Table 6.2. Note that the leaching requirement increases at least 10 percent for each 1000 parts per million of dissolved solids in the irrigation water. Hence, irrigation water having 4000 parts per million will necessitate that approximately 50 percent of the applied water passes through the root zone.

TABLE 6.2 Leaching Requirement Related to Quality of Drainage and Irrigation Water for an Equilibrium Condition, Ignoring Influence of Rainfall

Quality of Irrigation Water				Percentage Leaching Requirements for Given Qualities of Drainage Water			
Micromhos/cm $EC \times 10^6$	Approx. mg/l	MEQ/L	Tons/ha-m	5 millimhos/cm $EC \times 10^3$	10 millimhos/cm $EC \times 10^3$	15 millimhos/cm $EC \times 10^3$	20 millimhos/cm $EC \times 10^3$
100	63	1	.6	2	1	0.7	0.5
200	125	2	1.2	4	2	1	1
400	250	4	2.5	8	4	3	2
800	500	8	5.0	16	8	5	4
1600	1000	16	10.0	32	16	11	8
3200	2000	32	20.0	64	32	21	16
6400	4000	64	40.0	128	64	43	32

The leaching requirement represents, therefore, a necessary water use for a permanent irrigated agriculture. When water needs for a project are being planned, the leaching requirements should be considered.

Since drainage water carries salt downstream to the detriment of lower water users, the natural tendency of regulatory water quality management agencies is to reduce leaching to a minimum. The lower the design leaching fraction, the higher will be the requirements for water management control. Several practical factors such as the following make precise irrigation management nearly impossible: variability of soil and topography, inability to obtain uniform water distribution, inherent inaccuracy of water measurement, and the difficulty of obtaining competent field labor. Unless a reasonable margin of safety is maintained in the leaching requirement, salts will accumulate in portions of the root zone and production and profits will reduce significantly.

6.11 RECLAMATION AND MANAGEMENT OF SALINE AND ALKALI SOILS

Sometimes irrigation farmers who own alkali land on which the water table has been adequately lowered, and from which the excess salts have been leached, erroneously conclude that the task of reclamation has been completed. Reclamation is attained only when the lands are made to produce large crop yields. Usually, restoration to full productive capacity, or its establishment in virgin salty lands, requires both chemical and physical improvements in the soils.

Alkali soils usually contain excessive amounts of sodium, held by the finer soil particles, and therefore have a poor physical condition. To be reclaimed, these soils must be changed chemically and improved physically.

The chemical changes consist of exchanging calcium for sodium and then leaching away the sodium salts. Any soluble calcium compound will do, and the greater the concentration of calcium in the reclamation water, or in the soil solution bathing the soil particles during reclamation, the more rapidly this process will take place. If the soil does not contain gypsum, it can be added and dissolved by irrigation water. Other soluble calcium compounds such as calcium chloride are also very helpful but are expensive.

Some alkali and saline soils are low in available phosphorus and will give better crop yields if phosphate fertilizers are used. Superphosphate, treble superphosphate, and ammonium phos-

phate are among those generally recommended. Liberally applying barnyard manure, plowing under cover crops, the avoiding of
plowing or of other farm operations when the soil is too wet or
too dry are all helpful. Applying enough water to assure adequate penetration into heavy soils, keeping drains open and in
good repair, preventing excessive evaporation, and other important precautions are urgently essential to the management of
salty lands in order to assure permanent relief from waterlogging
and excess salts, and to assure perpetuation of soil productivity.

6.12 SALINE WATER FOR IRRIGATION

It is probable that population pressures and public welfare will
require the utilization of all available water in the arid regions of
the world. Domestic water supplies are of first importance, and
next in importance are the uses of water for irrigation and for
industrial purposes. Proper quality of water is essential for satisfactory domestic use and irrigation. The intensity of demand for
water is greater in some irrigated sections than in others, but in
only a very few of the arid regions of the world is the total water
supply, when fully stored and controlled, large enough to completely irrigate all the arable land. In some sections all the
available water is used, and the so-called saturation point of irrigation expansion has been reached.

As the saturation point in irrigation expansion is approached,
the need to use saline waters is increased. However, excessive
amounts of salts in irrigation water result in fatality to crops and
financial losses to water users. On the other hand, small
amounts of salts in water not only are harmless, but also actually stimulate crop growth under some conditions. A study of the
safe limits of salts in irrigation water, and of the conditions
under which saline water may best be used for irrigation, is essential to complete an intelligent utilization of arid-region resources.

Since water is a vital resource essential to agriculture, saline
irrigation water should not be pronounced unfit for irrigation
without careful consideration of all factors concerned. Frequently, the salty water is the only water available. Then it becomes a matter of necessity to accept its limitation, use it wisely,
and be grateful for a water supply even though it is not entirely
satisfactory.

Too often saline waters are pronounced unsuitable without
considering the need for water in the area. Permanent irrigated
agriculture can be maintained using rather salty waters, with
proper management. For example, the greater the salt content of

the water, the larger is the percentage of applied water that must be leached through the soil to maintain a favorable salt balance. Hence, the amount and kinds of salt present in the irrigation water should be a guide to necessary management if the water is to be used successfully.

6.13 TYPICAL ANALYSES OF IRRIGATION WATERS

The analyses of selected irrigation waters shown in Table 6.3 brings out the following points:

Sample 1, from the Rio Grande at Del Norte, Colorado, represents a mountain water low in total salts, whereas sample 2, taken from the same river in southern New Mexico, shows the marked increase in concentration between Colorado and New Mexico.

Sample 3, from the Pecos River near Barstow, Texas, is an example of one of the most concentrated waters used for irrigation purposes. The water is applied to a soil high in gypsum. Fair to good crops of cotton and alfalfa are obtained.

Sample 4, from the Snake River, Idaho, is representative of irrigation waters in the Northwest. They are, as a whole, low in total salt content.

Sample 5, from the Colorado River, Arizona, was selected to show the type of irrigation water used on fairly heavy soil in the Imperial Valley. The water contains 8.1 tons of salt per hectare-meter, which is a little more than average for irrigation water.

Sample 6, from an irrigation well, was selected as a water of high sodium percentage. Used on sandy soils it does some injury to grapes. The boron content is rather high.

6.14 STANDARDS FOR IRRIGATION WATERS

Plants in saline soils are adversely affected by high concentrations of salts in the soil solution and by poor physical condition of the soil. Both soil conditions are affected by the irrigation water. An irrigation water having a high sodium percentage will, after a time, give rise to a soil having a large proportion of replaceable sodium in the colloid, often designated as "black alkali" soil. Even on sandy soils with good drainage, waters of 85 percent sodium[2] or higher are likely to make soils impermeable after prolonged use. With higher total salt content, there is a floc-

[2]Percent sodium $= \dfrac{100\ Na}{Ca + Mg + Na + K}$

TABLE 6.3 Analyses of Selected Irrigation Waters

Sample	Source	Concentration			Constituents									
		Electrical Conductivity, EC × 10^6	Tons/ha-m	Total Solid (mg/l)	Cations (bases)				HCO₃	Anions (acids)			Sodium (%)	Boron (mg/l)
					Ca	Mg	Na	K		SO₄	Cl	NO₃		
1	Rio Grande, Colorado	85.0	0.7	81	0.56	0.27	0.30		0.70	0.32	0.12	(²)	26	
2	Rio Grande, New Mexico	870.0	5.9	641	3.76	1.34	4.03		2.97	4.69	1.53		44	
3	Pecos River, Texas	9150.0	61.7	6198	30.62	17.19	53.62			44.52	54.00		52	
4	Snake River, Idaho		3.7	380	2.60	1.80	1.74	0.14	3.28	1.69	1.18	0.04	28	
5	Colorado River, Arizona	1135.0	8.1	795	5.08	2.21	4.54		2.56	6.97	2.31	0.04	35	0.14
6	Well in Coachella Valley, California	1740.0	8.8	910	2.14	0.08	12.67		1.02	1.80	12.04	0.14	85	0.71

culating action that tends to counterbalance the poor physical condition caused by a high sodium concentration in the water. On a heavy soil already high in replaceable sodium the poorest irrigation water would be that which is low in total salts but which has a high sodium percentage.

The concentration of the soil solution also modifies plant growth, and is usually 2 to 100 times the irrigation water concentration and seldom more diluted. In heavily irrigated sandy soils, the soil solution will tend to approach the same concentration as that of the irrigation water. On heavy soils, where evaporation may greatly exceed drainage, the concentration of the soil solution may be 100 times that of the irrigation water, and may become too great for plant growth.

In most arid-region valleys neither the irrigation companies nor farmers can modify their irrigation water to any substantial degree. However, the standards for waters presented in Table 6.4 can be a guide to interested agencies in appraising and developing irrigation and drainage practices to harmonize with qualities of water.

In this evaluation shown in Table 6.4, water rated number 1 would be considered excellent for agricultural use. Water rated number 6 generally would not be usable for irrigation under any management or drainage condition. Each factor is important, and no attempt should be made to combine the ratings into a

TABLE 6.4   Evaluation of Irrigation Water Quality (Table of Limiting Values)

Evalua-tion Rating	EC (mil-limhos)	Na$^+$ (%)	SAR	Na$_2$CO$_3$ (MEQ/L)	Cl$^-$ (MEQ/L)	ES (MEQ/L)	Boron (mg/l)
1	0.5	40	3	0.5	3	4	0.5
2	1.0	60	6	1.0	6	8	1.0
3	2.0	70	9	2.0	10	16	2.0
4	3.0	80	12	3.0	15	24	3.0
5	4.0	90	15	4.0	20	32	4.0
6[a]							

[a]Values are higher than limits for 5.
Source: Christiansen, Jerald E., Olsen, Edwin C., and Williardson, Lyman S. "Irrigation Water Quality Evaluation," *Journal of the Irrigation and Drainage Diversion*, ASCE, no. IR2, proc. paper 13015, June 1977.

single index. Excessive boron might make a water unsuitable even though all other factors are favorable.

The suitability of a particular water for irrigation will depend upon (1) the adequacy of the drainage, (2) the method of irrigation, (3) the physical properties of the soils to be irrigated, (4) the salt tolerance of the crops to be irrigated, and (5) the management of the irrigation and drainage system. A single criterion cannot be used to evaluate irrigation water quality.

If the salts present are largely sulfates, the values for salt content in each class can be raised 50 percent. Some writers have indicated that waters of 70 percent sodium are unsuitable under most conditions, and a few have suggested an even lower limit. Yet, on sandy soils in the Coachella Valley, California, waters of more than 80 percent sodium are used and the farmers stay in business.

Some writers have selected a specific electrical conductivity ($K \times 10^5$) of 300 as being the upper limit for good production on most soils and waters. A marked exception occurs in the Pecos Valley, where a good growth of alfalfa and cotton is found on lands irrigated with water having a conductance of 915. The Pecos Valley may be considered special because of the heavy content of gypsum and lime in the soil.

The boron content of water is of great importance for many crops. Some crops, as beans, are very sensitive to an excess of boron; others, like sugar beets, will tolerate large quantities. Water containing more than 2.0 milligrams per liter will, in time, usually cause trouble with many of the crops grown.

6.15 SOURCES OF SALINITY IN WATER

Salinity in irrigation water that is obtained from gravity canals originates in one or more of three sources:

(a) In the natural drainage water yielded by watersheds that contain alkali salts in the soils and rocks; or

(b) In the conveyance of rivers or canals through soil or rock formations that are impregnated with alkali salts; or

(c) In the diversion of canals from the lower reaches of streams and rivers that receive seepage and return flow from irrigated lands.

The amounts of salts in natural drainage water near the stream sources are usually so small as to be of little concern. However some streams contain so large an amount of soluble salts that

TABLE 6.5   Increase in Salt Content of the Water in Typical Rivers Due to Seepage and Return Flow of Water from Irrigated Lands

River	State in U.S.A.	Salt Content (mg/l) Upper	Lower	Increase (mg/l)	Distance (km)	Increase per km (mg/l)
Colorado	Colorado	110	1178	1068	32	33
Jordan	Utah	890	1970	1080	23	47
Sevier	Utah	205	831	626	97	6
Sevier	Utah	205	1316	1111	241	5
Pecos	New Mexico	760	2020	1260	48	26
Pecos	New Mexico	760	5000	4240	290	15
Arkansas	Colorado	trace	2200	2200	193	11

the application of its water for irrigation is seriously harmful to trees and general farm crops. Some irrigation water is almost entirely free from salts at the diversion works but absorbs excessive amounts of salts as it is conveyed through canals constructed in saline formations.

The greater and more dangerous source of salts in irrigation water is from seepage and return flow. This fact is illustrated by a study of the salt content of water at different points along rivers that traverse irrigated lands and receive return seepage waters. Table 6.5, reporting some determinations of total alkali salts in five rivers at points separated by 23 to 290 kilometers, shows appreciable increases in alkali from the higher to the lower points on the river. The maximum increase per kilometer, 47 milligrams per liter, occurred in the Jordan River, Utah, and the minimum in the Sevier River, also in Utah. The owners of irrigation projects that divert water from the lower reaches of streams which receive seepage and return flow from upstream alkali lands should know the salt content of the water, and if necessary, take special precautions to avoid injury to crops and soils from this source of alkali.

The salt content of irrigation water in streams often varies appreciably from one part of the irrigation season to another, as is shown by the data of Figure 6.4.

6.16 TOLERANCE OF CROPS TO SALINITY

Some plants can survive in a waterlogged soil for short periods while others cannot survive under the same conditions. For soils having a high water table as well as considerable salinity,

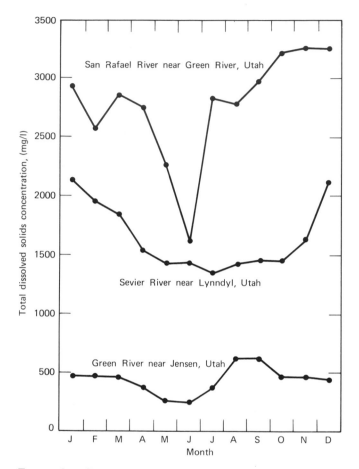

Figure 6.4. Seasonal variation of total dissolved solids of typical rivers.

plants should be selected which can tolerate the waterlogged soils as well as excess salts. Strawberry clover, Burmuda grass, and sweet clover owe part of their popularity to this characteristic.

Selection of salt-tolerant crops depends on the intended use of the crop, moisture conditions of the soil, climate, farm-management practices, and other local factors. Three general tolerance groups are listed in Table 6.6 with group I including those plants which have good tolerance; group II, moderately tolerant to moderately sensitive and group III, sensitive.

The plants listed in Table 6.6 have been classified into three broad divisions; namely, (1) common fruit and vine crops, (2) field

TABLE 6.6 Tolerance of Three Types of Crops for Salinity. Under each of the three types of crops the most tolerant crops are listed first and the least tolerant last.

Type of Crop	Tolerant (Group I)	Moderately Tolerant to Moderately Sensitive (Group II)		Sensitive (Group III)
		Salt Tolerance		
Fruit	Date palm	Pomegranate Fig Grape Olive		Lemon Grapefruit Pear Almond Apricot Peach Plum Apple Orange
Field and Truck	Sugar beet Garden beet Milo Rape Kale Lettuce Alfalfa	Alfalfa Flax Tomato Asparagus Foxtail millet Sorghum (grain) Barley (grain) Rye (grain) Oats (grain) Rice	Cantaloupe Sweet potato Sunflower Carrot Spinach Squash Sugar cane Corn Onion Pepper Wheat (grain) Cotton Potato	Vetch Peas Celery Cabbage Artichoke Eggplant Sweet potato Potato Green beans Black walnut Pecan
Forage	Alkali sacaton Salt grasses Nuttall alkali Bermuda Rhodes Rescue Canada wild rye Beardless wild rye Western wheat grass Barley (hay)	White sweet clover Yellow sweet clover Perennial rye grass Mountain brome Barley (hay) Birdsfoot trefoil Strawberry clover Dallis grass Sudan grass Hubam clover Alfalfa (California common) Tall fescue Rye (hay)	Wheat (hay) Oats (hay) Orchard grass Blue grama Meadow fescue Reed canary Big trefoil Smooth brome Tall meadow oat grass Cicer milk vetch Sour clover Sickle milk vetch	White Dutch clover Meadow foxtail Alsike clover Red clover Ladino clover Burnet

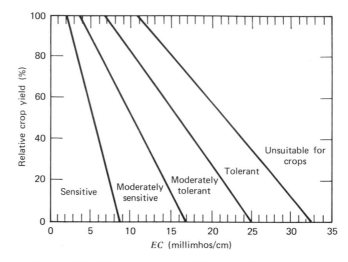

Figure 6.5. Relative crop yield related to salinity.

and truck crops, and (3) forage crops such as grasses, legumes, and cereals, which are used primarily for pasture or hay production. In each group the plants first named are considered to be more tolerant and the last name is more sensitive to salinity.

Figure 6.5 shows relative crop yield[3] related to salinity with categories from tolerant to sensitive as noted in Table 6.6.

Tolerance to Boron

Boron concentrations of as little as 2 to 4 milligrams per liter may be harmful to crops. Some crops are more tolerant to boron than others. Table 6.7 shows that beets have a relatively high tolerance whereas pear, apple, and peach trees have a low tolerance.

Availablility of Plant Nutrients

A healthy soil that is well drained and that has sufficient permeability to permit ready movement of air and water will have a higher percentage of soil nutrients available to plant roots. Figure 6.6 shows how acidity of the soil affects availability of plant nutrients.

[3]E. V. Mass and G. J. Hoffman, "Crop Salt Tolerance—Current Assessment," *Journal of the Irrigation and Drainage Diversion, A.S.C.E.,* June, 1977.

TABLE 6.7   Relative Tolerance of Some Crops to Boron

High	Medium	Low
Sugar beet	Sunflower	Pecan
Garden beet	Cotton	Black walnut
Alfalfa	Radish	Navy bean
Gladiolus	Field peas	Pear
Onion	Barley	Apple
Turnip	Wheat	Peach
Cabbage	Corn	
Lettuce	Milo	
Carrot	Oats	
	Pumpkin	
	Sweet potatoes	

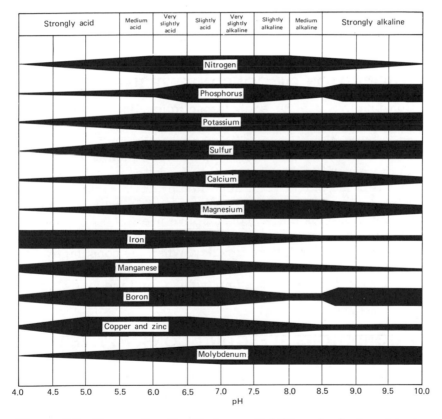

Figure 6.6. How soil pH affects availability of plant nutrients. Availability of nutrients to plants decreases as width of bar decreases. Soil pH is a key factor in regulating nutrient supply. (Courtest National Plant Food Institute.)

111

CONSUMPTIVE USE
OF WATER

Consumptive use of water involves problems of water supply, both surface and underground, as well as problems of management and economics of irrigation projects. Consumptive use has become a highly important factor in the arbitration of controversies over major stream systems where the public welfare of valleys, states, and nations is involved. Before available water resources of a drainage basin in arid and semiarid regions can be satisfactorily ascertained, careful consideration must be given to consumptive-use requirements for water in various subbasins.

7.1 DEFINITION OF CONSUMPTIVE USE

Consumptive use, or evapotranspiration, is the sum of two terms; (1) transpiration, which is water entering plant roots and used to build plant tissue or being passed through leaves of the plant into the atmosphere, (2) evaporation, which is water evaporating from adjacent soil, water surfaces, or from the surfaces of leaves of the plant. Water deposited by dew, rainfall, or sprinkler irrigation and subsequently evaporating without entering the plant system is part of consumptive use.

Consumptive use can apply to water requirements of a crop, a field, a farm, a project, or a valley. When the consumptive use of the crop is known, the water use of larger units can be calculated. Hence, the term "consumptive use" and the following discussion generally refer to the crop.

7.2 CONDITIONS AFFECTING CONSUMPTIVE USE OF WATER

Evapotranspiration is influenced by temperature, irrigation practices, length of growing season, precipitation, and other factors. The volume of water transpired by plants depends in part on the water at their disposal, and also on temperature and humidity of the air, wind movement, intensity and duration of sunlight, stage of development of the plant, type of foliage, and nature of the leaves.

Consumptive use by native vegetation of noncropped land is fairly constant from year to year where the water supply is adequate, but surface evaporation increases in wet years because of the expansion of water surface and moist soil areas. In some irrigated valleys irrigation is lavish when the water supply is abundant, and increased waste water and return flow are consequently available for nonproductive consumption. Also, evaporation is increased from the cropped land soils. Plant diseases and pests reduce consumptive use by inhibiting plant growth.

Size of a farm may affect water usage appreciably since borders, size of stream, and crop rotations may be different.

Evaporation from Soils

Where the water table is near the ground surface, evaporation from the soil is almost equal to evaporation from a free-water surface; whereas, with water levels at greater depths below the surface, evaporation from soil decreases until it becomes negligible when moisture no longer reaches the surface by capillary action.

Where irrigation water is applied by flooding methods, large amounts of water are lost by direct evaporation from soil surfaces without having passed through the roots, stems, and leaves of plants. After light showers during the growing season, considerable rainfall, if not all, is retained on the leaves where it subsequently evaporates without entering the roots of the plants. Showers of less than 5 millimeters seldom enter the soil to a sufficient depth to be used by the plants. Therefore, water received from light rain showers or from light sprinkler irrigations may not be used by the plants in the transpiration process. Subsequent discussion based upon the energy available for evaporation shows, however, that under normal conditions, water that evaporates directly from the soil or leaves is beneficially used, reducing by a like amount water that would otherwise have been transpired by the plants. This is true provided a normal mature crop exists over the soil surface. Early or late in the growing period, when the crop is not transpiring much water, evaporation from a wet soil surface may exceed normal consumptive use.

There are still decided differences in opinion on the effect of cultivation on direct evaporation of water from soils. Recent studies seem to throw considerable doubt on advantages of soil mulches for conserving water through reduction of evaporation losses from soils that are not excessively wet or in contrast with a water table at a shallow depth. Therefore, broad general conclusions concerning influence of cultivation on direct evaporation losses from soils may be misleading. The large number of variable factors involved—notably the differences in distances to free-water sources, in original moisture content of unsaturated soils, in texture, structure, and water conductivity—make it hazardous to generalize.

Transpiration

The process by which water vapor escapes from living plants, principally the leaves, and enters the atmosphere is known as transpiration. During the growing period of a crop, there is a continuous movement of water from the soil into roots, up stems, and out of leaves of the plants. Velocity of water flowing through plants varies widely from 25 to 175 centimeters per hour; but, under conditions of unusually high temperature, dry atmosphere, and wind, velocity of the stream may be greatly increased. A very small proportion of water absorbed by the roots is retained in the plants. If the rate of evaporation at the leaves exceeds the rate of absorption by the roots, wilting occurs and growth of the plant is impeded. On the other hand the available water is not used efficiently, if conditions are such as to stimulate excessive transpiration.

When water changes from liquid to vapor and passes from the plants into the atmosphere, about 590 calories of heat are required for every cubic centimeter being converted from liquid to vapor. Hence, if heat is not available, transpiration ceases; and conversely, as available heat increases, transpiration increases. Heat can come from the soil, the plant, the air, or from direct radiation. The majority of heat used for transpiration, even over short periods of time, comes directly from the sun as radiant energy.

Thus, a given increment of heat will vaporize a given quantity of water. If the available heat is used for vaporizing water from the ground surface, it is not available for vaporizing water within the cavities of the leaf. Therefore, a light shower of rain remaining on the surface of leaves or on the ground can evaporate, utilizing the bulk of available energy. When this occurs, the transpiration of water through the plant will be reduced accordingly.

For example, suppose a rainfall of only 3 millimeters occurred during the dry season and all the moisture was intercepted by leaves of the growing crop. Assume also that daily consumptive use was 6 millimeters per day. Experiments show that the bulk of consumptive use occurs during the day. Therefore, if consumptive use of 6 millimeters per day occurs over, say, 12 hours, the average use per hour would be 0.5 millimeters. During midday, the rate is likely to nearly double the average or 1.0 millimeters per hour. Hence, in 3 hours the rainfall of 3 millimeters would be evaporated and would have been fully utilized. Often overlooked is the fact that rainfall of 3 millimeters may represent only one-half of a normal daily consumptive use. Observations of the recovery of a wilted crop following a light shower will show that the rainfall is not without effect.

Rainfalls of less than 1 centimeter have frequently been neglected in project planning on the erroneous assumption that rain must get into the root zone to be fully effective. When the crop covers the ground surface, a light rain or a light irrigation will reduce by a like amount moisture removed from the soil by the roots, provided of course that an excessive wind does not exist.

7.3 DIRECT MEASUREMENTS OF CONSUMPTIVE USE

Various methods have been used to determine the amount of water consumed by agricultural crops and natural vegetation. Regardless of the method, the problems encountered are numerous. The source of water used by plant life whether from precipitation alone, irrigation plus rainfall, or ground water plus precipitation is a factor in selecting a method. Principal methods are: tank and lysimeter experiments, field experimental plots, soil moisture studies, integration, and inflow-outflow for large areas. These five methods of measuring consumptive use of water are described in the following articles.

Tank and Lysimeter Experiments

Reliability of consumptive-use determinations by means of tanks or lysimeters is dependent on nearness of reproduction of natural conditions. Artificial conditions are caused by limitations of soil, size of tank, regulation of water supply, and sometimes environment.

Tanks should be placed in surroundings of natural growth of the same species, that is, in their natural environment, so that consumptive use of water will presumably be the same as for similar growth outside the tank. It has been found that all tank vegetation must be protected from the elements by surrounding growth of the same species.

Lysimeters can be grouped into three categories: (1) nonweighing, constant water table type for use where a high water table normally exists, (2) nonweighing, percolation type often used in areas of high precipitation; and (3) weighing types, which provide the most accurate data for short time periods.

Soil Moisture Depletion

Consumptive use of water for various crops has been determined by intensive soil moisture studies. This method is usually suitable for areas where soil is fairly uniform and the depth to ground water is such that it will not influence soil moisture fluctuation within the root zone. Soil moisture is determined before and after each irrigation with some measurements between irrigations in the major root zone. The soil is usually sampled 2 to 4 days after an irrigation and again 7 to 15 days later or just before the next irrigation. At least six sampling sites representative of general field conditions are used. To minimize drainage and deep percolation errors, only those sampling periods where rainfall is light are used, and less water should be applied at each irrigation than can be retained by the soil within the root zone.

Soil water depletion can be measured by taking soil samples and gravimetric analysis. The neutron soil water probe has essentially replaced the gravimetrical procedure.

Water extracted per day from the soil is computed for each period. When rate of use is plotted against time, a curve can be drawn from which the seasonal use can be obtained.

Integration Method

The integration method is the summation of the products of unit consumptive use for each crop times its area, plus the unit consumptive use of native vegetation times its area, plus water surface evaporation times water surface area, plus evaporation from bare land times its area.

Before this method can successfully be applied, it is necessary to know unit consumptive use of water and the areas of various classes of agricultural crops, native vegetation, bare land, and water surfaces. By means of aerial maps and field surveys, areas of various types of native vegetative cover, bare land, and water surfaces can be determined.

Inflow-Outflow for Large Areas

Applying this method, valley consumptive use U is equal to the water that flows into the valley during a 12-month year I, plus yearly precipitation on the valley floor P, plus water in ground storage at the beginning of the year G_s, minus water in ground storage at the end of the year G_e, minus yearly outflow R.

Thus

$$U = (I + P) + (G_s - G_e) - R \qquad (7.1)$$

The difference between storage of capillary water at the beginning of the year and at the end of the year is usually considered to be negligible. It is assumed that stream measurements are made on bedrock controls, and that the subsurface inflow is about the same as subsurface outflow. The quantity $(G_s - G_e)$ is considered as a unit so that absolute evaluation of either G_s or G_e is unnecessary, only the difference being needed. This is the product of the difference in the average depth of water table during the year and multiplied by the specific yield[1] of the soil and area of the valley floor. The quantity P is obtained by multiplying the average annual precipitation in feet by the area of the valley floor. Unit consumptive use of the entire valley is obtained by dividing the total consumptive use by the area of the valley floor. Care must be taken to use the same dimensions in all terms of Equation 7.1.

Energy Balance and Heat and Mass Transfer

The energy balance method of determining consumptive use can be used for hourly values during daylight hours but accurate nighttime values are difficult to obtain. Eddy diffusion equations can be used and combinations of these procedures can be used also to calculate consumptive use. The method used is governed often by the data available, the accuracy needed, and the computational capability.

7.4 CONSUMPTIVE USE BY NATURAL VEGETATION

Water consumed by natural vegetation cannot be made available for irrigation of crops. Therefore, in considering the water supply of a region, water consumed by natural vegetation (such as salt grass, willows, cottonwoods, tamarisk, and tules) growing in irrigated valleys, moist areas, and along streams becomes of increasing importance as greater land areas are irrigated, especially during periods of drought. In planning new irrigation projects, consideration must often be given to differences in consumptive uses of water by irrigated crops and by natural vegetation replaced by the crops.

The relation of plant communities to moisture supply is one of the outstanding characteristics of the growth of natural vegetation. Whereas individual species are largely restricted to physical environments, the principal condition that governs distribu-

[1]Specific yield is defined as the total pore space of the soil less the moisture content at field capacity, both expressed as volume percentages of the total soil volume.

tion of vegetative groups in irrigated areas is the availability of water. Each species responds to water conditions for its most favorable growth and its widest distribution. Temperatures, moisture, and chemical and physical properties of soils are contributing factors in the distribution of natural vegetation. Quantity of water available for plant use and the effect of plant growth on supply are of great interest.

Measurements of consumptive use indicate that water-loving natural vegetation uses from 50 to 100 percent more water than most crop plants. Tules and cattails growing in or near irrigation canals and drainage ditches are exposed in narrow strips to sun and wind so that their consumption of water is high. Under such circumstances the natural vegetation along a kilometer of canal or ditch may consume enough water to irrigate 2 hectares of alfalfa or a greater area of other field crops or of fruit. Consumptive use of water by water-loving (phreatophytic) plants usually equals, or exceeds, the evaporation from a free-water surface.

7.5 CLIMATIC OBSERVATIONS AS AN INDEX TO CONSUMPTIVE USE[2]

How temperature, humidity, wind velocity, vapor pressure, and solar radiation influence consumptive use has been studied by several researchers. Penman, in England, has made the most complete analysis using several climatic variables, whereas temperature has been used as the principal variable to obtain an index to consumptive use by Blaney and Criddle in the arid western United States. Jensen and Haise used temperature and solar radiation to estimate the consumptive use of a reference crop (alfalfa) as did Hargreaves for grass as a reference crop. These methods and their significance are outlined in the following sections. An excellent discussion of factors affecting consumptive use along with a comparable evaluation of 15 or so different methods is presented in the report *Consumptive Use of Water and Irrigation Water Requirements*, M. E. Jensen, ed., American Society of Civil Engineers (A.S.C.E.), 1974. Other sources of detailed information are: "Crop Water Requirements," J. Doorenbos and W. O. Pruitt, FAO Irrigation and Drainage Paper no. 24 (Revised 1977), F.A.O., Rome.; "Irrigation Water Requirement," U.S.D.A., S.C.S., Technical Release no. 21, April

[2]Appreciation is extended to Dr. Robert W. Hill, Utah State University, for his contribution to this section which discusses climatic observations as an index to consumptive use.

1967 (Revised September 1970); and "Water Requirements Manual for Irrigated Crops and Rainfed Agriculture," G. H. Hargreaves, Utah State University, June 1977. An earlier digest of the Penman, Blaney-Criddle, and other methods was prepared by Criddle, in "Methods of Computing Use of Water," Proceedings Paper 1507, A.S.C.E., January 1958. Formulas and charts to follow were taken from these publications.

Modified Penman Method

Penman has made the most complete theoretical approach, showing that consumptive use is inseparably connected to incoming solar energy.

His formula representing the potential evapo-transpiration (consumptive use) is as follows in the modified form:

$$E_{tp} = \frac{\Delta}{\Delta + \gamma} (R_n + G) + \frac{\gamma}{\Delta + \gamma} 15.36 (w_1 + w_2 u_2) (e_s - e_a) \quad (7.2)$$

where E_{tp} = reference crop potential evapo-transpiration, well-watered alfalfa in cal/cm^2 per day (langleys/day).

   Δ = slope of saturation vapor pressure-temperature curve (de/dT) in mbar/°C

   γ = psychrometric constant

   R_n = net radiation in cal/cm^2 per day

   G = soil heat flux in cal/cm^2 per day

   u_2 = wind movement in km/day at 2 m

   e_s = saturation vapor pressure, mean of values obtained at daily maximum and daily minimum temperatures in mbar (This is a modification of the original Penman equation.)

   e_a = mean actual vapor pressure in mbar

w_1, w_2 = wind term coefficients, some empirically determined values are:

w_1	w_2	Location	Reference Crop
1.10	0.0106	Mitchell, Nebraska	alfalfa
0.75	0.0115	Kimberly, Idaho	alfalfa
1.00	0.0062	Penman	short grass

Judgment should be exercised in the use of any wind term coefficients if there is a different advective energy condition at night than during the day, such as experienced in some mountain valleys with canyon breezes or in coastal areas. The following additional equations are required for solution of Equation 7.2:

$$\gamma = c_p \ \frac{P}{(0.622\gamma)} \tag{7.3}$$

where $c_p = 0.240$

$P = 1013 - 0.1055 \ EL$, mbar. (*EL* is elevation in meters) (7.4)

λ = latent heat of water in cal/g; estimated by $\lambda = 595.9 - 0.55 \ T$, T in °C (7.5)

$\Delta = 33.86 \ [0.05904 \ (0.00738T + 0.8072)^7 - 0.0000342]$
for $T \geqslant 23$°C, Δ in mbar/°C (7.6)

$R_n = 0.77 \ R_s - R_b$ (7.7)

where R_s = incident solar radiation in cal/cm² per day.

The 0.77 value is obtained by assuming a reflectivity of 0.23 for a green growing crop.

$$R_b = R_{b0} \ [(aR_s/R_{s0}) + b] \tag{7.8}$$

where R_{s0} = clear day solar radiation in langleys/day. If actual records are not available, R_{s0} values may be estimated from Table 7.1.

a,b = empirical constants, see table following equation 7.9.

$$R_{b0} = (a_1 + b_1 \ \sqrt{e_a}) \ \ 11.71 \times 10^{-8} \ (T_a{}^4 + T_b{}^4)/2 \tag{7.9}$$

where a_1, b_1 = empirical constants, see following table

e_a = mean actual vapor pressure in mbar

T_a = maximum daily temperature in °K

T_b = minimum daily temperature in °K

Values of a, b, a_1 and b_1 have been determined for various locations as:

a	b	a_1	b_1	Location
0.90	0.10	0.37	−0.044	Mitchell, Nebraska[a]
1.35	−0.35	0.35	−0.046	Davis, California
1.22	−0.18	0.33	−0.044	Kimberly, Idaho
1.20	−0.20			Arid regions (suggested)
1.10	−0.10			Semihumid (suggested)
1.00	0.00			humid (suggested)
		0.39	−0.05	general

[a] The reported w_1, w_2, a, b, a_1, and b_1, values for Mitchell, Neb. are adapted from *Scheduling Irrigations Using a Programmable Calculator,* D. C. Kincaid and D. F. Heerman, U.S.D.A., ARS-NC-12, February 1974. Reported values for other locations are adapted from the A.S.C.E. report.

TABLE 7.1 Mean Daily Solar Radiation for Cloudless Skies, Langleys

Latitude °N	Month											
	Jan.	Feb.	Mar.	Apr.	May	June	July	Aug.	Sept.	Oct.	Nov.	Dec.
60	58	152	319	533	671	763	690	539	377	197	87	35
55	100	219	377	558	690	780	706	577	430	252	133	74
50	155	290	429	617	716	790	729	616	480	313	193	126
45	216	365	477	650	729	797	748	648	527	371	260	190
40	284	432	529	677	742	800	755	674	567	426	323	248
35	345	496	568	700	742	800	761	697	603	474	380	313
30	403	549	600	713	742	793	755	703	637	519	437	371
25	455	595	629	720	742	780	745	703	660	561	486	423
20	500	634	652	720	726	760	729	697	680	597	537	474
15	545	673	571	713	706	733	706	684	697	623	580	519
10	584	701	581	707	584	700	681	665	707	648	617	565
5	623	722	590	700	552	663	645	645	710	665	650	606
0	652	740	694	680	623	627	616	623	707	684	680	619
−5	648	758	690	663	590	587	577	590	693	690	727	677
−10	710	772	681	640	571	543	526	558	680	590	727	710
−15	729	779	665	610	516	497	497	519	657	587	747	739
−20	748	779	645	573	474	447	445	481	630	577	753	761
−25	761	779	626	533	419	400	406	439	600	665	767	777
−30	771	772	600	497	384	353	358	390	567	648	767	793
−35	774	754	568	453	335	300	310	342	530	629	757	806
−40	774	729	529	407	281	243	261	290	477	603	760	813
−45	774	704	490	357	229	183	203	235	447	571	747	843
−50	761	669	445	307	174	127	148	177	400	535	727	806
−55	748	630	397	250	123	77	97	123	343	497	707	794
−60	729	588	348	187	77	33	52	74	283	455	700	787

Source: From Table 3-1, Consumptive Use of Water and Irrigation Water Requirements, M. E. Jensen, ed. A.S.C.E. 1974.

An empirical equation for estimating the soil heat flux is:

$$G = [\overline{T}_{pr} - \overline{T}]\ 9.1 \qquad\qquad (7.10)$$

where $\overline{T}_{pr}$ = mean air temperature for a previous time period, usually the previous three days when daily estimates of E_{tp} are required

$\overline{T}$ = mean air temperature for the current time period, ie mean air temperature of the particular day for which E_{tp} is required.

In some locations wind movement may be available from an instrument at a height different than 2 meters. The following relationship provides a reasonable estimate of the wind movement at 2 meters when the anemometer is at height z:

$$u_2 = u_z \left(\frac{2}{z}\right)^{0.2} \qquad\qquad (7.11)$$

where u_z = wind movement at height z

z = height of anemometer above ground in m

Penman found that coefficients were necessary to reduce the potential rate of consumptive use to the actual water use of pasture growing in England. Very good correlation has been obtained between computed and measured consumptive-use values. The modified Penman equation has been used to estimate consumptive use for time intervals of 1 hour or less for research studies with comprehensive instrumentation. This method has been used extensively by government agencies and private companies who are providing computerized irrigation scheduling services for farmers. These groups usually set up their own weather stations in order to have daily data inputs for updating soil moisture status and estimating irrigation requirements on a field-by-field basis.

The principal limitation of the Penman approach is the lack of sufficient weather measurements in most localities. Only a few climatological stations record the needed data. Hence, the formula, even though quite reliable, does have some practical limitations.

Another fact to remember in use of the Penman equation is that the coefficients w_1, w_2, a, b, a_1, b_1, etc. should be calibrated for the specific location to better fit the relative importance of advective and solar energy in determining consumptive use. Oftentimes irrigated agriculture in arid regions is surrounded by vast desert areas with little vegetative cover and no free-water surfaces. This condition results in a greater quantity of energy being available for consumptive use than would be indicated by the incoming solar energy.

In the ASCE report comparing the various methods of calculating consumptive use, the modified Penman equation is ranked behind the Jensen-Haise method (which tied for first place in terms of reliability) for use in inland-semiarid to arid conditions. In coastal areas, Penman ranked fifth actually behind the Blaney-Criddle method. Apparently no site calibration was included in making the comparisons.

Jensen-Haise Method

Jensen and Haise used observations of consumptive use from the western United States to develop a linear relationship for estimating E_{tp} (defined in Equation 7.2) as:

$$E_{tp} = C_t \, (T - T_x) \, R_s \qquad (7.12)$$

where C_t = a temperature coefficient
$\quad T$ = temperature in °C
$\quad T_x$ = the intercept on the temperature axis
$\quad R_s$ = incident solar radiation in langleys/day

The values of C_t and T_x were determined as 0.025 and -3, respectively. These coefficients were further defined for other areas as:

$$C_t = (1/C_1 + C_2 C_H) \qquad (7.13)$$

and

$$C_H = 50 \text{ mbar}/(e_2 - e_1) \qquad (7.14)$$
$$C_1 = 38 - (2°C \times EL/305) \qquad (7.15)$$
$$C_2 = 7.6°C$$

where

$\quad e_1, \; e_1$ = saturation vapor pressures of water at the mean maximum and mean minimum temperatures, respectively, for the warmest month of the year in a given area

and

$$T_x = -\,2.5 - 0.14 \, (e_2 - e_1)°C/mbar - EL/550 \qquad (7.16)$$

Although the Jensen-Haise equation has been used to calculate E_{tp} on a day-to-day basis, it should be considered as reliably representing values for longer time periods (5 to 7 days), when effects of fluctuations in wind and humidity level are averaged out. This method has been widely used. However, there is some evidence that the elevation adjustments may be excessive and should be used with caution, particularly at higher elevations

(1500 to 2000 meters and above). This method performs quite reliably for semiarid to arid conditions as indicated by the comparison mentioned previously.

Estimates of crop consumptive use are calculated from E_{tp}, determined from either the Penman or Jensen-Haise method, by use of a crop coefficient which relates growth stage back to the reference crop. In equation form:

$$E_t = K_{co}E_{tp} \qquad (7.17)$$

where E_t = estimated crop consumptive use
$\quad\quad K_{co}$ = crop coefficient, see Table 7.2

Adjustment of the E_t value given by Equation 7.17 should be made for field situations where the soil water may drop to a level causing plant stress and subsequent reduction in consumptive use. The A.S.C.E. report and the paper "Model for Predicting Plant Yield as Influenced by Water Use," R. J. Hanks, Agronomy Journal, Vol. 65, 1974, each present acceptable techniques for making this adjustment.

TABLE 7.2   Experimental Crop Coefficients, K_{co}

| Crop | Planting to Effective Cover (%) | | | | | | | | | |
	10	20	30	40	50	60	70	80	90	100
Small grains	0.16	0.18	0.25	0.37	0.51	0.67	0.82	0.94	1.02	1.04
Beans	0.20	0.23	0.30	0.39	0.51	0.63	0.76	0.88	0.98	1.07
Peas	0.20	0.24	0.31	0.40	0.51	0.63	0.75	0.87	0.97	1.05
Potatoes	0.10	0.13	0.20	0.30	0.41	0.53	0.65	0.76	0.85	0.91
Sugar beets	0.10	0.13	0.20	0.30	0.41	0.53	0.65	0.76	0.85	0.91
Corn	0.20	0.23	0.29	0.38	0.49	0.61	0.72	0.82	0.91	0.96
Alfalfa	0.36	0.47	0.58	0.68	0.79	0.90	1.00	1.00	1.00	1.00[1]
Pasture	0.87	0.87	0.87	0.87	0.87	0.87	0.87	0.87	0.87	0.87

| Crop | Days After Effective Cover | | | | | | | | | |
	10	20	30	40	50	60	70	80	90	100
Small grains	1.04	0.94	0.74	0.49	0.19	0.10	0.10	0.10	0.10	0.10
Beans	1.02	0.96	0.85	0.73	0.59	0.45	0.31	0.19	0.10	0.10
Peas	0.98	1.02	0.99	0.76	0.20	0.10	0.10	0.10	0.10	0.10
Potatoes	0.90	0.85	0.75	0.60	0.38	0.10	0.10	0.10	0.10	0.10
Sugar beets	0.90	0.90	0.90	0.90	0.90	0.90	0.90	0.90	0.90	0.90
Corn	0.99	0.99	0.93	0.82	0.68	0.54	0.40	0.28	0.20	0.17
Alfalfa	0.75	1.00	1.00	1.00	1.00	1.00	1.00	1.00	1.00	1.00
Pasture	0.87	0.87	0.87	0.87	0.87	0.87	0.87	0.87	0.87	0.87

Source: From Table 6.5, Consumptive Use of Water and Irrigation Water Requirements, M. E. Jensen, ed., A.S.C.E., 1974.
[1]100% effective cover is at first cutting, K_{co} drops to 0.50 after each cutting and recovers to 1.00 in 20 days.

Hargreaves Equation

Hargreaves based his work on data from grass lysimeters. The method is similar in concept to Jensen-Haise and is subject to similar limitations. It has been used extensively in Latin America to estimate crop water requirements. In equation form:

$$E_{tg} = 0.0135 \, (T + 17.78) \, R_s \qquad (7.18)$$

where E_{tg} = reference crop potential consumptive use, well watered grass in langleys/day

   T = average daily temperature in °C

   R_s = incident solar radiation in langleys/day

Hargreaves indicates that a reference crop of alfalfa would have a consumptive use about 20 percent greater than grass, thus E_{tp} can be estimated from E_{tg} by

$$E_{tp} = 1.2 \, E_{tg} \qquad (7.19)$$

which allows use of the K_{co} values from Table 7.2.

This method does not have any adjustments for site specific conditions of elevation or humidity as does those previously presented. However, some unpublished comparisons have indicated that it performs as well as more complicated questions at many sites where lysimeter data were available. Hargreaves has also suggested that where measured R_s values are not available, a fair estimate is

$$R_s = 0.10 \, R_{so} \, (S) \, ½ \qquad (7.20)$$

where S = percent of possible sunshine

Conversion Factor for E_{tp} from Langleys per Day to Millimeters per Day

The preceding three methods for estimating reference crop consumptive use all present the E_{tp} or E_{tg} values in units of langleys per day. The units of a langley (calories per square centimeter) when multiplied by 10 and divided by the latent heat of water (λ of Equation 7.5) are converted to millimeters of water per day, which is a usual and more desirable unit for consumptive use calculations.

Blaney-Criddle Method

Blaney and Criddle developed a simplified formula using temperature and day-time hours for the arid western portion of the United States. Their formula has been used extensively by the

Latitude 0° North	January	February	March	April	May	June
0	8.50	7.66	8.49	8.21	8.50	8.22
5	8.32	7.57	8.47	8.29	8.65	8.41
10	8.13	7.47	8.45	8.37	8.81	8.60
15	7.94	7.36	8.43	8.44	8.98	8.80
20	7.74	7.25	8.41	8.52	9.15	9.00
25	7.53	7.14	8.39	8.61	9.33	9.23
30	7.30	7.03	8.38	8.72	9.53	9.49
32	7.20	6.97	8.37	8.76	9.62	9.59
34	7.10	6.91	8.36	8.80	9.72	9.70
36	6.99	6.85	8.35	8.85	9.82	9.82
38	6.87	6.79	8.34	8.90	9.92	9.95
40	6.76	6.72	8.33	8.95	10.02	10.08
42	6.63	6.65	8.31	9.00	10.14	10.22
44	6.49	6.58	8.30	9.06	10.26	10.38
46	6.34	6.50	8.29	9.12	10.39	10.54
48	6.17	6.41	8.27	9.18	10.53	10.71
50	5.98	6.30	8.24	9.24	10.68	10.91
52	5.77	6.19	8.21	9.29	10.85	11.13
54	5.55	6.08	8.18	9.36	11.03	11.38
56	5.30	5.95	8.15	9.45	11.22	11.67
58	5.01	5.81	8.12	9.55	11.46	12.00
60	4.67	5.65	8.08	9.65	11.74	12.39
South						
0	8.50	7.66	8.49	8.21	8.50	8.22
5	8.68	7.76	8.51	8.15	8.34	8.05
10	8.86	7.87	8.53	8.09	8.18	7.86
15	9.05	7.98	8.55	8.02	8.02	7.65
20	9.24	8.09	8.57	7.94	7.85	7.43
25	9.46	8.21	8.60	7.84	7.66	7.20
30	9.70	8.33	8.62	7.73	7.45	6.96
32	9.81	8.39	8.63	7.69	7.36	6.85
34	9.92	8.45	8.64	7.64	7.27	6.74
36	10.03	8.51	8.65	7.59	7.18	6.62
38	10.15	8.57	8.66	7.54	7.08	6.50
40	10.27	8.63	8.67	7.49	6.97	6.37
42	10.40	8.70	8.68	7.44	6.85	6.23
44	10.54	8.78	8.69	7.38	6.73	6.08
46	10.69	8.86	8.70	7.32	6.61	5.92

TABLE 7.3 Percent Daylight Hours

July	August	September	October	November	December
8.50	8.49	8.21	8.50	8.22	8.50
8.67	8.60	8.23	8.42	8.07	8.30
8.86	8.71	8.25	8.34	7.91	8.10
9.05	8.83	8.28	8.26	7.75	7.88
9.25	8.96	8.30	8.18	7.58	7.66
9.45	9.09	8.32	8.09	7.40	7.42
9.67	9.22	8.33	7.99	7.19	7.15
9.77	9.27	8.34	7.95	7.11	7.05
9.88	9.33	8.36	7.90	7.02	6.92
9.99	9.40	8.37	7.85	6.92	6.79
10.10	9.47	8.38	7.80	6.82	6.66
10.22	9.54	8.39	7.75	6.72	6.52
10.35	9.62	8.40	7.69	6.62	6.37
10.49	9.70	8.41	7.63	6.49	6.21
10.64	9.79	8.42	7.57	6.36	6.04
10.80	9.89	8.44	7.51	6.23	5.86
10.99	10.00	8.46	7.45	6.10	5.65
11.20	10.12	8.49	7.39	5.93	5.43
11.43	10.26	8.51	7.30	5.74	5.18
11.69	10.40	8.53	7.21	5.54	4.89
11.98	10.55	8.55	7.10	4.31	4.56
12.31	10.70	8.57	6.98	5.04	4.22
8.50	8.49	8.21	8.50	8.22	8.50
8.33	8.38	8.19	8.56	8.37	8.68
8.14	8.27	8.17	8.62	8.53	8.88
7.95	8.15	8.15	8.68	8.70	9.10
7.76	8.03	8.13	8.76	8.87	9.33
7.54	7.90	8.11	8.86	9.04	9.58
7.31	7.76	8.07	8.97	9.24	9.85
7.21	7.70	8.06	9.01	9.33	9.96
7.10	7.63	8.05	9.06	9.42	10.08
6.99	7.56	8.04	9.11	9.51	10.21
6.87	7.49	8.03	9.16	9.61	10.34
6.76	7.41	8.02	9.21	9.71	10.49
6.64	7.33	8.01	9.26	9.82	10.64
6.51	7.25	7.99	9.31	9.94	10.80
6.37	7.16	7.96	9.37	10.07	10.97

Soil Conservation Service of the United States Department of Agriculture (S.C.S.) wherein considerable data has been collected to determine the values of the coefficients to be used for various crops. The concept of a reference crop is not appropriate with the form of the equation presented here. Doorenbos and Pruitt, in the F.A.O. report, suggest a modification of the Blaney-Criddle formula for which the corresponding crop coefficients appear to be less dependent on climate.

By multiplying the mean monthly temperature t by the monthly percentage of day-time hours of the year p, there is obtained a consumptive use factor F. Expressed mathematically, it is:

$$U = 25.4 \, KF = 25.4 \, K \, \Sigma \, \frac{tp}{100} \qquad (7.21)$$

where the following quantities must be determined for the same period:

U = consumptive use of crop in mm for a given time period
K = empirical crop consumptive use coefficient (annual, irrigation season, or growing season)
F = sum of the consumptive use factors for the period (sum of the products of mean temperature and percent of annual day-time hours) $(t \times p)/100$
t = mean temperature in °F
p = percentage of day-time hours of the year, occurring during the period. (See Table 7.3 on pages 126–127).

For monthly calculations, lower case letters are frequently used for clarity as follows:

$$u = 25.4 \, kf = 25.4 \, k \, \frac{tp}{100} \qquad (7.22)$$

where u = monthly consumptive use, in mm
k = monthly crop coefficient
f = monthly consumptive-use factor, $(t \times p)/100$

The consumptive use of water by a particular crop in some areas being known, an estimate of the use by the same crop in some other area may be made by application of the formula $U = 25.4KF$.

Table 7.4 illustrates use of the method to compute monthly consumptive use by alfalfa in Salinas Valley, California. Monthly coefficients k were developed from measured consumptive use u and temperature t in San Fernando Valley, California.

Values of the normal seasonal consumptive-use coefficients for several crops are shown in Table 7.5. The last column also

TABLE 7.4   Computed Normal Unit Consumptive Use of Water by Alfalfa, Upper Salinas Valley, California

Month	Mean Monthly Temperature, (°F)	Daytime Hours (%)	Consumptive-Use Factor	Coefficient	Consumptive Use (mm)
	t	p	f	k	u
April	57.9	8.85	5.12	0.60	78
May	62.5	9.82	6.14	0.70	109
June	65.7	9.84	6.46	0.80	131
July	68.4	10.00	6.84	0.85	148
August	67.8	9.41	6.38	0.85	138
September	66.6	8.36	5.57	0.85	120
October	62.2	7.84	4.88	0.70	87
Total consumptive use for irrigation season					811

shows maximum monthly values that can be expected. Usually, irrigation systems must be designed to meet the maximum monthly water needs rather than the seasonal average.

The SCS Technical Release No. 21 presents a modified Blaney-Criddle equation for computing monthly consumptive use. The modification is made on the monthly k as:

$$k = k_c k_t \tag{7.23}$$

k_c = monthly crop coefficient, see SCS TR21
k_t = a climatic coefficient (calculated from
  $k_t = 0.0173\ t - 0.314$) subject to $k_t \geqslant 0.30$,
  where t is in °F

The Blaney-Criddle formula has a serious limitation in representing consumptive use for shorter time periods than 1 month due to the use of temperature as the only climatological variable. However, this formula has been used very extensively. In some instances, the monthly crop coefficients have been recalulated to better fit observed crop water use in a particular region. It is an empirical formula developed to fit arid conditions and will give good estimates of seasonal water needs under these conditions.

Computing Consumptive Use

Table 7.6 has been prepared to assist in the interpretation and use of the forgoing equations for computing consumptive use.

The computed consumptive uses from Table 7.6 arranged in order of increasing magnitude are as follows:

TABLE 7.5 Seasonal Consumptive-Use Crop Coefficients (K) for Irrigated Crops, for Use in Equation 7.21

Crop	Length of Normal Growing Season or Period[a]	Consumptive-Use Coefficient K[b]	Maximum Monthly k[c]
Alfalfa	Between frosts	0.80 to 0.90	0.95 to 1.25
Bananas	Full year	0.80 to 1.00	—
Beans	3 months	0.60 to 0.70	0.75 to 0.85
Cocoa	Full year	0.70 to 0.80	—
Coffee	Full year	0.70 to 0.80	—
Corn (Maize)	4 months	0.75 to 0.85	0.80 to 1.20
Cotton	7 months	0.60 to 0.70	0.75 to 1.10
Dates	Full year	0.65 to 0.80	—
Flax	7 to 8 months	0.70 to 0.80	—
Grains, small	3 months	0.75 to 0.85	0.85 to 1.00
Grain, sorghums	4 to 5 months	0.70 to 0.80	0.85 to 1.10
Oilseeds	3 to 5 months	0.65 to 0.75	—
Orchard crops:			
Avocado	Full year	0.50 to 0.55	—
Grapefruit	Full year	0.55 to 0.65	—
Orange and lemon	Full year	0.45 to 0.55	0.65 to 0.75[d]
Walnuts	Between frosts	0.60 to 0.70	—
Deciduous	Between frosts	0.60 to 0.70	0.70 to 0.95
Pasture crops:			
Grass	Between frosts	0.75 to 0.85	0.85 to 1.15
Ladino whiteclover	Between frosts	0.80 to 0.85	—
Potatoes	3 to 5 months	0.65 to 0.75	0.85 to 1.00
Rice	3 to 5 months	1.00 to 1.10	1.10 to 1.30
Soybeans	140 days	0.65 to 0.70	—
Sugar beet	6 months	0.65 to 0.75	0.85 to 1.00
Sugarcane	Full year	0.80 to 0.90	—
Tobacco	4 months	0.70 to 0.80	—
Tomatoes	4 months	0.65 to 0.70	—
Truck crops, small	2 to 4 months	0.60 to 0.70	—
Vineyard	5 to 7 months	0.50 to 0.60	—

[a]Length of season depends largely on variety and time of year when the crop is grown. Annual crops grown during the winter period may take much longer than if grown in the summertime.
[b]The lower values of K for use in the Blaney-Criddle formula, $U = KF$, are for the more humid areas, and the higher values are for the more arid climates.
[c]Dependent upon mean monthly temperature and crop growth stage.
[d]Given by Criddle as "Citrus orchard."
Source: From "Irrigation Water Requirements," Technical Release No. 21, Soil Conservation Service, U.S.D.A., September 1970.

TABLE 7.6 Computation Sheet for Calculating Consumptive Use

Location: Logan, Utah	Latitude: 41°45′
Crop: Spring Wheat, Seeding Date May 1	Elevation: 1430 meters
Heading Date July 15	

A. *Data, July 1977*
1. Air temperature
    mean maximum in °C   30.8
    mean minimum in °C   11.6
2. Average dewpoint °C   8.0
3. Average wind movement in km/day   23.7
4. Average clear day solar radiation
    in langleys/day   752
5. Average solar radiation in
    langleys/day   549

B. *Modified Penman Equation*

$\gamma = c_p P/(0.622\lambda)$ Eq. 7.3

1. c_p 0.240
2. $P = 1013 - 0.1055$ EL   Eq. 7.4   862
3. $\lambda = 595.9 - 0.55T$ Eq. 7.5   584
4. $1/(0.622\lambda)$ 0.00275
5. $\gamma = c_p P/(0.622\lambda)$ Eq. 7.3   0.569
    $R_b = R_{bo}[(aR_s/R_{so}) + b]$ Eq. 7.8
6. $T_a = $ °C + 273°C, °K (mean max)   304
7. $T_b = $ °C + 273°C, °K (mean min.)   285
8. a_1 0.39
9. a_2 −0.05
10. e_a (e_s at dew point) in mbar   10.7
11. $R_{bo} = (a_1 + b_1 \sqrt{e_a})\ 11.71 \times 10^{-8}\ (T_a{}^4 + T_b{}^4)/2$ Eq. 7.9   200
12. a 1.2
13. b −0.20
14. $R_b = R_{bo}[(aR_s/R_{so}) + b]$ Eq. 7.8   135
15. $R_n = 0.77\ R_s - R_b$ Eq. 7.7   288
    $G = [\bar{T}_{pr} - \bar{T}]\ 9.1$ Eq. 7.10
16. Assume $G = 0$
    $\Delta = 33.86\ [0.05904\ (0.00738T + 0.8072)^7 - 0.0000342]$
                                               Eq. 7.6
16. Δ Eq. 7.6   1.54

$$E_{tp} = \frac{\Delta}{\Delta + \gamma}\ (R_n + G) + \frac{\gamma}{\Delta + \gamma}\ 15.36\ (w_1 + w_2 u_2)\ (e_s - e_a)$$

                                               Eq. 7.2

17. $\dfrac{\Delta}{\Delta + \gamma}\ (R_n + G)$ 210
18. $\gamma/(\Delta + \gamma)$ 0.228
19. w_1 1.10
20. w_2 0.0062
21. $15.36\ (w_1 + w_2 u_2)$ 19.2

Table continues on next page

Table 7.6 Continued

22. e_s in mbar		29.4
23. $(e_s - e_a)$ in mbar		18.7
24. E_{tp}		307
25. K_{co}	Table 7.2	1.03
26. $E_{t\ crop}$ in langleys/day		316
27. $E_{t\ crop} \times 10/\lambda$ in mm/day		5.4

C. *Jensen-Haise Equation*
  Necessary additional data

1. T_2 (mean maximum for warmest month) in °C		30.8
2. T_1 (mean minimum for warmest month) in °C		15.6
$\quad C_t = 1/(C_1 + C_2\ C_H)$	Eq. 7.13	
$\quad C_H = 50\ mbar/(e_2 - e_1)$	Eq. 7.14	
3. $C_1 = 38 - (2°C \times EL/305)$	Eq. 7.15	28.6
4. $C_2 = 7.6°C$		7.6
5. e_2		44.5
6. e_1		17.7
7. $C_H = 50\ mbar/(e_2 - e_1)$	Eq. 7.14	1.87
8. $C_T = 1/(C_1 + C_2\ C_H)$	Eq. 7.13	0.0234
9. $T_x = -2.5 - 0.14\ (e_2 - e_1)$ °C/mbar $- EL/550$		
	Eq. 7.16	−8.85
10. $E_{tp} = C_t\ (T - T_x)\ R_s$	Eq. 7.12	386
$\quad E_{t\ crop} = K_{co}E_{tp}$	Eq. 7.17	
11. K_{co}	Table 7.2	1.03
12. $E_{t\ crop} = K_{co}E_{tp}$ in langleys/day	Eq. 7.17	398
13. $E_{t\ crop} \times 10/\lambda$ in mm/day		6.8

D. *Hargreaves Equation*

1. $E_{tg} = 0.0135\ (T + 17.78)\ R_s$	Eq. 7.18	289
2. $E_{tp} = 1.2\ E_{tg}$, in langleys/day	Eq. 7.19	347
3. K_{co}	Table 7.2	1.03
4. $E_{t\ crop} = K_{co}\ E_{tp}$ in langleys/day	Eq. 7.17	357
5. $10/\gamma$ in mm/day		6.1

E. *Modified Blaney-Criddle Equation*

$u = 25.4kf$	Eq. 7.22	
$k = k_t\ k_c$	Eq. 7.23	
1. $t°F = \dfrac{9}{5}\ T°C + 32°$		70.2
2. $k_t = 0.0173\ T°F - 0.314$		0.90
3. k_c (from SCS, USDA Technical Release no. 21, p76)		1.27
4. $k = k_t k_c$	Eq. 7.23	1.14
$\quad f = \Sigma tp$		
5. p (July, 42°N. Latitude)	Table 7.3	10.4
6. $f = \Sigma tp$		7.30
7. $u = 25.4\ KF$ in mm/month	Eq. 7.21	211
8. u in mm/day		6.8

Computional Method	Average Daily Use (mm/day)	Compared to Blaney-Criddle
Modified Penman	5.4	81%
Hargreaves	6.1	90%
Jensen-Haise	6.8	100%
Modified Blaney-Criddle	6.8	100%

Wherever possible water requirements should be evaluated after using more than one computional method to establish a range of probable values.

Since irrigation practices and resulting efficiencies are variable and difficult to assess, the variability in estimated rates of consumptive use shown above may be within the accuracy needed in normal irrigation practice.

Evaporation

Evaporation from a United States Weather Bureau pan has been found to be a good index to peak consumptive use. Some observations in low-humidity areas indicate that peak use may be as much as 10 to 20 percent higher than the evaporation from a Weather Bureau pan for a short period of time. Exposure of the pan and the crop vary the rate considerably.

Piche evaporometers, used extensively in many countries, also give good indices to maximum rates of consumptive use. The Piche unit is essentially a test tube filled with water, inverted with a blotter over the end, and installed in a conventional shelter with other weather instruments. Because of the small size of Piche units, rates of evaporation are in excess of the rates of water use by crops. Piche values are also larger than those obtained from a weather bureau pan. Multiplying Piche readings by 0.7 gives average comparable values, although the coefficient does change with climate, season, and exposure.

Atmometers, which are porous ceramic bulbs, have also been used successfully as indices to consumptive use. Differences between readings obtained from black and white bulb atmometers are indicative of maximum consumptive use rates that can be expected at the time.

Principal limitations of evaporation pans are that they are expensive and require considerable water. Piche evaporometers and atmometers have one drawback in common. The evaporating surface of each is subject to contamination by dust, oil, and other foreign material. Blotters for Piche units are inexpensive and can be changed readily. However, since atmometers are relatively more expensive than porous ceramic bulbs, they are frequently not replaced when dirty. Fingerprints, which leave oil on the surface, are especially harmful. Atmometers must be recalibrated when cleaned. All of these evaporative devices are independent, of course, of the physiology of the plant. They are indices to potential consumptive use. Admustment must be made for the kind of crop and stage of growth. The importance of these physiological plant factors is discussed next.

7.6 CLIMATE AND PLANT PHYSIOLOGY RELATED TO CONSUMPTIVE USE

Kind of crop and stage of growth certainly have an influence upon transpiration, and hence upon consumptive use. Typical consumptive use and evaporation rates during the growing season for barley are shown in Fig. 7.1. Note that the rate of water use increases to a peak rate of consumptive use and then diminishes. The maximum rate of use occurs in August in the northern hemisphere and in January in the southern hemisphere when days are hottest.

Two principal variables are not considered in Fig. 7.1. These variables are causing variation in rate from day to day and change during the season. The first is climate and the second is the physiology of plants.

Variations in consumptive-use rate occur from day to day because of changes in climate. A hot, dry, windy day will increase

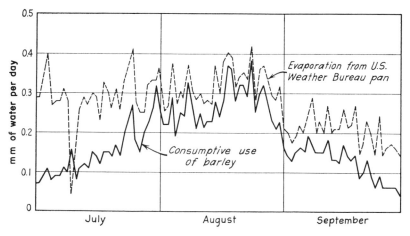

Figure 7.1 Typical consumptive use and evaporation rates during the growing season.

the rate; whereas a cool day will decrease the rate of consumptive use. Characterizing this weather influence is not easy. Several methods for estimating the influence of climate were outlined in the previous sections. Penman has used several variables, whereas Blaney-Criddle have relied upon temperature. Evaporation devices have been used as well to characterize climatic changes. All are helpful, but none are fully adequate. For example, the influence of weather upon a growing plant is not the same as the influence upon an open-water surface, for which the evaporation pan was developed. Exposure, height, and color of the leaf surfaces are all important. Much progress has been made in characterizing the influence of weather upon crop consumptive use.

Since evaporation integrates many of the weather factors, the influence of climate will be assumed to be represented by a suitable evaporation device. Then the ratio of consumptive use to evaporation should be very significant. Plotting this dimensionless ratio in Fig. 7.2 (using data from a US Weather Bureau Pan) as the ordinate, much of the variation disappears from the consumptive-use curve shown in Fig. 7.1. Note that the ratio approaches 1.0 as a peak value; i.e., the use is essentially the evaporation rate at that time. Remembering that the measured consumptive use and the evaporation index both involve errors, fluctuations in the curve shown in Fig. 7.2 are not unexpected.

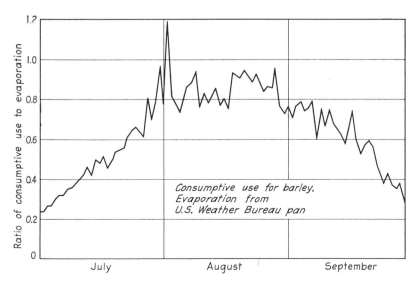

Consumptive use for barley. Evaporation from U.S. Weather Bureau pan

Figure 7.2 Typical variation of ratio of consumptive use to evaporation during the growing season.

Any suitable measure of weather, calculated or measured by evaporation devices, can be used.

Abscissa of both figures still has the dimension of time, and the water-use curve for each crop is unique. By changing the horizontal scale to a scale of maturity or stage of growth, much of this seasonal variation is eliminated. The curves are essentially superimposed.

Analyses of consumptive use and weather data for many crops grown in many countries of the world give strong support for the empirical curve shown in Fig. 7.3. Most crops which have rapidly expanding root systems or an expanding vegetative system follow this basic curve. The zero point on the scale of maturity is obtained by extrapolating the curves to zero.

The physiology of the growing plant can be characterized by flowering, fruiting, and other distinctive characteristics like tasseling that occurs in corn. Peak use comes at the beginning of flowering and at the end of the vegetative stage of growth. How irrigation and other management practices should be modified according to the state of growth is the subject of the next chapter.

For an example of the use of Fig. 7.3, to estimate actual consumptive use of the crop, suppose that at 30 percent of the growing period, evaporation was 7 millimeters per day. Since the ratio of consumptive use to evaporation at 40 percent growing season is 0.75 as shown in Fig. 7.3, the consumptive use will be 0.75 × 7 = 56 mm/day.

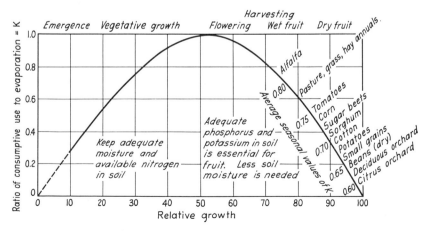

Figure 7.3. Generalized curve comparing consumptive use-evaporation ratio to relative growth of crop.

Variation of Crop Consumptive-Use Coefficient During the Growing Season

The change of crop coefficient during the growing season can best be illustrated using the Blaney-Criddle formula. Note that their formula $U = KF$ can be written as $K = U/F$, which is the ratio of consumptive use to the weather index spoken of previously. Hence, the Blaney-Criddle K is essentially the consumptive-use evaporation ratio shown in Fig. 7.3. Observe the seasonal values of K in Table 7.5 and the curve of Fig. 7.3. Dry beans, for example, are generally allowed to grow during most of the growing season (the average value of the ordinate K for beans is given in Table 7.5 as 0.65), whereas alfalfa is cut shortly after peak use is reached, resulting in an average ordinate for alfalfa of 0.85, which is considerably higher than for dry beans. Hence, the value of K is indicative of the point on the descending portion of the curve at which harvest occurs. Figure 7.3 shows these values on the descending portion of the curve for a series of typical crops. Values for other crops can be estimated from their similarity to the crops shown.

Naturally, crops like paddy rice and varieties of cacti do not have consumptive-use rates corresponding to those shown here, but the principles are very similar.

7.7 LENGTH OF GROWING SEASON

The length of the growing season, where temperature is not adverse, is governed by the variety of crop being grown and by the available energy during the growing period. Hotter days will shorten the growing period.

When frost is common, crops and crop varieties grown are determined by the available energy between killing frosts. The length of the growing season is generally considered to be the number of frost-free days. Some growth of perennial crops and some consumptive use of water occurs before and after the frost-free period. However, the number of frost-free days is a useful index to climate and water requirements.

Guy O. Woodward[3] has obtained an interesting and useful relationship between the average number of frost-free days in the growing season and the average temperatures during July, Fig. 7.4. This concept can be extended further to improve accuracy and can be applied to other areas. Such a relationship es-

[3]Guy O. Woodward, *Sprinkler Irrigation*, Sprinkler Irrigation Association, Washington, D.C., second ed., 1959.

Figure 7.4 Relation between frost-free days and July temperatures for arid western and humid eastern United States. (After Woodward, see footnote 3.)

tablished for areas of new development where limited data is available would allow estimates to be made as a basis for subsequent agricultural development. Furthermore, July temperatures in the Northern Hemisphere or January temperatures in the Southern Hemisphere can also be used to obtain estimates of peak consumptive uses.

7.8 CONSUMPTIVE-USE REQUIREMENTS OF TYPICAL CROPS

Seasonal consumptive-use requirements of common irrigated crops in relation to requirements for alfalfa are shown in Table 7.7 for arid regions in the United States. Seasonal use and peak daily use values have also been prepared as Table 7.8 for seven

TABLE 7.7   Seasonal Consumptive-Use Requirements of Common Irrigated Crops in Relation to the Requirements for Alfalfa—Western States (after Woodward, see footnote 3)

Crops	Relation of Seasonal Consumptive Use of Crops listed to the Seasonal Consumptive Use of Alfalfa— (Optimum Yield)
Alfalfa	1.00
[a] Almonds	0.42
[a] Apricots	0.50
Artichokes	0.50
[a] Avocados	0.52
Beans	0.42
Berries	0.40
Clover	0.90
Carrots	0.40
Corn	0.65
[a] Citrus	0.70
Cotton	0.65
Grain	0.50 (variable)
Grain-Sorghum	0.35 to 0.40
Grapes	(variable)
Hops	0.50
Lettuce	0.15 (Plant use only)[b]
Melons	0.40
Pasture	0.90 (variable)
[a] Peaches	0.65
Potatoes—Winter	0.50
Potatoes—Spring or Seed	0.30
Seed Peas	0.30
Sugar Beets	0.82
Strawberries	0.60
Tomatoes	0.45 to 0.50
[a] Walnuts	0.60

[a] Add 20–25% for permanent grass—legume cover.
[b] Water may be applied in addition to that required for growth for the purpose of quality improvement.

TABLE 7.8 Total Consumptive Use and Peak Daily Use, Western United States (Computed from Woodward, see footnote 3)

Crops	Southern Coastal[a] (1) 300 Days Plus Season Use (cm)	Daily Use (cm/day)	250–300 Days Season Use (cm)	Daily Use (cm/day)	South Pacific Coastal Interior and North Coastal (2) 250–300 Days Season Use (cm)	Daily Use (cm/day)	210–250 Days Season Use (cm)	Daily Use (cm/day)	180–210 Days Season Use (cm)	Daily Use (cm/day)	150–180 Days Season Use (cm)	Daily Use (cm/day)
Alfalfa	91	0.51	76	0.43	94	0.69	81	0.56	66	0.51	56	0.46
Pasture	85	0.51	71	0.43	84	0.69	76	0.56	61	0.51	51	0.46
Grain—small	41	0.46	36	0.41	43	0.56	37	0.51	31	0.51	25	0.46
Beets—sugar	74	0.51	64	0.46	76	0.69	66	0.56	—	—	—	—
Beans—field	31	0.46	25	0.41	33	0.56	28	0.46	—	—	—	—
Corn—field	—	—	—	—	—	—	48	0.64	41	0.56	36	0.51
Potatoes	—	—	—	—	61	0.51	51	0.46	—	—	—	—
Peas—green	25	0.46	21	0.41	28	0.64	23	0.46	20	0.41	18	0.41
Legume seed	64	0.51	56	0.46	66	0.51	56	0.56	51	0.51	46	0.46
Tomatoes	46	0.41	38	0.41	48	0.46	41	0.46	33	0.41	28	0.41
Vegetable seed	41	0.41	36	0.41	46	0.46	41	0.46	36	0.41	31	0.41
Beans—pole	—	—	—	—	46	0.51	41	0.51	36	0.46	31	0.41
Corn—sweet	41	0.46	36	0.41	41	0.51	36	0.46	31	0.46	28	0.41
Apples	—	—	—	—	61	0.51	56	0.51	51	0.46	46	0.46
Cherries	—	—	—	—	61	0.51	56	0.51	—	—	—	—
Peaches	—	—	—	—	61	0.51	56	0.51	—	—	—	—
Prunes	—	—	—	—	56	0.51	51	0.51	—	—	—	—
Apricots	—	—	—	—	56	0.51	51	0.51	—	—	—	—
Oranges	51	0.41	46	0.41	56	0.46	—	—	—	—	—	—
Avocados	46	0.41	—	—	—	—	—	—	—	—	—	—
Walnuts	56	0.51	46	0.46	61	0.64	56	0.56	51	0.51	—	—
Strawberries	56	0.51	46	0.46	58	0.64	46	0.56	41	0.51	36	0.46
Lettuce	10	0.41	10	0.41	15	0.46	13	0.46	—	—	—	—
Mint	—	—	—	—	58	0.61	53	0.56	48	0.51	43	0.46
Hops	—	—	—	—	51	0.56	51	0.51	—	—	—	—

[a] Fog belt.

TABLE 7.8 Total Consumptive Use and Peak Daily Use, Western United States (Continued)

| | Central Valley—California and Valleys East Side of Cascade Mountains (3) | | | | | | | | | | | |
| | 250–300 Days | | 210–250 Days | | 180–210 Days | | 150–180 Days | | 120–150 Days | | 90–120 Days | |
Crops	Season Use (cm)	Daily Use (cm/day)	Season Use (cm)	Daily Use (cm/day)	Season Use (cm)	Daily Use (cm/day)	Season Use (cm)	Daily Use (cm/day)	Season Use (cm)	Daily Use (cm/day)	Season Use (cm)	Daily Use (cm/day)
Alfalfa	102	0.76	86	0.71	76	0.64	66	0.56	51	0.51	36	0.46
Pasture	92	0.76	76	0.71	71	0.64	61	0.56	46	0.51	33	0.46
Grain—small	46	0.56	41	0.56	38	0.51	36	0.46	33	0.46	31	0.41
Beets—sugar	84	0.76	71	0.64	61	0.56	51	0.51	46	0.46	—	—
Beans—field	43	0.56	33	0.51	33	0.51	31	0.46	31	0.46	—	—
Corn—field	66	0.89	56	0.81	55	0.76	51	0.64	46	0.56	43	0.51
Potatoes—summer	31	0.41	—	—	—	—	—	—	—	—	—	—
Potatoes—fall	—	—	48	0.71	46	0.56	46	0.51	43	0.46	41	0.41
Peas—green	—	—	20	0.46	18	0.46	18	0.46	18	0.41	15	0.38
Peas—field	—	—	25	0.46	23	0.46	22	0.46	20	0.41	20	0.38
Tomatoes	51	0.51	46	0.46	46	0.46	43	0.43	41	0.41	—	—
Cotton	66	0.76	56	0.71	—	—	—	—	—	—	—	—
Grain—sorghum	38	0.51	33	0.46	31	0.43	25	0.41	23	0.38	20	0.38
Apples	—	—	66	0.51	58	0.51	53	0.46	—	—	—	—
Cherries	—	—	61	0.51	53	0.51	48	0.46	—	—	—	—
Peaches	56	0.56	56	0.51	51	0.51	46	0.46	—	—	—	—
Apricots	51	0.56	43	0.51	38	0.51	—	—	—	—	—	—
Oranges	71	0.46	—	—	—	—	—	—	—	—	—	—
Strawberries	61	0.51	51	0.51	46	0.46	—	—	—	—	—	—
Lettuce—winter	102	0.51	—	—	—	—	—	—	—	—	—	—
Mint	—	—	51	0.56	46	0.51	—	—	—	—	—	—
Hops	76	0.64	46	0.51	41	0.46	—	—	—	—	—	—
Grapes	—	—	64	0.56	56	0.51	—	—	—	—	—	—
Walnuts	61	0.56	51	0.51	—	—	—	—	—	—	—	—
Almonds	56	0.64	51	0.56	—	—	—	—	—	—	—	—

TABLE 7.8 Total Consumptive Use and Peak Daily Use, Western United State (Continued)

Intermountain, Desert, and Western High Plains (4)

Crops	250–300 Days Season Use (cm)	250–300 Days Daily Use (cm/day)	210–250 Days Season Use (cm)	210–250 Days Daily Use (cm/day)	180–210 Days Season Use (cm)	180–210 Days Daily Use (cm/day)	150–180 Days Season Use (cm)	150–180 Days Daily Use (cm/day)	120–150 Days Season Use (cm)	120–150 Days Daily Use (cm/day)	90–120 Days Season Use (cm)	90–120 Days Daily Use (cm/day)
Alfalfa	132	1.02	112	0.81	91	0.74	76	0.66	76	0.66	48	0.51
Pasture	121	1.02	102	0.76	84	0.71	71	0.64	71	0.64	41	0.51
Grain—small	53	0.64	46	0.56	41	0.51	41	0.51	41	0.51	36	0.46
Beets—sugar	94	0.76	81	0.76	76	0.71	66	0.64	66	0.64	46	0.51
Beans—field	56	0.64	43	0.51	36	0.51	36	0.46	36	0.46	31	0.38
Corn—field	—	—	76	0.89	66	0.76	61	0.71	61	0.71	—	—
Potatoes—fall	—	—	58	0.76	53	0.71	51	0.64	51	0.64	43	0.51
Peas—field	—	—	—	—	25	0.48	25	0.46	25	0.46	23	0.38
Tomatoes	—	—	51	0.56	46	0.51	43	0.46	43	0.46	—	—
Cotton	81	0.76	76	0.71	—	—	—	—	—	—	—	—
Grain—sorghum	48	0.64	46	0.51	41	0.51	36	0.46	36	0.46	—	—
Apples	—	—	—	—	71	0.56	61	0.51	61	0.51	—	—
Cherries	—	—	—	—	66	0.56	—	—	—	—	—	—
Peaches	—	—	74	0.64	69	0.56	—	—	—	—	—	—
Apricots	66	0.64	61	0.64	64	0.51	—	—	—	—	—	—
Almonds	56	0.64	51	0.64	—	—	—	—	—	—	—	—
Vineyards	102	0.69	81	0.64	66	0.56	—	—	—	—	—	—
Legume seed	—	—	—	—	—	—	—	—	—	—	36	0.41
Grass seed	—	—	—	—	—	—	—	—	—	—	31	0.36
Potatoes—seed	—	—	—	—	—	—	—	—	—	—	36	0.38
Grapefruit	114	0.51	—	—	—	—	—	—	—	—	—	—
Oranges	91	0.46	—	—	—	—	—	—	—	—	—	—
Lettuce—winter	15	0.46	—	—	—	—	—	—	—	—	—	—
Melons	56	0.64	51	0.56	46	0.51	41	0.46	41	0.46	—	—
Palm dates	152	0.76	—	—	—	—	—	—	—	—	—	—
Truck crops	51	0.64	46	0.56	36	0.51	31	0.46	31	0.46	25	0.38

regions shown in Fig. 7.4. These tables developed for the United States can be used in other areas by determining the length of the growing season and average July temperatures, or January temperatures in the southern hemisphere. Thus, if a given arid region had 160 frost-free days and an average July temperature of 25°C, it would be comparable to the Intermountain and High Plains, or area No. 4. Data for Table 7.8 for the Intermountain and High Plains should apply, therefore, to the new area in question with reasonable accuracy.

7.9 PRACTICAL CONSIDERATIONS

A hazard exists for water resource planners and managers looking at and working with the intricacies of consumptive determinations. Not infrequently they become lost in the details of the method and overlook fundamentals, obtaining results which are misleading. The prime concern of water resource management is to obtain useful data. A clear understanding of the pitfalls along the road to obtaining reliable data is essential to avoid becoming lost in a maze of detail associated with a method of calculating consumptive use. The accuracy required depends upon the use to be made of the answer. Are estimates needed of hourly, daily, monthly, or seasonal water use? What impact will the answer have on the project? These questions should be ascertained before a procedure is selected for arriving at the answer and before the accuracy required can be determined.

The most important part of the estimate is the first significant figure. Be sure this is correct. Complexity of formula or accuracy of measurement does not ensure reliability. The accuracy of the estimate is no better than the accuracy of the least reliable component used to obtain the answer. How accurate then are the component parts used to obtain the answer?

For research purposes, two or perhaps even three significant figures may be needed. Whereas, for actual management of a project, two significant figures are often more than is required. A research determination of consumptive use accurate to three significant figures has little real value, if it is divided by an estimated efficiency accurate only to one significant figure.

Sometimes accurate values of consumptive use are used which are not representative values. Care must be taken to ensure that the sample is representative of the population being measured. One sample from a given location accurate to ± 1 percent may not be as good as 20 samples from 20 locations individually accurate to only ± 20 percent.

Estimates of consumptive use must reflect spacial changes of geography, variation of climate, and changes in project management with time. Estimates of consumptive use should reflect expected variation of climate. Average climatic values are useful but not sufficient. Probability is a useful tool.

Changes in project management must be contemplated, since these influence the efficiency of water use on the project. Management practices have minor effect upon consumptive use, but management decisions do directly affect the acreages of various crops in agricultural production, the extent and duration of recreational uses, as well as efficiency of use, and thus determine the amount of water that is used consumptively. Sometimes estimates of project consumptive use are made without adequate consideration of geography, climate, or management.

In most project areas measured values of consumptive use can be obtained to compare with estimated values as a check of the coefficients and methods used. Sole reliance should not be placed upon estimates unless one understands clearly the limitations and unique characteristics of the method of estimating water use. Likewise, "measured" values need to be carefully appraised to ensure lack of serious bias. For example, a lysimeter is often used as a reasonably accurate measure of consumptive use. However, position of water table, fertility of soil, cultural practices, quality of water used, leaks in the system, errors in measurement, unrepresentative exposure, border effects, etc. can all drastically effect the "measured" values. Likewise, consumptive use obtained from measuring soil moisture depletion can be radically affected by techniques of measurement and errors of analysis. Thus, extreme care needs to be exercised to ensure that "measured" values are sufficiently accurate to be used to calibrate "estimated" values.

With an understanding of the bias and peculiarities of different methods of estimating consumptive use, these methods can be compared to real advantage. From such knowledgeable comparisons, justified confidence can be established in the final estimate. In some localities, one method will apply better than another because of natural environment. Also, available climatic data will generally vary from location to location. Utilizing the method that fits the climatic data and then comparing the results to obtain a continuous function of estimated consumptive use for the project area will give an overall reliability that cannot normally be obtained by using a single method.

Thus, a clear understanding of the characteristics and of the methods used to estimate consumptive use is of basic importance in wise water resource management.

CHAPTER
8
WHEN TO IRRIGATE— HOW MUCH WATER TO APPLY

Three major considerations influence the time of irrigation and how much water should be applied, namely, (a) water needs of the crop, (b) availability of water with which to irrigate, and (c) capacity of the root-zone soil to store water. Water needs of the crop are of paramount importance in determining the time of irrigation during the crop-growing season on irrigation projects which obtain their water supplies from storage reservoirs or from other dependable sources of water. Some irrigated areas have a limited water supply during the irrigation season, but an abundance of water during late autumn or winter and early spring. Irrigation farmers cannot always apply water when the crop is most in need; sometimes to save water they must apply it even though the crop does not need it, provided the soil has the capacity to store additional water. Therefore, crop needs, available water supply, and storage capacity of the soil must be considered in a discussion of the proper time to irrigate.

Growing crops use water continuously, but the rate of use varies with the kind of crop grown, age of the crop, and atmospheric conditions—all variable factors. At each irrigation a volume of water sufficient to supply the needs of the crop for a period varying from a few days to several weeks is stored in the unsaturated soil in the form of available soil water. How frequently water should be applied to soils of different properties in order to best supply crop needs is a question of real and practical significance. A factor of major importance in arriving at the desirable frequency and time of irrigation is the water need of the crop.

8.1 LIMITING SOIL MOISTURE CONDITIONS

Growth of most crops produced under irrigation farming is stimulated by moderate quantities of soil moisture and retarded by either excessive or deficient amounts. Air in the soil is essential to satisfactory crop growth; hence, excessive flooding and filling the soil pore spaces with water, thus driving out air, inhibits proper functioning of the plants even though supplying an abundance of available water. On the other hand, soils having deficient amounts of water hold it so tenaciously that plants must expend extra energy to obtain it. If the rate of intake by the plant is not high enough to maintain turgidity of the leaves, permanent wilting follows. At some soil moisture content between these two extreme moisture conditions, designated optimum moisture percentage, plants grow most rapidly (Fig. 8.1).

The objective of considerable research has been to extend knowledge of these two limiting soil moisture conditions; i.e., permanent wilting percentage and optimum plant growth percentage. Because of wide variations in physical properties of different soils, it is easy to understand that the moisture percentage in a clay soil at permanent wilting of a plant may be several times the moisture percentage in a sandy soil when the same plant wilts permanently.

8.2 APPEARANCE OF CROP

A light green color in alfalfa is generally indicative of adequate moisture and satisfactory growth, whereas a dark green color indicates lack of moisture. Among the root crops, sugar beets readily indicate need for water by temporary wilting, particularly during the warmest part of the day. Grain crops also indicate need for water by temporary wilting. In production of fruit crops, it is impractical to detect the need for water by the appearance of the leaves of the trees. It is, therefore, more essen-

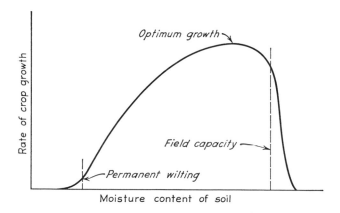

Figure 8.1.   Rate of growth to moisture content of the soil; optimum growth varying somewhat with aeration, water-holding capacity of soil, and crop growth.

tial to base the time of irrigation of orchards on observations of moisture content of the soil.

Crop growth should not be retarded by lack of available soil moisture; and the practice of withholding irrigation until the crop definitely shows a need for water is likely to retard growth. It is essential to maintain readily available water in the soil if crops are to make satisfactory growth.

8.3 SEASONAL USE OF WATER BY DIFFERENT CROPS

Irrigation farmers must select their crops, to some extent, on the basis of time at which water will be available. In valleys having no storage reservoirs, larger quantities of water are available early in the season. During the forepart of the crop-growing season, streams in mountain areas fed by melting snow banks and drifts are much larger than during the summer. Under such conditions, alfalfa, wheat, and oats may well be produced as each of these crops requires large amounts of early-season water. Canning peas may be matured before a water shortage begins. Alfalfa continues to grow throughout the late summer months, provided water is available. Sugar beets, potatoes, and corn require less water early in the season, but during the late summer months these crops need an abundance of water. Unless late-season water is assured, it is inadvisable to attempt to grow sugar beets and potatoes.

8.4 AVAILABLE WATER SUPPLY

Irrigation during the dormant or nongrowing season, in many localities, is an economical means of storing water in the soil for future use. Immediately after heavy rains which increase the flow in rivers and creeks, large quantities of water for irrigation may be available for short periods of time. It is desirable to build surface reservoirs—both large and small—in which to store water that becomes available from sudden torrential rains or that is available only during the fall, winter, or spring. As a general rule, water that is used in irrigation at times when not really needed by crops is used less efficiently than it would be if it were possible to apply the water to the soil when most needed by crops.

In some localities storage of water in surface reservoirs is impracticable because of high costs and lack of suitable natural facilities. In such cases, it is advantageous to use the soil as a storage reservoir and to apply water whenever available as a means of storing it for future use. Deep percolation from shallow soils underlain by coarse-textured sands and gravels may be excessive unless water is applied carefully. Some areas can be benefited by applying during the dormant season depths of water sufficient to saturate the subsoil gravels and cause the water table to rise to an elevation near the land surface. Where it is not feasible to drain ground water during the irrigation season, excessive depths applied during the dormant season may cause appreciable damage by waterlogging the soil and thus reducing the depth of the root zone.

8.5 FALL IRRIGATION

In many places fall and winter precipitation is insufficient to moisten the soil fully. Fall irrigation under such conditions is a means of saving water and of getting the land in favorable condition for germination of seeds and early growth of crops during the following season.

Alfalfa grown on well-drained soils may be irrigated in the fall; and fall irrigation of meadow forage crops and of pasture lands is usually desirable if the soils become very dry without irrigation. In the practice of irrigation during fall, winter, or early spring, it is important to guard against use of excessive depths of water. As shown in Chapter 3, there are definite limitations to the capacity of any soil to retain water. Permitting water to flow over the lands for many days, and sometimes weeks, is injurious both to land irrigated and to lower-lying areas to which means of the excess water seeps.

Fall irrigation can also be beneficial in land preparation for the subsequent spring planting. By plowing early in the fall and then irrigating, the water causes clods to crumble; also the freezing and thawing of wet soils reduces the need for harrowing, etc., to prepare a good seed bed. The cost of land preparation is reduced thereby, and usually better soil structure is obtained. Thus, fall irrigation may be very helpful in minimizing tillage and improving or maintaining good soil structure.

8.6 WINTER IRRIGATION

In mild climates where the soil does not freeze, winter irrigation may be practiced advantageously as a means of saving water that would otherwise be lost.

At higher elevations and in colder parts of irrigated regions, winter irrigation is of little, if any, practical importance. Frozen soils absorb water slowly, if at all, and it is difficult to spread water over the snow-covered fields effectively. Furthermore, some crops are injured by winter irrigation in cold climates.

8.7 EARLY SPRING IRRIGATION

Some arid-region lands need irrigation during the early spring months in order to supply the moisture essential to satisfactory germination and early growth of annual crops. Arid-region streams usually have ample water to meet the needs for early spring irrigation. Even where the discharge of streams at high mountain elevations is held in storage reservoirs, enough water is frequently available from rains and melting snows on lower elevations to supply the needs for early spring irrigation. The value of early spring irrigation as a means of storing available water in root-zone soil is often not fully realized. Irrigators are frequently misled when spring rains moisten the soil to a depth of 20 to 40 centimeters. Some consider the soil "wet enough" when, in fact, there may be 1 to 2 meters of dry soil below the moist surface soil.

Use of a soil auger or of a soil tube will enable irrigators to decide intelligently needs of soils for early spring irrigation. Although it is highly desirable to save water by applying it to inadequately moistened soils, it is undesirable early in the spring to irrigate soils that are already moistened to field capacity.

Fully moistened soils in which the water table is at a shallow depth may be injured rather than benefited by early spring irrigation. Under such conditions it is better to waste water than to apply it to the soil. Also, early spring irrigations may retard growth if irrigation cools the ground considerably. Damage to crops may result from an early irrigation in soils of low soluble nitrogen content, particularly if roots are shallow at the time of irrigation, or on soils susceptible to limited aeration. When nitrogen deficiency and limited aeration occur, the crop may show a yellow, unhealthy appearance. The nitrogen problem can be avoided by maintaining a higher nitrogen level in the soil, or by adding soluble nitrogen to water during spring irrigation.

8.8 SOIL MOISTURE REMOVAL BY PLANT ROOTS

When to irrigate and how much to apply are affected considerably by where and when water is removed from the soil by the plant roots. Shallow-rooted crops will require more frequent irrigation than deep-rooted crops. Soil conditions which restrict the growth of roots will likewise influence irrigation practice. Figure 8.2 showing shallow alfalfa roots resulting from a water table at 75 centimeters is in sharp contrast to Fig. 8.3 which shows that less water was extracted from the surface 25 centimeters than from the second and third increment of soil. However, in Fig. 8.4 more water was extracted by cotton from the first 25 centimeters

Figure 8.2   Distribution of alfalfa roots with a shallow water table at 75 cm below ground surface.

than from any succeeding 25 centimeters of soil, even though the alfalfa shown in Fig. 8.3 and the cotton in Fig. 8.4 were both grown in the same valley with similar soils. Rooting data from irrigated crops grown in semihumid regions shows considerably more water removed from the first 25 centimeters of soil than from any succeeding depth, while data from hot arid regions generally shows less water removed from the first 25 centimeters

Depth (cm)	Water used	
	cm/yr	% of total
0–25	17	⑬
25–50	18	⑭
50–75	19	⑭
75–100	17	⑬
100–125	14	⑪
125–150	12	⑨
150–175	10	⑧
175–200	9	⑦
200–225	7	⑥
225–250	6	⑤

Figure 8.3   Use of water by alfalfa from each meter of the root-zone soil in an arid climate with a deep water table.

Depth (cm)	Water used	
	cm/yr	% of total
0–25	22	㉚
25–50	19	㉖
50–75	13	⑱
75–100	9	⑬
100–125	6	⑧
125–150	3	⑤

Figure 8.4   Use of water from each meter of soil growing cotton in an arid climate with a deep water table.

than from the next 25 centimeters of soil. This difference is caused by two factors: first, the depth to which the applied water penetrates, and second, the moisture content of the soil during the growing period. For example, in humid and semihumid areas several rainfalls occur during the growing season. However, seldom do these rains penetrate beyond the first foot of soil. In the very arid regions, essentially no rainfall occurs during the growing season. Likewise, the cotton shown in Fig. 8.4 is a more shallow-rooted crop than the alfalfa shown in Fig. 8.3. For this reason the cotton is irrigated more frequently, especially during the early growing period. Thus, more water is extracted from the first foot of soil whenever the first foot is kept more moist either by frequent rains or by frequent irrigations.

8.9 EFFECT OF MOISTURE CONTENT ON RATE OF REMOVAL OF SOIL MOISTURE

Other conditions being equal, roots of a plant in moist soil will extract more water than the roots of the same plant growing in dryer soil. Figure 8.5 illustrates the fact that when the soil is wet, most of the crop's moisture is withdrawn from soil near the sur-

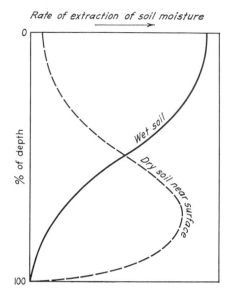

Figure 8.5. Variation of extraction of soil moisture with moisture content and depth.

face. This is because more roots are normally growing near the surface. However, as the moisture content of soil near the surface decreases, more moisture is extracted from lower depths. Finally, as the moisture content of the soil near the surface approaches permanent wilting percentage, essentially all moisture removed by the roots comes from deeper soil. Since fewer roots exist in the lower portion of a soil profile, more energy must be expended in extracting required moisture, and frequently insufficient moisture is available to the plant to prevent wilting. Moisture may be available in the soil, but the plant roots may not be able to extract this moisture at a sufficient rate to meet transpiration requirements of the plants.

When the upper portion of the root zone is kept moist, most water used consumptively by the plant will be removed from the soil near the surface. However, when infrequent irrigations are applied, and essentially no rainfall occurs, less water may be used from the surface foot than from succeeding depths. Since the soil is wet following a good irrigation and normally dry on and near the soil surface before another irrigation is applied, the rate of removal will be a composite between the two qualitative curves of Fig. 8.5. Normal arid-region irrigation practice generally results in an average extraction pattern shown in Fig. 8.6 which is essentially triangular in shape. The distribution of active roots in a normal soil is also approximately triangular in shape, the greatest concentration being near the soil surface. Note that a crop extracting moisture to a depth of 100 centimeters

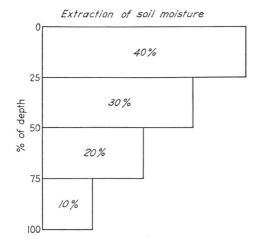

Figure 8.6. Average extraction of soil moisture by plant roots between irrigations in an arid region.

would remove four times more moisture from the surface 25 centimeters than from the fourth 25 centimeters of soil.

8.10 INFLUENCE OF A RESTRICTING SOIL LAYER

Any restricting layer will cause a concentration of roots above that layer. Frequently, roots are restricted by a compacted layer resulting from faulty tillage practices. A 10 percent reduction in soil voids will generally prevent root growth. This much reduction in voids can easily occur using rubber-tired vehicles, disks, and vibrating tillage implements. Restricted roots have less soil available to extract needed soil moisture and plant nutrients. Plants with restricted root systems show symptoms of low fertility and drought, and have deformed tap roots.

If the restricting layer results in a temporarily-perched water table, the saturated soil will cause an aeration problem. Therefore, a restricting layer is particularly harmful. Good farming practices will eliminate and prevent compacted layers, but silty clay layers and other restricting layers are not easily avoided or eliminated. Whenever they occur, heavy irrigations should be given in the dormant season to store water in the deeper soils, and lighter, more frequent irrigations during the active growing period will be beneficial.

Remember also that roots cannot penetrate a dry soil. Special precautions should be taken to be sure that the soil moisture is favorable throughout the entire depth of rooting. For a young crop, it is especially necessary to maintain moisture ahead of the expanding roots so that they will not be inhibited in their growth and expansion.

8.11 STAGE OF GROWTH AFFECTS IRRIGATION PRACTICE

Growth of all plants can be divided into three stages with regard to irrigation practice: vegetative, flowering, and fruiting. The relationship of these three stages of growth to the consumptive use was shown in Fig. 7.3. During the vegetative stage consumptive use continues to increase. Flowering occurs near and during the peak of consumptive use. The fruiting stage is accompanied by a decrease in consumptive use until the transpiration essentially ceases during the latter part of the formation of dry fruit.

The fruiting stage can well be considered in two parts: the wet-fruit stage which follows flowering, and the dry-fruit stage following wet-fruit growth. Dry-fruiting is accompanied by a decrease in consumptive use until transpiration ceases and the plant is dormant.

The amount of water applied and the frequency of irrigation must be adjusted to the actual consumptive use of the crop, water-holding capacity of the soil, and depth of rooting. Naturally, a shallow sandy soil will require quite different scheduling of irrigation than will be required for a deep clay loam soil.

Examples of crops harvested in different stages of growth are shown in Table 8.1.

TABLE 8.1.   Typical Crops Harvested During the Different Stages of Growth

		Fruiting	
Vegetative	Flowering	Wet	Dry
Lettuce	Flowers	Tomatoes	Dry peas
Alfalfa	Cauliflower	Green peas	Dry beans
Grasses	Broccoli	Green beans	Potatoes
Cabbage		Sweet corn	Mandioca
Asparagus		Deciduous fruits	Seeds and grains
Mint		Sugar cane	Field corn
		Artichokes	Cotton
		Sugar beets	Rice
		Berries	Onions
		Bananas	Nuts
		Cantaloupes	
		Citrus	
		Watermelons	

Irrigation During the Vegetative Stage of Growth

A good moisture supply normally should be available to plants at all stages of vegetative growth. When vegetation is being produced, nitrogen is particularly essential. Light, frequent irrigations are generally desirable because of the need to keep ample moisture in the soil for the relatively shallow root system. For perennial crops like alfalfa with a deep root system, less frequent but heavier irrigations are necessary. Whenever excessive temperatures exist, frequent irrigations may be desirable for cooling the soil and the plants. When a vegetative crop like lettuce is grown in areas where peak consumptive use may be as high as 10 millimeters per day, benefit will result from frequent irrigations for cooling as well as for maintaining ample moisture in the soil.

Irrigation Practice During the Flowering Stage of Growth

Since consumptive use is at or near a maximum during the flowering stage, care must be exercised to ensure adequate moisture in the root zone. However, the increased consumptive use is offset by the normally increased depth of roots. Deeper roots have a greater root-zone depth and hence a larger water supply available. In fact, the ratio of consumptive use to depth of extraction for a normally expanding root system remains remarkably constant. This ratio for green peas, potatoes, sugar beets, and grain grown in Logan, Utah, was about 1.5 millimeters depth of water each day for each 30 centimeters of root extraction depth. Thus, when grain roots were 90 centimeters deep, the consumptive use was 1.5×3 or 4.5 millimeters per day. Hence, when the increasing depth of rooting is considered, the frequency of irrigation for most crops remains remarkably constant during the vegetative and flowering periods.

Actually, best production is obtained if the crop is kept adequately irrigated during both vegetative and flowering stages. However, harvesting procedures and early frosts do modify irrigation practice. Recall that a crop consists of thousands of individual plants, with some more mature than others. The best yield can be obtained when the plant has reached a certain stage of maturity. However, it is not practical to harvest each plant at its optimum stage of growth. Economics generally requires harvesting all plants at a given time; this is true in the harvesting of grain, hay, and sugar cane. Even though tomatoes are harvested by handpicking several times during the season, it is still an advantage to minimize the number of pickings. Also, frost may occur before some of the fruit has matured. Therefore, harvesting methods and frost may justify withholding irrigation for a short period during flowering to force the less mature plants out of the vegetative stage and into the flowering and fruiting stages. Reducing the amount of water available to the plant at this stage of growth will reduce the yield of more mature plants, but this decrease may be offset by an increased harvest from less mature plants. Normally, ample water should be available during fruiting unless maturing is not uniform or unless damaging frosts occur before harvesting is completed.

Special precautions should be taken to prevent the soil from becoming too dry, causing the flowers to fail to mature into seed.

Irrigation Practice During the Fruiting Stage of Growth

The root system is essentially extended to its maximum depth by the time fruiting occurs and the consumptive use has begun to decrease, reducing the water requirements of the crop and the frequency of irrigation. During production of dry fruit, irrigation has essentially ceased—the slight water requirements of the crop are met usually from stored water in the soil. The last heavy irrigation should normally be given during the wet-fruiting stage so that deep roots will have water available during final development of the fruit.

Tubers like potatoes and peanuts require adequate moisture during the entire growth period. Allowing the soil to become too dry between irrigations causes uneven, nobby tubers. Ample aeration is also essential. For this reason, good potatoes are generally produced on sandy loam soils which can be irrigated frequently without creating a problem. Heavier soils offer increased resistance to the growing tubers and thereby restrict growth and result in nonuniform tubers. For this reason, in addition to aeration problems, tuberous crops do better on sandy soils than on clay soils. Another unique consideration during the fruiting stage of apples is that color of red delicious apples can be enhanced by sprinkling to reduce the temperature of the fruit by evaporation.

It is well to remember that sufficient moisture must be available to fully develop wet fruit. For example, soft fleshy fruits, green peas, and grains will not be firm and fully formed unless ample moisture is available.

Excessive irrigation during the fruiting period will stimulate vegetative growth for some crops and result in a reduction of fruiting. For example, cotton, which is normally planted yearly, is a perennial which will start new vegetative growth during the time the cotton is ready to pick if excess moisture is available in the soil.

Phosphorus and potassium are especially essential during the fruiting stages of growth, and excessive nitrogen may prolong vegetative growth.

8.12 DEPTH OF ROOT ZONE

Assuming a favorable unrestricted root zone, the depth of rooting increases during the vegetative and flowering period. The hotter the climate or the longer the growing period, the deeper the roots will penetrate. Crops requiring only 2 months to mature generally do not penetrate more than 30 to 60 centimeters. Crops requiring 3 to 4 months to mature penetrate 90 to 150 centimeters. Crops requiring 6 months to mature may penetrate 180 to 300 centimeters or more. As a rough guide, depth of rooting varies from 30 to 45 centimeters depth per month of active growth, depending upon the crop and climate. There exists a tendency to underestimate the depth of rooting.

8.13 FREQUENCY OF IRRIGATION

How often to irrigate depends upon several factors. A number of management factors have been discussed in previous sections of this chapter. The importance of optimum moisture content for crops at a given stage of growth also has been discussed. The available water supply has been shown to be important, making it necessary to irrigate at times to store surplus water in the soil. Also, the need to leach out excess salts and the need to cool the soil may govern the frequency of irrigation.

During the active growing period of the crop, however, the most important factor governing frequency of irrigation is the need to keep adequate moisture in the soil for the crop. Two approaches can be made based upon the need for water by the root-zone soil. Both depend upon the capacity of the soil within the root zone to store water and the amount of water to be depleted. The available water per increment of soil depth multiplied by the depth to which moisture will be depleted will give the total water-holding capacity. However, maximum production can be obtained on most crops if not more than 50 percent of the available water is removed during the vegetative, flowering, and wet-fruit stages of growth. For some crops, removing not more than 25 percent of the available water will produce maximum yields. Yet, the cost of irrigating frequently usually makes the 50 percent depletion more economical than irrigating when only 25 percent has been depleted. At least 75 percent of the available moisture can generally be removed during the dry-fruit stage without detrimental results.

Frequency of irrigation can be determined by dividing the amount of moisture to be depleted from the soil by the consumptive use per day. Care should be taken to be sure the amount of moisture and the consumptive use are measured in similar units. Frequency of irrigation can also be determined by noting when the amount of moisture consumptively used is equal to the amount to be depleted from the soil.

8.14 IRRIGATION EFFICIENCIES

In a world where water is such a precious resource, no man has the right to waste water which another man needs. Efficient use of irrigation water is an obligation of each user. However, efficiency of use will vary from locality to locality. In areas where water is scarce and costly, available water is generally used carefully. Whereas, in areas of abundant water, the value is less and the tendency is to waste water. Also efficiency is influenced by cost and quality of labor, ease of handling water, crops being irrigated, and soil characteristics. For these reasons irrigation efficiency is a broad general term which can be applied to irrigation practices in a qualitative manner. To describe segments of the over-all efficiency picture quantitative evaluations can be made. The objective of these efficiency concepts is to show where improvements can be made which will result in more efficient irrigation. Adequate control and management of irrigation water requires that methods be available to evaluate irrigation practices from the time water leaves the point of diversion until it is utilized by the plants.

Water-Conveyance Efficiency

The earliest irrigation efficiency concept for evaluating water lossers was water-conveyance efficiency. Most irrigation water then came from diversions from streams or reservoirs. Losses which occurred while conveying water were often excessive. Water-conveyance efficiency formulated to evaluate this loss can be stated as follows:

$$E_c = 100 \ \frac{W_f}{W_r} \tag{8.1}$$

where E_c = water-conveyance efficiency
W_f = water delivered to the farm
W_r = water diverted from the river or reservoir

Water-Application Efficiency

Having conveyed the available water to the farm through costly diversions and conveyance structures, the need was apparent to apply the water efficiently. Often considerably more water was applied to the soil than it could possibly hold. The following concept of water-application efficiency was developed to measure and focus attention upon the efficiency with which water delivered was being stored within the root zone of the soil, where it could be used by plants.

$$E_a = 100 \; \frac{W_s}{W_f} \qquad (8.2)$$

where E_a = water-application efficiency
W_s = water stored in the soil root zone during the irrigation
W_f = water delivered to the farm

The concept of water-application efficiency can be applied to a project, a farm, or a field to evaluate irrigation practice. Irrigation efficiencies can vary from extremely low values to values approaching 100 percent. However, in normal irrigation practices, surface irrigation efficiencies of application are in the range of 60 percent, whereas well-designed sprinkler irrigation systems are generally considered to be approximately 75 percent efficient.

Common sources of loss of irrigation water from the farm during water application are represented thus:

R_f = surface runoff from the farm
D_f = deep percolation below the farm root-zone soil

Neglecting evaporation losses during the time water is being applied and immediately after, it follows that

$$W_f = W_s + R_f + D_f \qquad (8.3)$$

therefore

$$E_a = 100 \; \frac{W_f - (R_f + D_f)}{W_f} \qquad (8.4)$$

At each irrigation the farmer applies to his land a given volume of water. His irrigation problem is to store this water in the root zone of his soil. He cannot store all the water as soil moisture, for some loss of water is unpreventable. He should store in his root-zone soil the maximum percentage of the water that he applies consistent with good irrigation practice and economy. The most common losses of irrigation water are represented by runoff and deep percolation. Irregular land surfaces, shallow soils underlain by gravels of high permeability, small irrigation streams, nonattendance of water during irrigation, long irrigation runs, excessive single applications—all these factors contribute to large losses and low efficiency. Also excessively large irrigation streams, improper preparation of land, compact impervious soils, steep slopes of land surfaces, and nonattendance contribute to inefficiency.

The depth of water applied in each irrigation is a dominant factor influencing efficiency of application. Even if water were spread uniformly over the land surface, excessive depths of application would result in low efficiencies. Many variable factors such as land uniformity, irrigation method, size of irrigation stream, length of run, soil texture, permeability, and depth influence the time the irrigator keeps water running on his farm and hence the depth he applies.

Water-Use Efficiency

Having conveyed water to the point of use and having applied the water, the next efficiency concept of concern is the efficiency of water use. What proportion of the water delivered was beneficially used on the project, farm, or field can be calculated using the following formula:

$$E_u = 100 \; \frac{W_u}{W_d} \tag{8.5}$$

where E_u = water-use efficiency
W_u = water beneficially used
W_d = water delivered

Water-Storage Efficiency

The increasing scarcity of water and the growing realization of the value to be gained from irrigation has resulted in high-priced water, discouraging excessive use of water. Low financial returns from irrigation often occur not because of excessive water application but because of insufficient application. In many cases only a small fraction of the needed water is being applied. The water application efficiencies under such practices are essentially 100 percent, and yet the irrigation practice is poor. To assist in the evaluation of this problem, the concept of water-storage efficiency is useful. This concept directs attention to how completely the needed water has been stored in the root zone during the irrigation.

$$E_s = 100 \ \frac{W_s}{W_n} \qquad (8.6)$$

where E_s = water-storage efficiency
W_s = water stored in the root zone during the irrigation
W_n = water needed in the root zone prior to the irrigation

Research has shown that improved storage efficiency will triple production in the lower Rio Grande Valley of Texas where insufficient water is normally available. Similar research in western Kansas where water must be pumped at considerable cost produced double the production from better storage efficiency. High water application efficiencies and low production measured in both Texas and Kansas gave positive proof that for these conditions water-storage efficiency is the important irrigation concept.

Water-storage efficiency becomes important whenever insufficient water is stored in the root-zone soil during an irrigation. This condition may occur because of high water costs, because of a scarcity of water, or because of excessive time required to secure adequate penetration. The existence of a salt problem may require that water-storage efficiency be high in order to keep salts washed out of the surface soil. Often a high water-application efficiency indicates that water-storage efficiency may be the important index to better irrigation practice.

Water-Distribution Efficiency

Another important characteristic of irrigation is uniform distribution of irrigation water throughout the root zone. Under most conditions, the more uniformly water is distributed, the better will be the crop response. Uneven distribution has many undesirable characteristics. Drought areas appear in a field which is not irrigated uniformly unless excess water is applied, which in turn results in a waste of water. Whenever a tendency exists for salt accumulation, those areas receiving less than the desired depths of water will show the greatest accumulation of salt.

The formula for water-distribution efficiency, which evaluates the extent to which water is uniformly distributed, is shown below.

$$E_d = 100 \ \ 1 - \frac{y}{d} \qquad (8.7)$$

where E_d = water-distribution efficiency
y = average numerical deviation in depth of water stored from average depth stored during the irrigation
d = average depth of water stored during the irrigation

To illustrate this concept used as an efficiency, suppose two sprinkler systems were each delivering average water applications of 7.6 millimeters per hour. However, system A had a water-distribution efficiency of 90 percent and system B had a distribution efficiency of 70 percent. Assume the owner of system B cannot afford to improve his system and hence will continue to use it as it is. Would you recommend identical operation of both systems to grow similar crops on similar soils? Or would you find it advisable for system B to apply more water in order to obtain results approximately comparable to system A? If you consider that system B must apply more water per irrigation than system A to produce satisfactory results, how much more water would you recommend? (Obviously system B is not as efficient as system A.)

When the variation in distribution is considered in terms of water distribution efficiency, the answer to the question of how much to increase the application from system B is obvious. The depth to be applied is divided by the water-distribution efficiency, thus

$$\text{Depth of application} = \frac{\text{Rated depth}}{\text{Water-distribution efficiency}}$$

Therefore, system A would be applying 11 percent more water and System B 43 percent more water to produce comparable results. Obviously the less efficient system will cost much more to operate.

On the average 75 percent of each area irrigated in this example will have received an amount of water equal to or in excess of the rated amount, and 25 percent will still be deficient. It is undoubtedly impractical and uneconomical to apply sufficient excess water to wet all of the area to the desired depth.

Water-distribution efficiency will provide the measure for comparison between systems, whether it be sprinkler compared to sprinkler, surface compared to surface, or sprinkler compared to surface irrigation.

Consumptive-Use Efficiency

Water stored in the soil during an irrigation may not remain in the soil for use by the crops. A wide furrow spacing and considerable exposed ground surface may result in excessive surface evaporation and continual significant downward movement of moisture beyond the root zone. Loss of water by deep penetration and by excessive surface evaporation following an irrigation can be evaluated by the following concept of consumptive-use efficiency:

$$E_{cu} = 100 \ \frac{W_{cu}}{W_d} \tag{8.8}$$

where E_{cu} = consumptive-use efficiency
$\quad\quad\ W_{cu}$ = normal consumptive use of water
$\quad\quad\ W_d$ = net amount of water depleted from root-zone soil

This concept may be extremely important, particularly where high moisture, high permeability, wide spacing, and ridges are combined. When irrigating potatoes under moist conditions in ridges of permeable soil with widely spaced rows, consumptive-use efficiency may be in the order of 50 percent. Efficiency can be materially increased if ridges are not used. Efficiency can also be increased if row spacing is decreased, or if good sprinkler irrigation is used to more uniformly distribute water. Hence, if better irrigation practice is followed, consumptive-use efficiency is increased. Consumptive-use efficiency is another measure of irrigation efficiency which may be very useful under certain conditions.

This concept also involves the efficiency with which roots are able to utilize moisture stored in the soil during an irrigation. Consumptive-use efficiency is affected by soil texture, amount and distribution of vegetation, profile of the soil surface, distribution of roots within the soil, and variation of soil moisture within the root zone. These factors affect the water lost by excessive evaporation from the soil surface and the water lost from percolation downward beyond the root zone. The water lost by excessive evaporation is increased as the amount of bare ground increases, and decreases as the foliage increases.

Consumptive-use efficiency is useful in explaining the differences in crop response from different methods of irrigation. The irrigation research literature is full of field tests comparing yields obtained from surface irrigation. Sometimes differences are significant and other times on the same crops, differences are not significant. Frequently variability lies in the differences in boundary conditions which are reflected in consumptive-use efficiency. When considering which irrigation method to use, the important aspect is which method will place water where it can be most efficiently utilized by plant roots. This concept may not be important for alfalfa, with its well-distributed root system, but may be very important for potatoes, corn, or melons.

Since losses from evaporation and deep percolation following an irrigation are often small, consumptive-use efficiency can best be evaluated on a seasonal basis, so that the effects of these factors are summated for the entire growing season.

Irrigation-Efficiency Examples

Figure 8.7 contains nine illustrative profiles of common irrigation practices. Examples (a), (b), and (c) are cases of heavier water application near the upper end of the field and illustrate how the three efficiencies vary as the amount of water applied increases. Examples (d), (e), and (f) illustrate a common case where heavier applications of water are applied near each end of the field. Examples (g), (h), and (i) illustrate the type of moisture distribution which frequently occurs under sprinkler irrigation. Example (g) is common sprinkler-irrigation practice when insufficient water is applied. The low-water storage efficiency E_s, and the low-water distribution efficiency E_d, show vividly why this is poor irrigation practice. Example (i) is also an undesirable irrigation practice and is well marked by the low-water application efficiency E_a.

Note, however, that efficiencies of 100 percent are not always desirable. Maximum net profit can often be obtained by filling the root zone only every second, third, or fourth irrigation, de-

Figure 8.7   Application, E, AROEFW E_s, and distribution, E_d, efficiencies and the effect upon corn production illustrated by two-dimensional profiles of the soil. Note: Water-application efficiency estimates are made assuming no runoff.

pending upon the soil and crop. This is primarily because of the capacity of plants to extract more water from the upper portion of the root zone than from the lower portions. Alternating the depth of application also promotes better bacterial growth in fine-textured soils. Thus, a water-storage efficiency of less than 100 percent periodically might produce the best results.

Also, water-application and -distribution efficiencies of nearly 100 percent are not always desirable or practical. The expense necessary to secure high application and distribution efficiencies is often in excess of the economic returns. Another important reason for not always desiring water-application efficiencies of nearly 100 percent is that one or more excessive applications of water per year, depending upon salt content of water and soil, is usually desirable to flush out excess salts in the root zone, thereby keeping the soil in a good productive condition.

8.15 EXAMPLE OF WHEN TO IRRIGATE AND HOW MUCH TO APPLY

Following is a digest of a detailed report[1] which shows the pattern that can be followed in arriving at a guide for management to estimate when to irrigate and how much to apply. Facts observed in this summary are: water-holding capacities of soil, depth of rooting, water use by crops, peak consumptive use, and variation in consumptive use during the season. These factors are all utilized to ascertain the actual water use and the frequency and amount of water to be applied by irrigation.

Maximum Available Moisture

The soils were divided into three convenient groups according to texture as coarse, medium, and fine. After determining the field capacity and permanent wilting point of each group of soils, the maximum depth of available water per foot of soil was determined to be as follows:

Texture	Maximum Depth of Available Water cm/m of Soil
Coarse	10
Medium	14
Fine	18

[1] Taken from "Design of Irrigation System for Plantacoes do Mucoso, Lda. with an Evaluation of Natural Resources Governing the Feasibility of Irrigated Agricultural Development and Recommended Management Practices," by Vaughn E. Hansen and Jay M. Bagley.

Depth of Rooting

Consideration was then given to the depth from which water was extracted from the soil. The average extraction pattern, shown in Fig. 8.6, was considered and then Fig. 8.8 was developed showing the depth of rooting during the growing period.

Water Requirements

Analysis of water use in the basins of the Cuanza, Lucala, and Bengo Rivers indicated that average evaporation from a Piche-type evaporometer at Dondo during the dry season was about 3.6 millimeters per day. Observations both in Angola and in the United States indicated that the peak consumptive use of water was about 0.8 of the Piche evaporation, or 2.8 millimeters per day. A balanced crop rotation will permit a reduction in design capacity to approximately 0.75 of peak design use or 2.1 millimeters per day.

Since not all water which is put into the irrigation system is efficiently used by the plants because of irregular placement of the water, etc., it is necessary to apply more water than will be needed by the plants. Thus, it was assumed that the average efficiency for application of water would be 70 percent for sprinkling and 50 percent for surface irrigation. Also, approximately 10 percent of the water was assumed to be lost in conveyance to the field when surface irrigation was used.

It was also necessary to consider that the system probably would not be operated more than 12 hours per day on the average, considering Sundays, etc. Training and lack of dependability of laborers prevent extensive irrigating at night. When all

Figure 8.8   Depth of rooting during the growing period.

these factors were considered, the design requirements which determined the size of the pipe lines and pumps were:

	l/s per ha	ha/l per s
Sprinkler irrigation	0.7	1.4
Surface irrigation	1.1	0.9

After detailed consideration, surface irrigation was deleted in favor of steel pipe conveyance lines and sprinkler irrigation. One of the factors considered in the decision to use sprinkler irrigation was that surface irrigation may require that 50 percent more water be pumped from the adjacent river than would be necessary if sprinklers were used. Hence, the 0.7 liters per second per hectare was used to estimate water needs.

The concepts embodied in the generalized consumptive-use curve were used to obtain Fig. 8.9 showing the amount of water used by the crops during the growing period.

Table 8.2 was then prepared from Fig. 8.9 to show the accumulated use of water during the growing period. Growing periods from 2 months to 6 months were considered. For example, cotton, which has a 6-month growing period, requires a total of almost 37 centimeters of water. If cotton is planted about March 15 after the rains have stored 7 centimeters of water in the soil, it will

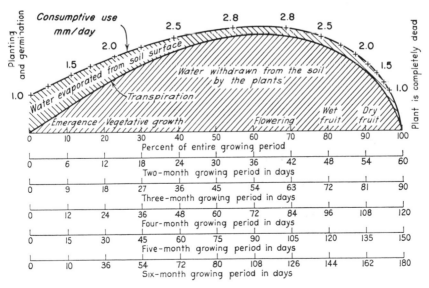

Figure 8.9   Consumptive use of water for various periods of growth.

TABLE 8.2 Accumulated Use of Water for Crops with Various Periods from Planting to Complete Maturity

Percentage of Growing Period [a]	Accumulated Consumptive Use of Water in Percentage of Total Use	Total Period of Growth									
		2 Months		3 Months		4 Months		5 Months		6 Months	
		Days Since Planting	Accumulated Water Use (cm)	Days Since Planting	Accumulated Water Use (cm)	Days Since Planting	Accumulated Water Use (cm)	Days Since Planting	Accumulated Water Use (cm)	Days Since Planting	Accumulated Water Use (cm)
10	6.0	6	0.76	9	1.0	12	1.5	15	2.0	18	2.3
20	14	12	1.80	18	2.5	24	3.6	30	4.3	36	5.1
30	24	18	2.80	27	4.3	36	5.8	45	7.4	54	8.9
40	35	24	4.30	36	6.6	48	8.6	60	10.9	72	13.0
50	47	30	5.80	45	8.6	60	11.7	75	14.5	90	17.3
60	59	36	7.40	54	11.2	72	14.7	90	18.5	108	22.1
70	72	42	8.90	63	13.5	84	18.0	105	22.6	126	26.9
80	84	48	10.40	72	15.7	96	21.1	120	26.2	144	31.5
90	95	54	11.90	81	17.8	108	23.6	135	29.5	162	35.3
100	100	60	12.40	90	18.8	120	24.9	150	31.2	180	37.3

[a] Growing Period is used herein to refer to the entire time from planting to the time the plant dies, which is usually longer than the period from planting to harvesting. Flowering will occur at about 50 to 60 percent of the growing period and fruiting after 60 percent.

Figure 8.10   Amount of water that must be applied by sprinkler irrigation in order to store the require amount of water in the soil.

require an average of about (37−7) or 30 centimeters of irrigation water.

Frequency and Depth of Water to Be Applied by Irrigation

The depth of water to be stored in the soil during irrigation affects the efficiency with which it can be applied. Figure 8.10 was prepared considering the irrigation efficiency to show how much water had to be applied by irrigation to store different depths of water in the soil. Note that the efficiency is less when smaller amounts of water are applied. If 100 centimeters are to be stored, 135 centimeters must be applied because the water is not uniformly distributed over the field.

Combining the growing period, rate of consumptive use, depth of rooting, efficiency of irrigation, and the water-holding capacity of the soils. Table 8.3 was prepared. Similar tables for medium- and fine-textured soils and also for 4-, 5-, and 6-month growing periods were prepared but are not included herein. These tables, together with the preceding material, contain data which are necessary to carefully regulate irrigation practice for different crops grown on either coarse-, medium-, or fine-textured soils. The report also contained recommended management practices similar to the recommendations presented in preceding portions of this book.

TABLE 8.3 Irrigation Frequency and Amount of Water to be Applied by Sprinkling Irrigation When Varying Amounts of Available Water Remain on the Soil. Length of Growing Period = 3 months. Soil Texture = Coarse.

Time Since Planting in Days	Consumptive Use of Water (mm/day)	Average Depth of Rooting (cm)	Maximum Depth of Available Water in Root Zone (cm)	Interval in Days Between Irrigations and Depth in cm to be Applied When Different Amounts of Available Water Remain in the Soil							
				75%		50%		25%		0%	
				Days	Depth (cm)	Days	Depth (cm)	Days	Depth (cm)	Days	Depth (cm)
0–9	1.3	5	1	1	1	2	1	3	1	4	2
9–18	1.8	15	2	2	1	5	2	8	3	10	3
18–27	2.0	28	3	3	2	7	3	10	4	13	5
27–36	2.3	41	4	4	2	8	4	13	5	17	6
36–45	2.5	53	5	5	3	10	4	15	6	20	8
45–54	2.8	66	7	6	3	12	5	17	7	24	9
54–63	2.8	79	8	7	3	14	6	21	8	28	11
63–72	2.5	89	9	8	4	17	7	25	9	34	12
72–81	2.3	97	10	11	4	21	7	33	10	44	13
81–90	1.3	104	11	20	4	41	8	61	11	82	14

CHAPTER
9
SPRINKLER AND TRICKLE IRRIGATION

The method of applying water to the surface of the soil in the form of a spray, somewhat as in ordinary rain, is known as sprinkling. This method of irrigation was started about 1900. The first agricultural sprinkler systems were an outgrowth of city lawn sprinkling. Before 1920 sprinkling was limited to truck crops, nurseries, and orchards.

It was used in humid regions as a supplemental method of irrigation. Most of these systems were stationary overhead perforated pipe installations, or stationary overtree systems with rotating sprinklers. These systems were expensive to install but often fairly inexpensive to operate. Portable sprinkler systems developed with the introduction of lightweight steel pipe and quick couplers in the early 1930's. This reduced the equipment cost for sprinkler systems and resulted in an increased number of sprinkler installations. The number of sprinkler installations has increased rapidly since 1950 owing to the development of more efficient sprinklers, lightweight aluminum pipe, more efficient pumps, and to the widespread distribution of low-cost electrical power and fuels for internal combustion engines. Sprinklers have been used on all soil types and on lands of widely different topography and slopes, and for many crops.

9.1 CONDITIONS FAVORING SPRINKLER IRRIGATION

When considering where sprinkler irrigation can be used to greatest advantage, the normal requirement of uniform distribution of water is of utmost importance. That method of irrigation which can most economically distribute the required water uniformly is generally the best method. Some of the conditions which favor sprinkler irrigation are as follows:

1. Soils too porous for good distribution by surface methods.

2. Shallow soils the topography of which prevents proper leveling for surface-irrigation methods.

3. Land having steep slopes and easily erodable soils.

4. Irrigation stream too small to distribute water efficiently by surface irrigation.

5. Undulating land too costly to level sufficiently for good surface irrigation.

6. Labor available for irrigating is either not experienced in surface methods of irrigation or is unreliable; good surface irrigation requires trained reliable labor.

7. Land needs to be brought into top production quickly. Sprinkler systems can be designed and installed quickly.

The following points should also be considered when comparing sprinkler and surface methods of irrigation.

1. Water measurement is easier with sprinkler than with surface methods.

2. Sprinkler systems can be designed so that less interference with cultivation and other farming operations occurs and less land is taken out of production than with surface methods.

3. Higher water-application efficiency can normally be obtained by sprinkler irrigation.

4. When water is already being pumped to the point of use, the pressure needed for sprinkling can be obtained with a minimum of additional capital investment.

5. When domestic and irrigation water comes from the same source, a common distribution line can frequently be used.

6. For areas requiring infrequent irrigation, sprinkler irrigation can be provided at a lower capital investment per acre of land irrigated than can surface irrigation.

7. Whenever water can be delivered to the field under gravity pressure, sprinkler irrigation is particularly attractive.

8. Frequent and small applications of water can be applied readily by sprinkler systems.

9.2 OTHER USES OF SPRINKLER SYSTEMS

Sprinkler systems have several secondary agricultural uses which are important in addition to the primary use for distributing irrigation water to be stored in the soil.

Light frequent irrigations, so easily managed by using

sprinklers, are helpful in many situations, such as the following: shallow-rooted crops, germination of new plants, control of soil temperature on certain crops such as lettuce, and control of humidity on such crops as tobacco. It is possible that light frequent irrigations may produce favorable results for coffee culture in lieu of shade trees which are commonly used in some parts of the world to control temperature and humidity.

Frost Protection

Frost protection can be provided by sprinklers. Sprinkler systems have been used for frost protection for several years on cranberry bogs, blueberries, strawberries, almonds, citrus, vegetables, fruits, and flowers. Crops have been saved in temperatures as low as $-8°C$ by sprinkler irrigation. Fine misty sprays are desirable. Caution should be taken not to break off tree limbs with ice accumulation from overhead sprinklers. Sprinkling should be sufficient to keep the ice which forms wet and slushy. Spray should be started at or above $30°C$ for best protection. Water should be applied until the ice is melted. A suggested irrigation application rate for different temperatures and wind velocities is shown in Table 9.1.

Application of Fertilizers, Pesticides and Soil Amendments

Several fertilizers, pesticides, and soil amendments can be applied quickly, economically, easily, and effectively through sprinkler-system spraying. The equipment is simple and the labor requirement is slight. By controlling the time of application most of the materials can be leached to the desired depth. Being in solution, these materials are distributed over the surface of the land quite uniformly. Fertilizer elements already in solution be-

TABLE 9.1 Irrigation Application Rate for Frost Protection at Different Temperatures and Wind Velocities

Plant Temperatures (°C)	Wind Velocity (KW/hr)		
	0–0.05	1.0–2.5	3.0–5.0
	Irrigation Rate (mm/hr)		
−3	2.5	2.5	2.5
−4	2.5	3.5	5.5
−5	3.0	5.0	
−6	3.5	6.5	
−8	5.0		

come available to the plant roots faster than when placed in the soil dry.

Fertilizers, pesticides, and amendment materials can be injected into the water in several ways. One of the simplest methods is to connect a barrel containing the dissolved material to the suction side of the pump. The barrel can also be connected to the throat of a venturi tube. Small high-pressure pumps are also used to inject the solution into the high-pressure side of the water line.

Preferred practice is to operate the sprinklers sufficiently before applying fertilizer to thoroughly wet the foliage and soil. Then the fertilizer should be applied for a sufficient time to provide good distribution over the land being irrigated. Application times of at least 30 to 60 minutes are required. Following application the sprinklers should be operated from 30 to 90 minutes depending upon the rate of application. Any corrosive material is thus washed out of the applicator pump and sprinkler system and toxic material is washed off the foliage of the plants. Also applying water after the application of fertilizer or soil amendment moves the material into the soil.

Application procedures for pesticides will vary with the material being applied, depending upon whether it should be left on the foliage or penetrates the soil.

Ammonium nitrate, phosphoric acid, and ammonium sulfate are quite corrosive. Brass and bronze impellers in pumps are strongly attacked by solutions containing phosphorus especially when ammonium salts are also present. Hence, special care should be exercised to be sure that solutions entering the system on the suction side of the pump are fairly dilute and that the rate of application of the material is low.

Crop Cooling

Irrigation can be used to improve both quality and yield of some crops by reducing the temperature on hot days. Potatoes, tomatoes, grapes, apples, strawberries, and similar wet fruit crops have responded favorably to evaporative cooling by sprinkling. Sprinkling of apples has resulted in improved color and marketability.

In some climates, early frost frequently occurs during flowering of fruit trees, reducing yield drastically. Sprinkler irrigation has been used successfully in the spring to cool the trees and soil, thereby delaying flowering and reducing the hazard of damage from an early frost.

9.3 TYPES OF SPRINKLERS

Three general types of sprinklers are used: fixed nozzles attached to the pipe, perforated pipe, and rotating sprinklers.

The earlier sprinkler systems were "fixed nozzle pipe" types, illustrated in Fig. 9.1. Parallel pipes are installed about 15 meters apart and supported on rows of posts. Water is discharged at right angles perpendicularly from the pipe line. The entire 15-meter width between pipe lines may be irrigated by turning the pipes through about 135°.

Perforated sprinkler lines illustrated in Fig. 9.2 are used more extensively in orchards and nurseries. Generally rates of application are in excess of 20 millimeters depth per hour and pressureheads are less than 25 meters, often as low as 7 meters. They do not cover a very wide strip. The rotating sprinklers illustrated in Fig. 9.3 are being used very extensively. The advantage of this sprinkler over other types is its ability to apply water at a slower rate while using relatively large nozzle openings. This factor is particularly favorable in waters containing silt and debris since

Figure 9.1.   An overhead sprinkling irrigation system used for frequent applications of water. (Courtesy Skinner Irrigation Co.)

Figure 9.2. Perforated sprinkler line in citrus orchard. (Courtesy W. R. Ames Col.)

Figure 9.3. Sprinkler system on a field of tomatoes. (Courtesy ShurRane Irrigation Co. and E. C. Olsen Co.)

less stoppage of sprinklers is experienced. Application rates less than 2.5 millimeters per hour are possible with these sprinklers. This slow rate is desirable on soils having low infiltration rates and advantageous to the small farmer doing his irrigation along with his field work. This low rate of application makes it necessary to move the sprinklers only once or twice a day on soils with low water-holding capacity. Pressure heads for rotating sprinklers normally range from 20 meters for the smaller sprinklers to over 70 meters for the large units.

9.4 PRESSURE HEADS

Sprinkler systems operate under a wide range of pressure heads from 3.5 meters to over 70 meters. The desirable pressure head depends upon power costs, area to be covered, type of sprinkler used, sprinkler spacing, and crop being irrigated. Low pressure heads range from 3.5 to 10 meters; medium from 10 to 20; intermediate from 20 to 40, and high pressure heads from 40 to 70 meters. Giant sprinklers usually operate at pressure heads in excess of 55 meters. Sprinklers in the low-pressure head range have small areas of coverage and relatively high sprinkling rates for the recommended spacings of the sprinklers. Their use is generally confined to soils having infiltration rates of more than 12 millimeters per hour during the irrigation.

Medium-pressure head sprinklers cover larger areas and have a wide range of precipitation rates, and water drops are well broken up.

High-pressure head sprinklers cover large areas, and precipitation rates for recommended spacing are higher than for the medium pressures. Distribution patterns are usually good but are easily disrupted by winds because of higher water trajectories. They have high-application rates, above 20 millimeters per hour. The wetted diameter of the circle is from 60 to 120 meters. Distribution patterns are very good in calm air but are easily disturbed by winds.

In general, sprinklers can be obtained which will fit practically any operating condition in the field.

9.5 TYPES OF SPRINKLER SYSTEMS

The sprinkler system includes the sprinkler, the riser pipe, the lateral distribution pipe, the main line pipe, and, often, the pumping plant. Sprinkler systems may be classified as semipermanent or portable according to the make-up of the components. Figure 9.4 illustrates three typical types of sprinkler irrigation systems; multiple sprinklers on a line, a single large sprinkler, and a boom sprinkler system with several sprinklers.

Semipermanent Installations (often referred to as solid set systems)

There are semipermanent installations wherein the main lines are buried and the lateral lines and sprinklers remain fixed during the irrigation season. Labor is reduced to a minimum, and quality and yield of crops are generally maximal, provided the system is operated properly. One man can irrigate 40 to 80 hectares per day compared to approximately 20 hectares with the movable system.

When light frequent irrigations are desirable and when the farmers are totally dependent upon irrigation to supply the crop moisture needed during the season, semipermanent systems are finding favor. The most extensive use has been on potatoes which show marked increases in yield and quality with light frequent irrigations. Shallow-rooted crops, like radishes, are also responding well to permanent systems. For years nurseries have

Boom—Sprinkler

Single—Sprinkler

Figure 9.4. Typical types of sprinkler irrigation systems.

found them useful for maintaining a misty moist condition over plants.

Since permanent sprinkler systems are often used to apply small quantities of irrigation water, it is well to remember that wind may be a serious problem. Short periods of application do not allow for natural shifts in wind direction and variations in wind velocity which produce a more even distribution of water.

A variation of this system has sprinklers mounted on quick-coupling risers so that they may be moved along the buried lateral lines. This reduces the number of sprinklers and increases the labor needed for operation of the system.

Portable Systems

Semiportable sprinkler systems consist of buried main pipe lines, portable lateral pipe lines and sprinklers, and a fixed pumping plant. The portable low-angle sprinkling system, illustrated in Fig. 9.5, is made up of a portable main pipe line, portable lateral pipe lines and sprinklers, and fixed pumping plant.

Multi—Sprinkler

Figure 9.5.   Low-angle sprinklers in orchards keep water off soft fruit and leaves. (Courtesy W. R. Ames Co.)

The fully portable system has portable lateral pipe lines with sprinklers and portable pumping plant. A fully portable system can be moved readily from field to field, thus extending its utility.

Large rotating booms carrying combinations of rotating and fixed sprinklers are illustrated in Fig. 9.6. The size of pipe, length

Figure 9.6. Rotating boom-type sprinkler. (Courtesy W. R. Ames Co.)

of the boom, nozzle sizes, and pressures can be varied to cover the desired area at each setting.

9.6 MOVING LATERAL LINES

Lateral lines containing sprinklers can be moved by several methods. The cheapest method in terms of capital equipment cost is the hand-move system where the pipe is uncoupled, moved a length at a time to the new location, and recoupled. Drag- or tow-type laterals are being used wherein the entire line is pulled by a tractor or other vehicle to its new location on skids provided at each coupling. Wheels can be secured at the couplings in place of the skids, permitting easier transport. Both the skid and wheel arrangement allow movement parallel to the line with slight curvature being possible to displace the line laterally.

The wheel-move system uses the pipe as an axle and by mounting large diameter wheels the pipe line can be moved perpendicular to the direction of the lateral lines. The wheel-tube units can either be hand powered or powered by a small auxiliary internal combustion engine, an electric motor, or a hydraulic motor.

The lateral system may also be moved by rotating the line in a circle about one end as a pivot. The rate of application is adjusted to compensate for different rates of travel along the line. The wheels are powered by hydraulic cylinders, designed to produce a rate of travel proportional to the distance from the pivot point.

9.7 GRAVITY SPRINKLER SYSTEMS

Often the most economical sprinkler system is one utilizing a source of water adjacent to and above the area to be irrigated. When the source of water is high enough on an adjacent hill to produce sufficient pressure at the sprinkler nozzles without use of a pump, the system is called a gravity sprinkler system.

There is no expense for a pump and motor, fuel or electricity. However, more main-line pipe must be purchased to convey the water from the source to the field than is required normally for a pumped sprinkler system. Larger main-line pipe produces more pressure at the sprinklers, but also increases the cost of pipe. The most economical design is obtained considering not only design requirements but also seasonal irrigation costs and expected production.

9.8 SOURCES OF WATER AND METHODS OF DEVELOPING PRESSURE

Sprinkler systems require sources of water free of debris that would clog the sprinklers. The best sources of debris-free water are wells and lakes. Screening devices are usually necessary to remove the debris when water is taken from irrigation canals or rivers.

Pressure for operation of sprinklers is obtained from gravity when practical and supplied by pumps when the source of water is located at such a level that gravity pressure will not operate the sprinklers. In some systems, combinations of both gravity and pump power are used to obtain the desired pressure.

When pumping from lakes, streams and irrigation canals, centrifugal pumps are generally used to develop pressure. When pumping from wells either centrifugal, submersible, or turbine pumps are used. Turbine pumps or submersible pumps are better adapted to wells in areas having a variable water table during the irrigation season.

9.9 REMOVAL OF SILT AND DEBRIS

Silt and debris are very detrimental to sprinkler equipment. The abrasive action of silt causes excessive wear on pump impellers and sprinkler nozzles and bearings. Debris plugs nozzles and thus interferes with uniform application of water to the land. Clogged sprinkler nozzles cannot be tolerated. The method of removal of debris and silt will need to be adapted to site conditions, but a few general considerations will usually apply and should be carefully considered.

Silt is heavier than water and will settle out if given an opportunity. Hence, silt removal can best be accomplished by creating a quiet condition where silt can settle and subsequently be flushed from the container. Basins of various shapes can be used effectively.

Since silt is heavier than water it will tend to be carried along the bottom of the stream, provided the stream is not too turbulent. The intake of the pump should be located where the silt content is a minimum. Never place the intake in position where the silt will be carried directly into the intake. Frequently the approach conditions can be adjusted so that the portion of the flow with the greatest silt concentration is deflected past the intake. Screens can also be used to prevent silt from entering the intake to the pump.

In some regards debris presents more of a problem than does silt. Debris can be water soaked and float beneath the surface or can be carried on the bottom of the stream while lighter material will be carried on the surface. Screening devices clog easily. Sprinklers are easily clogged with debris and prevented from functioning properly. The structure used to remove debris from the water should be designed specifically to combat the type of debris present in the water.

Floating debris can be skimmed off the surface with a fixed or rotating screen and the subsurface waters permitted to proceed into the pipeline. However, cleaning the screen presents a problem. If sufficient water is available to allow some to pass by the intake, the excess can be used to flush over the screen and remove the collected debris. In this case a horizontal screen just below the surface or a screen gently sloping up so that it is below the surface at the upstream end, and above or near the surface at the overflow, will clean reasonably well. Revolving screens have also been used with reasonable success.

Removal of debris such as moss and organic matter that is suspended in the flow presents more of a problem. Screens covering the entire cross section of flow are effective provided the size is small enough to remove the material and provided they are cleaned sufficiently often to prevent excessive resistance to flow. Two, or even three, screens used in sequence are helpful in reducing the tendency for clogging. The first should be coarser to retain the larger debris, and the final screen should be small enough to retain the smallest objectionable debris.

9.10 OPERATION OF SPRINKLER SYSTEMS

A sprinkler system may be well designed for the crop and field, but if it is not efficiently operated the result will be disappointing. A correctly designed sprinkler system will supply adequate water during periods of maximum water demand by the crop. Over-irrigation will result if the system is operated at full capacity when the water demand of the crop is less than maximum. Excessive application will cause leaching of soluble plant food, low water-application efficiencies, reduction in quality and quantity of crops, and, ultimately, a drainage problem.

9.11 COST ANALYSES

In considering the economics of any irrigation system, all costs and benefits should be included. Initial costs are important because the purchaser must finance this usually heavy capital investment. However, annual costs per hectare compared to annual returns per hectare are the best measure of the economics of an irrigation system. Annual costs per hectare should include the following:

1. Annual depreciation of irrigation system.
2. Interest on investment for the system.
3. Water costs.
4. Power costs.
5. Repair, operation, and maintenance.
6. Labor.
7. Taxes.

When considering the return to be expected from a new irrigation system, credit should be shown for any savings resulting from the following:

1. Increased yield and quality.
2. Less land out of production.
3. Reduction in land preparation, tillage, and harvesting costs.
4. Saving in labor, repair and maintenance, and operating cost.
5. Saving in water and power costs.

Actual costs vary widely. Initial cost will increase as labor-saving devices are added. Permanent systems where the main and lateral lines are not moved will cost more than portable lines. The average total cost per hectare per year includes initial investment and operating costs. The life expectancy of a sprinkler system varies with treatment, use, and storage but will average about 15 years.

The average annual cost of equipment is normally 40 percent of the total annual cost. Fuel and repair would average about 30 percent and labor 30 percent.

The sprinkler system should be designed to apply water at the lowest annual cost. A balance should be sought between pipe size and pumping costs in a system operated by pumping. Careful analysis should be made to arrive at a reasonable balance between equipment costs and power costs.

9.12 SPRINKLER SYSTEM DESIGN REQUIREMENTS

The important factors in the success of sprinkler irrigation systems are first, the correct design, and second, the efficient operation of the designed system. The basic information necessary for the design of a farm irrigation system is obtained from four sources, namely, the soil, the water supply, the crop to be irrigated, and the climate.

Information concerning soils includes the soil type, depth, texture, permeability, and available water-holding capacity of the root zone. Necessary water supply concerns the location of the water delivery point in relation to the fields to be irrigated, the quantity of water available, and the delivery schedule. The maximum consumptive use of water per day, the root-zone depth, and the peculiarities of irrigation necessary to be taken into account in the irrigation system are obtained from a knowledge of the requirements of the crop to be grown. Climatological information includes the natural precipitation and wind velocities and direction. All this information must be compiled in one form or another before starting to design a sprinkler system.

The performance requirements of a sprinkler system include applying water at a rate that will not cause runoff from the area irrigated. Water should be applied at such a rate that high water-application efficiency is obtained and in such a manner that high water-distribution efficiency results.

The sprinkler system must have the capacity to meet the peak water-use demands of each crop during the irrigation season. Allowance in capacity must be made for unavoidable water losses by evaporation, interception, and some deep percolation.

When a system is designed for supplemental irrigation or protective purposes, it should have a capacity to apply the necessary depth of water to the design area in a specified time. The cost of the system should be consistent with the insurance values involved.

There should not be more than a 20-percent variation in the depth of water applied to any part of the design area. To obtain reasonably uniform distribution the pressure in the lateral lines should not vary more than 20 percent. Variations in pressure occur as the result of friction loss in pipes and elevation changes in main lines or lateral lines. It may be necessary to control pressures with valves.

A sprinkler system must apply water so that it will not cause physical damage to the crop. In orchards, high-velocity streams of water from sprinkler nozzles have bruised growing apples and peaches when sprinklers have been placed too close to the trees. Also, in crops having fine seedling plants, a fine spray must be applied, or the plants will be beaten into the ground. Such a spray requires high pressures to break up the water drops at the nozzle.

Droplets must be small enough so that the soil is not damaged. Some soils will puddle under the impact of large droplets, causing the soil to crust. Smaller nozzles operated at higher pressures reduce drop size. Acceptable droplet size can be obtained using the following combination of nozzle sizes and pressure heads.

$$(3.175 \text{ to } 4.75 \text{ mm})—25 \text{ to } 35 \text{ m}$$
$$(4.75 \text{ to } 6.35 \text{ mm})—30 \text{ to } 40 \text{ m}$$
$$(6.35 \text{ to } 9.5 \text{ mm})—35 \text{ to } 50 \text{ m}$$

Two nozzle sprinklers will produce acceptable breakup of drops at 3.5 meters less pressure head than single nozzle sprinklers.

Wind can be critical, interfering excessively with distribution of the water. It is difficult to obtain good distribution of water in winds in excess of 15 kilometers per hour. Lateral lines should be set at an angle to the prevailing wind. It may be necessary to decrease the spacing between sprinklers and between lateral lines as much as 40 percent to obtain satisfactory distribution in the windy condition.

When used in practical field spacing with selected operating pressures, the sprinkler chosen must give satisfactory moisture distribution.

If a pump is necessary it must be picked on the basis of the maximum operating conditions of pressure head and flow and must not overload at minimum operating conditions.

Careful attention should be given to the planning of the sprinkler system. To select the best combination of power plant, sprinkler system, and operating schedule requires skill and experience.

9.13 PLANNING FARM SPRINKLER SYSTEMS

Adequate planning for sprinkler irrigation includes an inventory of the resources of the farm and integration of the operation of a system with other required farm or ranch operations. The system planner needs to consider the limitations of the farm resources as well as those factors of economics and farming operations which may influence the success or failure of the proposed irrigation enterprise.

In many cases, the initial installation of a sprinkler system is only a beginning step in the development of a complete farming sprinkler system. Adequate planning for sprinkler irrigation should recognize the need for future expansion. The initial system when installed should, insofar as is practical, fit with future expansions and, finally, into the ultimate farm irrigation system.

A factor often overlooked, both by the planner and the prospective user, is the cropping pattern which will develop from full use of the system. Often the system capacity, layout, and its operation are planned around one cash crop. Soil-building crop rotations will be required to maintain optimum production as the yields of this crop are stepped up with irrigation and the other necessary good farming practices. The requirements for moisture for the resulting crop rotation may be entirely different. With a crop rotation, it may be possible to reduce the required system capacity where peak moisture requirements for different crops occur at different times during the irrigation season. With the development of a balanced crop rotation, mechanized equipment may be adapted for use on both row crops and close growing crops. Recognition of this feature may lead to reduced labor costs.

To prepare a successful sprinkler system plan there should be complete understanding between the planning technician and the prospective owner or user. When the inventory of farm or ranch resources is made, the planner and owner or user should work closely together. Differences in soils should be understood. Quite often the landowner or operator can give clues as to the intake rate of soils in the design area. If he has had previous irrigation experience, he has probably observed differences in ability of the same soils to hold water for crop use.

A complete understanding is needed concerning the labor requirements for irrigation operations. Without a full discussion of this phase of sprinkler irrigation, misunderstandings can arise, leading either to improper operation of the system or disappointment on the part of the user. Labor should be included in the annual irrigation costs.

The supplier-planner, who can reach these understandings and plan the irrigation facilities and operations accordingly, is going to be in a better position to maintain a profitable and reputable longtime career in sprinkler irrigation. In addition, he will have the satisfaction of playing an important role in building the agricultural production in his territory. Table 9.2 is a guide to the selection of a sprinkler irrigation system considering several of the pertinent factors. The tabulated values and guidelines should be used as an indication of normal practice.

TABLE 9.2 Factors Affecting the Selection of Sprinkler Irrigation Systems

Type of System	Maximum Slope (%)	Water Application Rate (mm/hr) Min.	Water Application Rate (mm/hr) Max.	Shape of Field	Field Surface Conditions	Maximum Height of Crop (m)	Labor Required (hr/ha per irrigation)	Size of Single System (ha)	Approx. Relative Cost[a] (dollars/ha)	Cooling and Frost Protection	Adaptable To: Pesticide Application	Adaptable To: Fertilizer Application	Adaptable To: Liquid Animal Waste Distribution
Multi-Sprinkler													
Hand-moved:													
Portable set	20	2.5	50	Rectangular	No limit	No limit	1.2–3.7	1–15	50–200	No			Not Recommended
Solid set	No limit	1.3	50	Any shape			0.5–1.2	1 or more	400–900	Yes			
Tractor-moved:													
Skid-mounted	5–10	2.5	50		Smooth enough for safe tractor operation		0.5–1.0	10–15	100–300				
Wheel-mounted	5–10	2.5	50				0.5–1.0	10–15	100–300				
Self-moved:													
Side-wheel-roll	5–10	2.5	50	Rectangular	Reasonably smooth	1.2	0.2–0.7	10–30	100–300	No	Yes	Yes	Yes
Side move	5–10	2.5	50			1.2–1.8	0.2–0.7	10–30	125–300				
Self-propelled:													
Center-pivot	5–15	5.0	25	Circular	Clear of obstructions, path for towers	2.4–3.0	0.1–0.4	15–65	125–250				
Side-move	5–15	5.0	25	Square or Rectangular		2.4–3.0	0.1–0.4	30–65	125–250				
Single-Sprinkler													
Hand-moved	20	6.3	50	Any shape	No limit	No limit	1.2–3.7	10–15	50–200	No			Not Recommended
Tractor-moved:													
Skid-mounted	5–15	6.3	50		Safe operation of tractor		0.5–1.0	10–15	100–300		Yes	Yes	
Wheel-mounted	5–15	6.3	50		Lane for winch and hose		0.5–1.0	10–15	100–300				
Self-propelled	No limit	6.3	25	Rectangular			0.2–0.7	15–40	120–250				
Boom-Sprinkler													
Tractor-moved	5	6.3	25	Any shape	Safe operation of tractor	2.4–3.0	0.5–1.2	10–15	200–300				
Self-propelled	5	6.3	25	Rectangular	Lane for boom and hose	2.4–3.0	0.2–1.2	15–40	200–300				Yes
Permanent													
Manual or Automatic	No limit	1.3	50	Any shape	No limit	No limit	0.1–0.2	1 or more	400–1000	Yes			

[a] Does not include the cost of water supply, pump, power unit, and main line.

9.14 DESIGN OF A SPRINKLER SYSTEM

Proper sprinkler design involves careful consideration of essentially all of the concepts presented in earlier chapters. Soils, topography, crops, water supply, management, consumptive use, and hydraulics are involved in addition to labor, equipment, and power costs.

Table 9.3 shows how these factors can be organized in a systematic manner to arrive at a good sprinkler design. Technical data is available on sprinklers, pipe, pumps, and power units from the companies who sell sprinkler irrigation equipment.

Design should begin with the farm-resource inventory. Tabulation of location, water supply, crops, desired irrigation operations, power source, soils, and consumptive-use information, as well as a sketch of the farm, are essential before any detailed calculations can be made.

With this information the system capacity computations can be made. Note that allowance is made for variation from field to field, and yet one set of system-design values must be selected which will give the sprinkler system the capacity to meet the varying operating conditions.

Knowledge of the required capacity of the system and the operating conditions allows selection of an application rate, spacing, and distance between lateral lines. A particular sprinkler nozzle size and operating pressure can be selected and pipe lengths calculated. Power requirements and economic pipe sizes should be determined by studying several combinations. From these possibilities the most desirable pump and power unit can be selected.

The feasibility of irrigating by sprinklers cannot be fully ascertained until a cost analysis is completed showing the expected profits when production costs are subtracted from gross returns. Alternate designs should be considered in arriving at the best design.

The selection of sprinkler, nozzle size, pressure, and general layout should be based upon good engineering concepts tempered by field experience with sprinkler irrigation. Rate of application and droplet size are determined by considering several variables, but the feasible rate of application and droplet size will be restricted to a fairly narrow range of possiblities. Spacing of sprinklers along the lateral line, and spacing between lateral lines are also functions of sprinkler selection and pressure. Certain combinations work best for given field conditions of soil, crop, and wind. Much helpful information to guide designers is

TABLE 9.3 Sprinkler Irrigation Technical Data Sheet

FARM INFORMATION

Name of Owner _____ Address _____ Date _____

Section _____ Range _____ Township _____ County _____ State _____

IRRIGATION SYSTEM DESIGN AREA:

Topography—Include profile of proposed main pipeline locations. Give elevations of points around field boundaries, water level, pump locations, highest and lowest points in design area, and natural gas or electrical power line location.

Soils—Show predominating soil types and location on map.

	Field Nos.				
	1	2	3	4	5
Available moisture-holding capacity cm/m					
Intake rate (mm/hr)					
Effective depth of soil (m)					

WATER SUPPLY

Source _____ Amount available _____ l/s _____ m³

Seasonal variation (l/s) _____ to _____

Delivery scheduling _____

POWER SOURCE

Electrical _____ Internal combustion engine _____ Fuel type _____ Other _____

If electrical give power phase _____ Voltage _____ hp limitations _____

LABOR AVAILABILITY

Hours of operation per day _____ Operation days per week _____

IRRIGATION REQUIREMENTS

	DESIGN AREA			REQUIREMENTS		
Crops	Field No.	Hectares	Effective root zone depth (m)	Moisture to be replaced each irrigation (cm)	Peak moisture use rate (mm/day)	Peak use period irrigation frequency (days)

DESIGN AND SPECIFICATION INFORMATION

SYSTEM CAPACITY REQUIREMENTS

Field Nos.

1	2	3	4	5

Minimum system capacity ——————— l/s

Application rate (mm/hr) ....

Time of lateral operation per set (hr) ....

Depth applied per set (cm) ....

Number of lateral sets per day (no.) ....

Operation period to cover area (days) ....

System capacity required (l/s) ....

Water application efficiency (%) ....

Depth moisture replaced each irrigation (cm) ....

SPRINKLER SELECTION

Application rate _____ mm/hr. Sprinkler spacing on lateral _____ m.

Lateral spacing on main line _____ m. Sprinkler discharge _____ l/s.

Diameter of circle covered _____ m. Type _____ Nozzle sizes _____ mm × _____ mm.

Required operating pressure head _____ m.

LATERAL DESIGN

Lateral spacing on main line _____ m. Sprinkler spacing on lateral _____ m.

Lateral length _____ m. Area covered per lateral setting _____ ha.

No. sprinklers per lateral _____ Lateral discharge _____ l/s.

No laterals required _____ , to cover design area in _____ days.

Pipe size required: Length _____ m. Diameter _____ cm. Gage pipe _____

Length _____ m. Diameter _____ cm. Gage pipe _____ m.

Rise or fall (circle applicable condition) in lateral _____ m.

Pressure head loss in lateral due to friction _____ m. or _____ m.

Pressure head required (at main line) to operate lateral _____ m.

MAIN LINE DESIGN

Portable _____ or Permanent _____ Material: Steel _____ Aluminum _____ or Other _____

Main line length _____ m. Discharge capacity _____ l/s.

Rise or fall (circle one) in main line _____ m.

Allowable head loss due to friction _____ m.

Pipe sizes required _____ m of _____ cm pipe _____ gauge.

_____ m of _____ cm pipe _____ gauge.

_____ m of _____ cm pipe _____ gauge.

Head loss due to friction _____ m. Outlet spacing on main line _____ m.

SUPPLY LINE DESIGN (That portion of main line outside design area)

Supply line length _____ m. Discharge capacity _____ l/s.

Pipe size required _____ cm. Friction head loss _____ m.

TOTAL DYNAMIC HEAD REQUIREMENTS

Pressure head required at lateral _____ m.
Friction head loss in main line _____ m.
Friction head loss in supply line _____ m.
Friction head loss on suction line _____ m.
Elevation difference between pump and highest point of lateral line _____ m.
Elevation difference between water source and center of pump _____ m.
Miscellaneous friction head loss in special valves and fittings _____ m. Total

PUMP REQUIREMENTS

Capacity _____ l/s at head _____ Size _____ cm.

PUMP SPECIFICATIONS

Type _____ Capacity _____ l/s. Efficiency _____ %, at _____ m head
rpm @ required discharge _____ Brake hp @ required discharge

POWER REQUIREMENTS

Water hp _____ Efficiency _____ %. Brake hp _____ rpm

POWER UNIT SPECIFICATIONS

Type _____ Make _____ Model _____ Size _____ rpm.
Continuous hp rating _____ at _____ rpm. cm.
displacement _____ Stroke _____ cm.
Piston speed at design load _____ Bmep at design load _____ m.
Type of power conversion _____ Speed ratio _____
Electric motor: Voltage _____ Phase _____ rpm

193

ECONOMIC ANALYSIS OF IRRIGATION

SYSTEM COST

ITEM	Quantity and unit	Unit cost	Total cost	Annual fixed cost
Well unit				
Pumphouse unit				
Power unit				
Pump unit				
Power line extension				
Main pipe line (complete with valves)	m			
Lateral pipe lines (complete)	m			
Sprinklers and risers	m			
Special equipment:	no.			
List:				
Total Material Costs				
Cash outlay for installation				
Farm costs, Labor				
Tractor use				
Other				
Total Farm Costs				
Total Cost				

LABOR COSTS

Moving laterals (per irrigation) _____ hr. Moving mainline (per irrigation) _____ hr.

Starting and stopping pump (per irrigation) _____ hr.

Other time required (per irrigation) _____ hr.

Total labor (per irrigation) _____ hr. @ $ _____ per hour

Seasonal cost = Cost per irrigation _____ × No. of irrigations _____

INVESTMENT COST

Water development—(well, pond, sump, etc.) .... $ _____

Land development—(leveling, clearing) .... _____

Equipment .... _____

Materials .... _____

Labor .... _____

Other—(rights of way, legal and technical costs) .... _____

Total investment .... $ ....

ANNUAL FIXED COSTS

Annual depreciation of investment .... $ _____

Interest on average investment .... _____

Taxes—(irrigation equipment) .... _____

Insurance .... _____

Total fixed cost .... $ ....

ANNUAL OPERATING AND MAINTENANCE COST

Hours of operation per year _____

Fuel or electricity .... $ _____

Lubricating oils, grease, attendance .... _____

Labor, irrigating .... _____

Maintenance irrigation system .... _____

Total operating and maintenance cost .... $ ....

Total cost of irrigation = Fixed cost plus operating and maintenance costs .... $ _____

Cost per ha-mm = Total cost of irrigation divided by tota ha-mm applied .... $ _____

Cost per irrigated ha = Cost per ha-mm multiplied by mm applied .... $ _____

Remarks: _____

MAP OF AREA
TO BE IRRIGATED
(Use paper with coordinate grid lines)

Made by _____

Title _____

Date _____

195

available from commercial sources as well as from public agencies.

9.15 TRICKLE IRRIGATION

Trickle irrigation,[1] also referred to as drip irrigation, consists of an extensive network of pipes usually of small diameter that deliver filtered water directly to the soil near the plant. The water outlet device in the pipe is called an "emitter" discharging only a few liters per hour. From the emitter, water spreads laterally and vertically by soil capillary forces augmented in the vertical movement by gravity. The area wetted by an emitter depends upon the flow rate, soil type, soil moisture, and the vertical and horizontal permeabilities of the soil.

The "control head" usually located at the source of the water, consists of flow control valves, measuring devices, pressure controls, and filters. Usually a fertilizer injection system is located also at the control head. The filter system must remove essentially all debris, sand, and clay to reduce clogging of the emitters.

Flow may be controlled manually or may be set to automatically deliver (1) desired volume of water, (2) water for a predetermined time, or (3) water whenever soil moisture decreases to a predetermined amount.

Lateral lines are generally flexible PVC or polyethylene pipe 12 to 32 millimeters in diameter. Emitters are inserted into the lateral lines at a predetermined spacing chosen to fit crop and soil conditions. Twin bore pipe, porous pipe, and pipe with small perforations is used in some installations to function both as a lateral conveyance pipe and as an emitter system.

Emitters must produce a relatively small flow and yield a nearly constant discharge. The flow cross section needs to be relatively large to reduce clogging of the emitters.

Benefits of Trickle Irrigation

Often greater crop yields and better quality are obtained, probably the result of the ability to maintain a nearly constant moist soil in the root zone. Fruit that contains considerable moisture when harvested (such as tomatoes, citrus, deciduous fruits, grapes, and berries) respond well to trickle irrigation. Trickle irrigation is not practical or economical for closely planted crops such as cereal grains and alfalfa.

[1]Jack Keller and David Karmeli, *Trickle Irrigation Design*, Rainbird Sprinkler Manufacturing Corporation, 1975.

Irrigation water can be applied very efficiently to small trees and widely spaced plants, where adequate water can be placed in the root zone without wetting the soil where no roots exist, as shown in Fig. 9.7. In arid regions with good management the ratio of transpired to applied water is usually at least 0.90. Water application efficiencies approaching 100 percent and water savings of 30 to 50 percent over other irrigation methods are obtained for those crops and conditions favoring trickle irrigation.

Insect, disease, and fungus problems are reduced by minimizing the wetting of the soil surface. Fewer weeds, less soil crusting, reduced cultivation, and thus less soil compaction and less interference with harvesting are benefits of trickle irrigation.

Potential Problems of Trickle Irrigation

Clogging of the small conduits and openings in the emitters is the most serious problem. Sand and clay particles, debris, chemical precipitates, and organic growth can block flow from emitters. The clogging often occurs gradually, reducing flow and causing poor water distribution along the lateral with a reduction in crop growth.

Dissolved salt is left in the soil as the water is used by the plants. The greatest deposition is near the edge of the wetted zone at the soil surface. If a rain flushes the salts near the sur-

Figure 9.7.   Trickle irrigation of a newly planted field where only the root zone is supplied water.

face down into the active root zone, severe damage to the crop may result. The more saline the irrigation water, the more severe the salt deposition problem becomes.

As with all types of irrigation the buildup of excess salts must be periodically flushed out of the potential root zone. Application of water by sprinkler or surface irrigation may be necessary. The cost of equipment, labor, and water to remove excess salts must be considered when analyzing the cost and efficiency of trickle irrigation.

Since only part of the soil is wetted, care must be taken to see that root growth is not inhibited by supplying moisture to too small a soil volume. Roots will not penetrate a dry soil.

In windy areas, movement of dry cultivated surface soil between emitters may be a hazard to growing crops.

CHAPTER 10

SURFACE
AND SUBSURFACE
IRRIGATION

Irrigation water is supplied to land by four general methods: namely, (a) surface, by flooding or furrows; (b) subsurface, in which the surface is wetted little if any; (c) trickle from pipes near the plant; and (d) sprinkling, in which the soil surface is wetted much as it is by rainfall. Sprinkler and trickle irrigation is the subject of Chapter 9. Surface and subsurface methods are further subdivided as follows:

1. Surface:

   a. Uncontrolled or "wild" flooding.

   b. Flooding controlled with corrugations, borders, checks, or basins.

   c. Furrows.

2. Subsurface:

   a. Controlled by lateral supply ditches.

   b. Uncontrolled, from excess application of water to adjacent or higher lands.

Irrigation methods vary in different parts of the world and on different farms within a community because of differences in soil, topography, water supply, crops, and customs. Forage crops such as alfalfa, clover, hay, and pastures in some areas are irrigated by use of corrugations. Flooding methods of irrigation as well as border strips and basins are suitable for forage crops and rice. Row crops are irrigated by furrows. Any one or a combination of several methods may be best suited to one farm, as illustrated in Fig. 10.1.

10.1 UNCONTROLLED OR "WILD" FLOODING

In the early irrigation of centuries past, throughout Asia and southern Europe, water was applied by flooding extensive areas of rather smooth, flat land. In Egypt especially, the flooding

Measuring and turnout structure

Canal or lateral

Division box

Drop structure

Drop structure

Drop structure

Furrow irrigation row crops

Border irrigation of row crops

Border Alfalfa or small grains

Corrugations Alfalfa or small grains

Basins level ground for flooding

Garden

Border irrigation grains

Contour ditches

Border ditches

Irrigated pasture

Waste ditch

Irrigation methods vary in different parts of the country and on different farms within a community because of various differences in soil, topography, water supply, crops, and customs. Close-growing crops such as alfalfa are irrigated by use of corrugations, borders, border ditches, basins, and contour ditches while row crops are irrigated by furrows and borders. Any one or a combination of several may be best suited to one farm.

Figure 10.1. Various methods of applying irrigation water to field crops.

200

method was of general adoption, the water being forced to spread over vast tracts, during the season of high stream flow.

In modern irrigation, several improved flooding methods have been developed. Brief descriptions of these methods follow.

Where water is applied from field ditches without any levees to guide its flow, or otherwise restrict its movement, the method is designated uncontrolled or "wild" flooding. It is practiced largely where irrigation water is abundant and inexpensive.

If the water is made to flow over the surface too quickly, an insufficient amount will percolate into the soil. Whereas, if water is kept on the soil surface too long, waste will result from percolation beyond the root zone. It is an important and difficult problem to apply water efficiently by uncontrolled flooding methods. The size of stream used, the depth of water as it flows over the soil surface, and the rate of intake of water into the soil all influence application efficiency. In uncontrolled flooding, the smoothness of the land surface and the attention and skill of the irrigator are also important.

Water is brought to the field in permanent supply ditches and distributed from ditches built across the field. The spacing of the ditches is determined by the grade of the land, the texture and depth of the soil, the size of stream, and the nature of the crop. The distances between the diversions from ditches down the steepest slope are similarly determined.

Flooding from field ditches is well adapted to some lands that have such irregular surfaces that the other flooding methods are impractical. However, even on lands that may advantageously be irrigated by the other flooding methods, irrigators often continue to use uncontrolled flooding because of the low initial cost of preparation of land for this method. The extra labor cost in the application of water and the greater losses of water by surface runoff and deep percolation usually offset the apparent advantages of low initial cost of preparation of land.

Where land, water, or labor is expensive, where soil is deep and not likely to crust badly, and where the land is not too rough or steep, it is generally advisable to prepare for controlled flooding in border strips or level or contour basins.

10.2 BORDER-STRIP FLOODING

Dividing the farm into a number of strips, preferably not over 10 to 20 meters wide and 100 to 400 meters long, separated by low levees or borders, is designated the border-strip method. Water is turned from the supply ditch into these strips along which it flows slowly toward the lower end, wetting the soil as it advances.

The surface is essentially level between levees, so that the advancing sheet of water covers the entire width of land strip; but lengthwise of the levee the surface slopes somewhat according to the natural slope of the land. It is desirable, although not urgently essential, that the slope be uniform within each levee. If practical, it is best to make the border slope from 0.2 to 0.4 percent, but slopes as low as 0.1 percent and as high as 8 percent may be used where it is impractical to obtain the more appropriate slopes. Special care is essential to prevent erosion of soil on the steeper slopes.

The size of stream turned into a single border strip varies from 15 to 300 liters per second, depending on the kind of soil, the size of border, and the nature of the crop.

Because of the relatively high initial cost of preparing land for the border method, it is desirable to plan the location of the levees and strips so that different forage and grain crops may be irrigated with the same borders. Crops which are to be furrow-irrigated, such as sugar beets, potatoes, and corn, may be grown on land on which the forage crops have been irrigated by the border method. Provided the soil conditions are favorable to lateral water movement underneath the low, broad border levees, it is practical to plant and mature crops on the levees. It is difficult to furrow the levees satisfactorily and to keep irrigation water in furrows on the levees.

The border method is suitable to soils of wide variation in texture. It is important, however, to study the physical soil properties in advance of preparing land for border irrigation. Rather impervious subsoils overlain by compact loams permit long border strips, whereas open soils having highly permeable, gravelly subsoils necessitate short narrow strips.

At the head of each border strip a gate is placed in the supply ditch for convenience in turning water into and out of the strip. Borders being strengthened and prepared for planting and irrigation are shown in Fig. 10.2.

10.3 CHECK FLOODING

The check-flooding method consists of running comparatively large streams into relatively level plots surrounded by levees. This method is well suited to very permeable soils which must be quickly covered with water in order to prevent excessive losses near the supply ditches through deep percolation. It is also

Figure 10.2.   Borders being strengthened and prepared for planting and irrigation. (Courtesy Soil Conservation Service, U.S.D.A.)

suited to heavy soils into which water percolates so slowly that they are not sufficiently moistened during the time a sheet of water flows over them, making it necessary to hold the water on the surface to assure adequate penetration.

Checks are sometimes prepared by constructing levees along contours having vertical intervals of 5 to 15 centimeters and connecting them with cross levees at convenient places. These are called contour checks and are formed by building longitudinal levees approximately parallel to the contours, and connecting them at desirable places with levees at right angles.

The check method of irrigation for grain and forage crops is advantageous in localities where large irrigation streams are available and also on projects which depend on direct flow from widely fluctuating streams. In some areas, torrential summer rains suddenly make swollen streams which must be quickly applied to the land to prevent loss of the water. On land of very small slopes the area of each check may be several acres.

The levees should be about 2 meters wide at the base and not over 25 centimeters high because it is essential to avoid obstruction to farm machinery, and also to assure satisfactory growth of crops on the levees.

10.4 BASIN FLOODING

Basin flooding is used extensively to irrigate rice. Water can be held in the basin at the desired depth for the required time.

The basin method of flooding is frequently used to irrigate orchards. On some farms a basin is made for each tree, but under favorable conditions of soil and surface slope from two to five or more mature trees are included in one basin. From the supply ditch the water is conveyed to the basin, either by flowing through one basin and into another, or preferably by small ditches constructed so that the water may be turned directly from a ditch into each basin.

10.5 THE FURROW METHOD

In the irrigation methods thus far described, almost the entire land surface is wetted in each irrigation. Using furrows for irrigation, as shown in Fig. 10.3, necessitates the wetting of only a part of the surface (from one-half to one-fifth) thus reducing evaporation losses, lessening the puddling of heavy soils, and making it possible to cultivate the soil sooner after irrigation. Nearly all row crops are irrigated by the furrow method, rather than by flooding.

Grain and alfalfa crops are irrigated often by means of small furrows designated corrugations. These corrugations are advantageous when the available irrigation streams are small, and also for land of uneven topography. Furrow irrigation is adapt-

Figure 10.3.   Four corrugations supplied from single outlet. (Courtesy U.S.D.A.)

able to a great variation in slope. It is customary to run the furrows down the steepest slope to avoid overflowing the banks of the furrows.

Length of Furrows

On some soils, furrows having slopes of 10 to 15 percent are successfully used by allowing only very small streams to enter the furrows and by careful inspection to control erosion. Slopes of 0.5 to 3 percent are preferable, but many different classes of soil are satisfactorily irrigated with furrow slope from 3 to 6 percent.

The length of furrows varies from 25 meters or less for gardens to as much as 500 meters for field crops. Furrow lengths of 100 to 200 meters are common. Excessive deep percolation losses and soil erosion near the upper end of the field result from use of very long furrows. However, reduction of waste land for ditches and turning of machinery favor longer rows.

Spacing and Depths of Furrows

Spacing of furrows for irrigation of corn, potatoes, sugar beets, and other row crops is determined by the proper spacing of the plant rows, one irrigation furrow being provided for each row. In orchard irrigation, furrows may be spaced from 1 to 2 meters apart. Soils having unusually favorable capillary properties, or impervious subsoils, may permit orchard furrows 3 to 4 meters apart. With the greater spacing it is essential to check on the moisture distribution after each irrigation by making borings with a soil auger or tube to find whether or not the lateral moisture movement from furrows is adequate.

Furrows from 20 to 30 centimeters deep facilitate control and penetration of water into soils of low permeability. They are well suited to orchards and to some furrow crops. Other furrow crops, such as sugar beets, are best irrigated with furrows from 8 to 12 centimeters deep. It is highly desirable in irrigating sugar beets and similar root crops to have the furrows deep enough, and the stream in each small enough, so that the water cannot come in contact with the plant.

Water Distribution to Furrows

Water is distributed to the furrows from earth-supply ditches or from wood or concrete flumes or metal or concrete pipe placed underground. The earth-supply ditch is very common. Small openings are made through the bank, and the water flows into one or more furrows. Figure 10.3 shows four small ditches or corrugations supplied from a single outlet. This method necessitates careful supervision to avoid erosion of the supply-ditch openings, and consequent excess flow in some and inadequate flow in others. On the other hand, it provides flexibility, permitting a large stream in each furrow when the water is first turned in, thus wetting the furrow through its entire length quickly, and then decreasing it so that just enough water enters the furrow to keep it wet, thereby reducing to a minimum the runoff from the lower end of the furrow.

The use of small-diameter 100-centimeter length curved pipe, made of lightweight plastic, aluminum, galvanized iron, or rubber, enables the irrigator to siphon water from the ditch to the furrows as shown in Fig. 10.4 and keep ditch banks solid. This method permits easy and frequent change of water from furrow to furrow.

Irrigators can increase the uniformity of application of water to their furrow-irrigated crops by frequent regulation of the size of stream flowing into the furrow. For this purpose, gated pipe shown in Fig. 10.5 is especially helpful. Small, easily adjusted gates in the pipe facilitate control of the size of stream delivered to the furrow. Streams as small as 0.1 liter per second or as large as 1 liter per second or more can be delivered. The lightweight aluminum or galvanized gated pipe is easily placed, easily connected, and easily moved after irrigation.

Figure 10.4. Plastic siphon tubes used in furrow irrigation.

Figure 10.5.   Portable, gated aluminum pipe delivering water to furrows. (Courtesy Hastings Irrigation Pipe Co.)

10.6 SUBIRRIGATION

In a few localities, natural soil and topographic conditions are favorable to the application of water to soils directly under the surface, a practice know as subirrigation. An impervious subsoil at a depth of 2 meters or more, a very permeable subsoil, a permeable loam or sandy loam surface soil, uniform topographic conditions, and moderate slopes favor subirrigation. Under such conditions, proper water control to prevent alkali accumulation or excess waterlogging usually results in economical use of water, high crop yields, and low labor cost in irrigation.

California has several large tracts of low-lying lands in the Sacramento-San Joaquin Delta that are successfully subirrigated. Before being reclaimed some of these tracts were flooded every year by the overflow waters of the Sacramento and San Joaquin rivers. Reclamation was made possible by building large levees around tracts of several thousands of hectares, followed by installing drainage systems and by pumping the water discharged from the drains over the levees into the river channels. The soils are composed largely of decayed organic matter and are known as peat, tule, or muck soils. During several months of each year, the water in the river channels, now controlled by artificial levees, is 0.5 to 4 meters or more higher than the land surface. In order to obtain water for irrigation, siphons are built from the channels over the levees and the water is thus siphoned to the lands. It is distributed in a series of ditches from 0.6 to 0.9 meters deep and 0.3 meter wide having vertical sides. These ditches, spaced from 45 to 90 meters apart, provide adequate distribution of the water to irrigate small grains and root crops.

Some of these subirrigated lands have been injured and made less productive by saline and alkali conditions developed by the upward capillary water flow from the shallow water table. This reduction in productivity has made it necessary to discontinue subirrigation and irrigate large areas by the sprinkling head.

On the Egin Bench in Idaho, 28,000 acres are irrigated principally by subirrigation.

The land slopes uniformly about 0.2 percent. Surface loams and gravelly loams from 0.5 to 2 meters in depth are underlain by more permeable soil materials which rest on impervious lava rock at depths up to 25 meters.

The flow of water in the main canals is regulated by a ditch-rider, whose job is very different from that of his co-worker on a conventional irrigation system. Essentially he is operating a huge groundwater reservoir, the water level of which is controlled by the rate of inflow and outflow. The inflow is from controlled water supplies and precipitation; the outflow, consumptive use by plants, evaporation from the soil, and seepage. Excess inflow means waterlogged soils, flooded fields, farmyards, and roads, while excess outflow results in wilted crops.

Irrigation does not interrupt cultivation of fields. Farm machinery can be operated in the fields at any time, even though the water table is near the ground surface.

The principal crops grown on the Egin Bench are, in order of percentage of cropped area, potatoes, alfalfa and clover hay, small grains, sugar beets, and field peas. Early in the agricultural development of the area, an attempt was made to irrigate the land by the usual flooding methods. Excessive deep percolation losses resulted, and frequent irrigation was found essential to ordinary crop yields. The gradual rise of the water table convinced the irrigators that smaller quantities of water would suffice under more favorable irrigation methods. Irrigation water is applied in shallow ditches about 1 meter wide and spaced from 30 to 90 meters apart. In general, these ditches do not exceed 400 meters in length. A stream from 7 to 14 liters per second is run into each ditch, from which it sinks to the ground water, causing the water table to rise high enough to moisten the root-zone soil by capillary action that subirrigation methods may be advantageously employed.

10.7 SUBIRRIGATION AND DRAINAGE

In some localities, natural drainage is insufficient to carry away the excess water and salts applied in subirrigation. It has been found necessary in some projects to construct large open drains to prevent excessive waterlogging and salinity of soils. In certain parts of an irrigated valley where deep percolation from irrigation has resulted in the rise of the ground water to the extent that drainage is essential to maintain the soil productivity, it is possible that subirrigation methods may be advantageously employed.

10.8 HYDRAULICS OF SURFACE IRRIGATION

The hydraulics of surface irrigation is rather complex and, consequently, not well understood. Nevertheless, good irrigation design depends upon understanding its principles. Figure 10.6 shows a schematic view of border irrigation, illustrating the basic variables involved in the hydraulics of surface irrigation. They are as follows:

Size of stream.	Slope of land surface.
Rate of advance.	Surface roughness.
Length of run and time required.	Erosion hazard.
Depth of flow.	Shape of flow channel.
Intake rate.	Depth of water to be applied.

It should be clearly understood that the hydraulics of surface irrigation has all the complexities of unsteady open-channel flow plus the added major complication of a variable intake. Hence, direct solution of this complex problem is not simple.

Figure 10.6.   Schematic view of border irrigation illustrating the basic variables involved in the hydraulics of surface irrigation.

10.9 DESIGN OF SURFACE IRRIGATION SYSTEMS

At least ten principal criteria are important in the design of a surface irrigation system. All ten are in turn governed by the over-all economics of the entire farming operation. Thus the design of an irrigation system is complex and not readily subject to quantitative analysis.

1. *Store the required water in the root zone of the soil.* The amount of water to be stored varies with the crop and length of the growing season. Development of an irrigation system flexible enough to meet the varying conditions is important, but sometimes difficult. A compromise design is usually the most economical design.

2. *Obtain reasonably uniform application of water.* It is almost impossible to design a surface irrigation system that will provide satisfactory control under all conditions. Size of stream may vary, and intake rate does vary from crop to crop. It varies also with seasons and the time since irrigation was commenced. The more elaborate and costly the irrigation control devices, usually the more nearly a uniform irrigation can be obtained. For good distribution, the water should reach the end of the field in about one quarter of the total irrigation time.

3. *Minimize soil erosion.* Erosion cannot be eliminated, but it can be minimized. Observation of amount of erosion occurring with different size streams is about the only way to evaluate the permissible size stream. Observation will indicate the size stream which can be used without excessive erosion.

4. *Minimize runoff of irrigation water from the field.* Sizable quantities of water are wasted from the end of most fields during irrigation. This waste water is referred to as runoff. One of the most effective ways of minimizing runoff is to reduce the flow entering the field when the water nearly reaches the end of the field. The amount of wetted area will reduce somewhat, causing a slight decrease in intake, while the velocity will reduce considerably. Whenever possible, runoff should be reused on lower lands.

5. *Provide for beneficial use of runoff water.* The beneficial use of waste water should not be overlooked, since sizable quantities of runoff usually occur. Wherever possible, runoff water should be reapplied to lower lands or recirculated to upper or adjacent lands.

6. *Minimize labor requirement for irrigation.* Labor requirements should be held to a minimum. It is generally necessary to have well-trained and responsible labor for irrigation. Good

preparation of the land, good control, and good irrigation layout will minimize labor requirements.

7. *Minimize land used for ditches and other controls to distribute the water.* Normally 5 percent of the land is used for surface irrigation ditches and controls. Thoughtful layout can minimize the amount of waste land. Sometimes it is advisable to use longer irrigation runs than are best for uniform water distribution to reduce the amount of waste land and to facilitate cultivation.

8. *Fit irrigation system to field boundaries.* The design of the irrigation system is often controlled by the size and layout of the farm. Frequently the length of an individual field is controlled by legal boundaries rather than desirable irrigation design.

9. *Adapt system to soil and topographic changes.* Considerable difficulty will result if a furrow traverses soils of markedly different texture, slope, or depth. Intake rate will be different in each portion and thus the flow rate cannot be adjusted to fit both types. Also the water-holding capacity will require a different irrigation frequency. Irrigation layout should be such that a minimum variation in these factors occurs in a given field.

10. *Facilitate use of machinery for land preparation, cultivating, furrowing, harvesting, etc.* Allowance must be made for machinery. As the degree of mechanization increases, wider borders and longer fields will be desirable.

Table 10.1 summarizes the factors affecting the selection of surface irrigation systems. Maximum slope, water application rate, shape of field, crops to be grown, labor required, and cost are all critical factors. The tabulated values and guidelines of Table 10.1 should be used as an indication of normal practice. Local conditions may necessitate design of a particular irrigation system outside of the recommended normal values.

TABLE 10-1 Factors Affecting the Selection of Surface Irrigation Systems

Type of System	Maximum Slope (%) Humid Areas — Nonsod Crops	Humid Areas — Sod Crops	Arid Areas — Nonsod Crops	Arid Areas — Sod Crops	Water Application Rate of Intake Family[a] (mm/hr) Min.	Max.	Shape of Field	Adaptable to — Row Crops (Row or Bedded)	Sown, Drilled, or Sodded Crops	Weed Control in Rice	Orchards and Vineyards	Labor Required (hr/ha per irrigation)	Approximate Relative Cost[b]
Level:													
Level border	Nearly Level				2.5	50	Any shape	Yes	Yes	Yes	Yes	0.25–1.2	100–250
Contour levee	0.1				2.5	15	Any shape	Yes	Yes	Yes	Yes	0.12–1.2	50–150
Level furrow	Nearly level				2.5	50	Rows should be of equal length	Yes	Yes	No	Yes	0.50–1.7	100–250
Graded:													
Graded border	0.5	2.0	2.0	4.0	7.5	50	Rectangular	No	Yes	No	Yes	0.5–2.5	100–300
Contour ditch	NA	4.0	4.0	15.0	2.5	75	Any shape		Yes	No	Yes	2.5–5.0	25–50
Graded furrow	0.5	NA	3.0	NA	2.5	75	Rows should be of equal length	Yes	Yes	No	Yes	1.0–3.0	100–300
Corrugation	NA	NA	4.0	8.0	2.5	38	Rectangular	No	No	No	Yes	1.0–3.0	50–100
Contour furrow	Cross slope 3.0		Cross slope 6.0		2.5	50	Rows should be of equal length	Yes	No	No	Yes	1.2–3.7	75–150

[a] Intake family is a grouping of soil by the Soil Conservation Service. It is based on the ability of the soil to take in the required amount of water during the time it takes to irrigate.

[b] Does not include the cost of water supply, pump, power unit, and main line.

Adapted from "Planning for an Irrigation System," published by American Association for Vocational Instructional Materials in cooperation with the Soil Conservation Service, U.S.D.A. June 1971.

CHAPTER
11

IRRIGATION IMPLEMENTS AND STRUCTURES

Extensive use of well-designed modern implements for leveling irrigated lands, making borders, construction and cleaning of ditches, and for making corrugations and furrows is contributing to irrigation advancement. It is essential to efficiency and economy that the lands of each irrigated farm be well prepared for irrigation and also be provided with structures that facilitate easy control and regulation of the irrigation water during its application to the land. It is essential also that the irrigator be able to control the stream at his disposal and to spread the water uniformly over the land surface in order to moisten the soil to the desired depth without sustaining excessive losses of water. Some of the modern implements used in irrigation farming and farm structures that facilitate the control of water are described in this chapter.

11.1 IMPLEMENTS

The farm implements of first importance in irrigation are plows, spike-tooth harrows, disk harrows, scrapers, and levelers. Good plows and good plowing contribute greatly to the possibility of uniformity in distributing irrigation water. Lands that are irrigated by the ordinary flooding methods especially require good plowing because, owing to the lack of specially prepared levees, there is no means of crowding water over the higher portions of poorly plowed fields. Careless plowing of lands that are to be irrigated by flooding, or plowing with dull, poorly kept plows, is followed by inefficient irrigation.

11.2 IMPLEMENTS FOR LEVELING LANDS

Scrapers and soil movers like the unit shown in Fig. 11.1 have contributed greatly to advancement in land leveling. After the large cuts and fills are accomplished with the scraper the lands are leveled and smoothed with large automatic levelers like the land plane shown in Fig. 11.2.

Figure 11.1. Farm tractor and carry-all scraper used for land leveling (Courtesy Eversman Manufacturing Co.)

Figure 11.2. Land plane for automatic leveling. (Courtesy The Farmhand Co.)

Flooding and furrow methods of irrigation requires use of good land leveling and tillage implements in order to obtain a smooth land surface in which there are few, if any, small depressions or elevations.

11.3 BORDER-MAKING IMPLEMENTS

In addition to the use of well-built scrapers and other land-leveling implements on the larger farms, special border-making scrapers drawn by tractors have proved to be economical. For smaller farms adjustable bordering machines like the one illustrated in Fig. 11.3 are very helpful. The frame attached to the rear end of the implement smooths and grades the top of the levee in the same operation which pushes the soil together to make a levee. Commercial equipment to construct borders is available also.

Figure 11.3.   Preparing land for the border method of irrigation. (Courtesy Union Pacific Railroad Co.)

11.4 IMPLEMENTS FOR MAKING AND CLEANING DITCHES

The era of pioneer methods of digging irrigation ditches with hand labor and the pick and shovel is largely past in many countries. Power-drawn ditching machines, illustrated in Fig. 11.4 have greatly reduced the costs and the time required for construction of canals and ditches. Where ditches with steep banks are feasible and desired, carry-type scrapers drawn by tractors are efficient and economical. In the construction of new canals as well as in the repairing of old ones it is often advantageous to compact the soils well in order to reduce seepage losses and also to add to the stability of the soil and thus reduce erosion of the bed and the banks of the canal. The sheepsfoot roller, for many years always used in the construction of earth dams, is useful also on canals.

Ditchers of the type shown in Fig. 11.5 are especially useful for cleaning the banks of canals.

11.5 CORRUGATION IMPLEMENTS

Shallow furrows are designated as corrugations. Several types of homemade corrugators are used: one, a roller around which collars of the desired thickness and depth are built; another, a drag having runners as corrugators. The roller type compresses

Figure 11.4.   Ditching machine, hydraulically operated for excavating ditches. (Courtesy Eversman Manufacturing Co.)

Figure 11.5.   This ditcher with 5.5-m blade for clearing canal banks weighs 3 tons and will operate in 1.5 m-depth ditches having steep banks. (Courtesy Robinson Machinery Co.)

and compacts the soil as a means of making furrows; the drag type crowds the soil to both sides of the runner. These corrugators are limited in use to newly plowed land. For old alfalfa land, clover land, or other land having a compact surface, heavy, well-constructed steel shovels are needed to make satisfactory furrows.

11.6 DEEP-FURROW IMPLEMENTS

Potatoes, sugar cane, corn, asparagus, celery, and orchards on some soils are best irrigated by using deep furrows, especially in heavy soils. Many orchard crops, such as apples, peaches, lemons, olives, and almonds are also best irrigated by means of deep furrows. Common shovel plows are sometimes used for making the furrows. For orchard irrigation on land having a steep sidehill slope a standard moldboard plow may be used by throwing the soil downhill so as to avoid overflowing of the furrows. A two-way sulky plow saves time in making deep furrows on sidehill land.

11.7 FARM IRRIGATION STRUCTURES

The word structure is applied to dams, head gates, sluices, flumes, inverted siphons, chutes, and drops which are built to divert water from natural sources and convey it to the farms for irrigation. The devices and pieces of equipment used by the individual irrigator to divert water from a large canal into his ditch and convey it to the several parts of his farm are here designated farm irrigation structures. In some communities rather crude farm irrigation structures are made to suffice even though the labor cost required in the use of such structures is sometimes very high. As a rule it is economical, and it is always most satisfactory to the irrigator, to build structures that have the required capacity and the strength to control the water. Many irrigation canals are built along the rims of the valleys immediately above the irrigated lands so that each irrigator obtains water directly from the main canal which conveys water during the entire irrigation season. On such canals satisfactory take-out structures are especially necessary.

Farm irrigation structures include two general classes, namely, permanent and temporary structures. No structure is truly permanent, in the strict sense of the word, but the term is applied to those structures which remain in place during one or more irrigation seasons. Temporary structures are those that are moved from place to place during an irrigation, or those that are built for only one season's use. The irrigation devices here classed as "temporary farm irrigation structures" are described by some authorities as "irrigation equipment." A further desirable classification of structures is based on the function of the structure and includes: diversion, conveyance, and distribution structures.

11.8 PERMANENT DIVERSION STRUCTURES

A check gate is a gate placed across a stream from which it is desired to divert water. The function of the check gate is analogous to that of the dam or the diversion weir on the rivers at the heads of canal systems. Check gates are built across laterals and ditches for the purpose of diverting part or all of the stream. The take-out gate is a part of the farmer's "diversion works" and is analogous to the head gate of the main diversion works on the river system. Its function is to regulate the quantity of water flowing into the small lateral, the field ditch, or the furrow. Typical wooden check gates are illustrated in Figs. 11.6 and 11.7. These gates may also be used as take-out gates, although pipe or culvert take-outs are commonly used, especially to take small streams out of large canals. For check gates in large canals in which the quantity of water fluctuates appreciably, it is desirable to place flashboards on the bottom of the canal to raise the water level so that more water flows out of the canal. A well-made concrete check gate with wooden flashboard is illustrated in Fig.

Dimensions (cm)

Flows (l/s)	A	B	C	D	E	LUMBER THICKNESS
25-50	240	60	92	30	74	2.5
50-150	275	110	92	60	92	4
150-225	305	120	122	70	92	4

Figure 11.6. Standard single-wing wooden check gate.

Dimensions (cm)

Flows (l/s)	A	B	C	D	E	F
75-170	275	110	91	60	60	92
170-280	366	122	122	70	122	75
280-425	435	140	155	80	140	75

Figure 11.7. Standard double-wing wooden check gate.

11.8. A concrete structure that facilitates the diversion of the inflowing stream to either of three outflowing streams is shown in Fig. 11.9.

Permanent diversion structures can be made of wood as illustrated by Figs. 11.6 and 11.7, or by concrete as illustrated by Figs. 11.8 and 11.9. However, increasing use is being made of bricks and concrete blocks to form irrigation structures. Assembly costs in some countries are less than when concrete is used.

Prefabricated structures are beginning to be used widely. Mass production of simple shapes which can be combined into complex structures is reducing the cost of the material and the cost of installation. A prefabricated low-cost steel structure is shown in 11.10.

11.9 HYDRAULIC PRINCIPLES OF DIVERSION STRUCTURES

In taking water out of a canal, it is desirable that the farmer obtain a flow as nearly constant as possible. Sudden increases in the quantity of water flowing in the canal, which occur as a re-

Figure 11.8. Concrete turnout gate and wooden flashboard.

Figure 11.9. Concrete three-way outlet box on irrigation lateral. Water can be diverted to any one or more of the outlets. (Courtesy Soil Conservation Service, U.S.D.A.)

Figure 11.10. Prefabricated low-cost steel water-control structure. (Courtesy United States Steel Co.)

sult of storms or from closing off take-out gates, should be permitted to flow down the canal with as little obstruction as possible. These two conditions, i.e., approximately constant flow for the irrigator and a minimum of obstruction in the main canal, in general, may be provided by using submerged pipes or culverts as take-outs and overflow flashboards as checks in the main canal to cause the water to rise high enough to submerge the take-out gate and divert the desired quantity of water.

To double the quantity of water flowing through a submerged culvert take-out the effective head, h, must be increased four times the original head. To double the quantity of water flowing over flashboards as checks, the depth need be increased only 1.6 times. Therefore, streams through submerged take-outs are subject to much less variation than those through overpour take-outs.

11.10 TEMPORARY DIVERSION STRUCTURES

In order to divert water from the small ditches on the farm many irrigators use only temporary earth dams. They make each dam at the time and place desired by means of an ordinary shovel. In some soils that erode easily it is helpful to use a little partly rotted straw or weeds temporarily held in place by means

Figure 11.11.   A portable irrigation control gate. (Courtesy Page Metal Products Corp.)

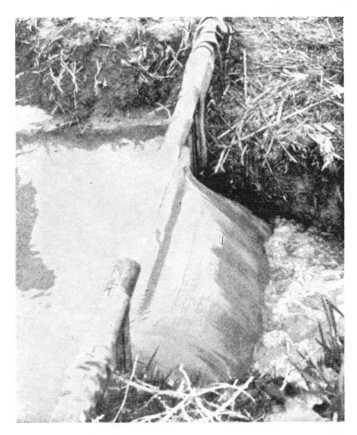

Figure 11.12.   Portable, rectangular weir made from canvas or plastic, utilizing sheet metal for the notch. (Photo by James R. Barker.)

of wooden stakes driven into the soil at the bottom of the ditch. Temporary earth dams are unsuited to streams of more than 50 liters per second and in some soils are very difficult to maintain with a stream of 25 liters per second or more.

The labor requirement of temporary dams is greatly reduced by using portable dams of either metal, canvas, or plastic. Portable irrigation dams are illustrated in Figs. 11.11 and 11.12. Both dams can be used with fair accuracy to measure the flow since the opening through which the water passes has fixed dimensions. Portable metal dams are suited to streams smaller than those diverted by canvas dams. For streams of 125 liters per second or more the metal dam required may be so large as to become burdensome to move or carry about the field. Well-built canvas or plastic dams are used to divert streams as large as 200 liters per second or more, although streams of 75 liters per sec-

Figure 11.13.   Using an adjustable canvas-dam stick to control the amount of water passing the dam. (Photo by James R. Barker.)

ond are more commonly diverted by using these portable dams. A heavy, durable, closely woven canvas is necessary to stand the water pressure and prevent excessive leaking.

An adjustable canvas or plastic dam is shown in Fig. 11.13. The dam stick hinges at the center and is held in place by a wing nut. Adjusting the wing nut will allow varying quantities of water to pass over the dam.

11.11 WATER CONVEYANCE STRUCTURES

The term structure as used herein applies quite as fully to ditches, levees, etc., which are made of earth as it does to devices built of wood, concrete, or metal. Conveyance of water under pressure through underground concrete pipe to various points on the farm is becoming increasingly popular. On excessive slopes caution must be exercised against erosion, and on small slopes and in wet and humid climates it is important to guard against growth of weeds and grasses in order to maintain a satisfactory discharge capacity.

Structures used for water conveyance are varied to meet conditions of terrain and flow requirements. To convey water to the point of use for irrigation in the most efficient manner is a challenge to the irrigation engineer. A few structures commonly used to complement canals and pipe lines are discussed.

Flumes

For crossing natural depressions or narrow canyons, and for conveying irrigation water along very steep sidehills, flumes are constructed either of wood or metal, or both, as shown in Figs. 11.14 and 11.15. Concrete is also used for flumes. To attain economy in the application of materials for flumes, it is desirable to give the flume sufficient slope to assure water velocity appreciably higher than in earth canals, thus making possible a proportional reduction in flume cross section.

Tunnels

To shorten the length of a diversion canal, to avoid difficult and expensive construction on steep, rocky hillsides, and to convey irrigation water through mountains from one watershed to another, many tunnels have been constructed. It is usually economical to line the bottoms and sides of tunnels through rock formations as a means of decreasing seepage losses and lessen-

Figure 11.14.   Small flume crossing over arroyo. This structure saved 500 m of ditch length. (Courtesy Soil Conservation Service, U.S.D.A.)

Figure 11.15.   Large metal flume over wash. Steel substructure set on concrete piers. (Courtesy Soil Conservation Service, U.S.D.A.)

ing frictional resistance. An irrigation tunnel constructed through loose material is lined with concrete as the boring of the tunnel progresses.

Drops and Chutes

Wood, concrete, or masonry bulkheads are used in places where the natural slopes down which canals must flow are so high as to cause excessive water velocities and erosion of the canal banks and bed. Water is dropped several feet, and the energy of a flowing stream is dissipated in the stilling basin without causing erosion. Fig. 11.16. A number of closely spaced drops may be used, as shown in Fig. 11.17. Chutes built of wood, concrete, or steel, Fig. 11.18, are serviceable where it is necessary to convey water down relatively steep hills which would require many drops closely spaced to control water velocity, and in which water would cause serious erosion if not controlled. Chutes may well be considered in three sections: (1) the section of transition and accelerating velocity; (2) the section of uniform high velocity; and (3) the stilling basin. In the first section the velocity is increased from approximately 1 meter per second up to 6 meters or more per second, and the cross-sectional area of the water is proportionately decreased. In the second section, because of very high velocity, the retarding forces are essentially equal in magnitude and opposite in direction to the driving forces; hence the water velocity remains constant. To dissipate

Figure 11.16 A masonry canal drop. Long apron reduces erosion below the structure. Suitable for small canals and low drops. (Courtesy Soil Conservation Service, U.S.D.A.)

Figure 11.17. Series of check drops to prevent high canal-water velocities and erosion. Flashboards can be placed in notch to raise water level so that farmer can irrigate fields on both sides of the canal. Structure in foreground appears to have apron below the check set too high. (Courtesy Soil Conservation Service, U.S.D.A.)

Figure 11.18.   Conveyance flumes of concrete in the foreground, and aluminum to the left, to be removed for harvest of sugarcane in Hawaii. (Courtesy Hawaiian Commercial and Sugar Co.)

the kinetic energy at the lower end of a chute, it is necessary to provide a deep stilling basin.

Inverted Siphons

For crossing wide deep hollows, depressions, or canyons it is customary to build pipe lines and convey water through them under pressure. The cost of flumes for crossing wide depressions is often so high as to prohibit their construction. Pipes by which irrigation water is conveyed across canyons usually built on or near the ground surface are known as inverted siphons.

The velocity of water flowing through a siphon of given diameter is fixed by the hydraulic slope and by the roughness of the inside of the pipe. The velocity is not influenced by the total water pressure inside the pipe.

11.12 REMOVABLE CONCRETE STRUCTURES

Several unique methods of water control have been developed on the large sugarcane plantations of Hawaii. Figure 11.19 shows thin-wall concrete sections with metal slide gates in the fore-

Figure 11.19.   Conveyance flumes of concrete in the foreground, and aluminum to the left, to be removed for harvest of sugarcane in Hawaii. (Courtesy Hawaiian Commercial and Sugar Co.)

ground and to the left an aluminum flume diverting water from the concrete flume. All concrete and aluminum sections shown will be removed from this field for harvesting and will be replaced after harvest. Figure 11.20 shows an aluminum flume conveying water down a steep hillside without loss of water or erosion. Note the plastic spill for discharging the water into the furrow. Several unique gates particularly adaptable for steep slopes have been developed by the Hawaii plantations. Thin-walled rubber pipe is being used in Fig. 11.21, and portable gated pipe is supplying water to furrows in Fig. 11.22.

11.13 UNDERGROUND PIPE

Irrigators are relying on buried concrete, plastic, and steel pipe more and more for distributing irrigation water. From the underground pipe water flows upward through riser pipes, and through low-cost irrigation valves to an irrigation ditch, a check, or a basin. To prevent erosion of the soil around the riser pipe, top boxes of concrete or wood are sometimes used. Water flows from the pipe riser into the distribution ditch. Plastic or metal

Figure 11.20. Portable aluminum flumes used to irrigate sugarcane growing in steep land. (Courtesy Hawaiian Commercial and Sugar Co.)

Figure 11.21. Flexible rubber pipe with a thin-walled, removable concrete flume behind. (Courtesy Hawaiian Commercial and Sugar Co.)

Figure 11.22. Fourteen furrows are being supplied controlled streams by this gated pipe. (Courtesy New Mexico State College.)

pipe sections about 1 meter long and ranging from 2 to 5 centimeters or more in diameter are often placed on the ditch bank to convey water from the ditch to the furrow.

Lightweight surface-gated pipe, shown in Fig. 11.22, is sometimes used with underground concrete pipe and connected to the concrete-pipe riser, thus at once solving the soil-erosion problem around the riser outlet and providing for complete control of the quantity of water flowing into each furrow. The convenience of the water control and the ease of regulating the quantity that is delivered to each furrow from the underground pipe are also illustrated in Fig. 11.23, showing the owner operating a valve to assure the best flow to the furrow.

11.14 SIPHON TUBES

Siphon tubes convey the water over the ditch bank into the furrow. They are especially useful with new ditch banks and are available in plastic, aluminum, galvanized iron, and rubber. Figure 11.24 shows water being conveyed in a concrete ditch and diverted by siphons to the furrows. Note that no gates are necessary and leakage from the ditch is thereby nullified.

The flow from the siphon is directly proportional to the size of the siphon and also proportional to the square root of the differ-

Figure 11.23.   A Washington farm owner operating a valve to release water from underground concrete pipe for irrigation of his orchard by the furrow method. (Courtesy Bureau of Reclamation.)

Figure 11.24.   Siphons diverting water from a concrete-lined ditch. (Courtesy Portland Cement Association.)

ence in elevation between the water in the ditch and the outlet, or the elevation of the downstream water surface if the siphon is submerged. Flow into borders or furrows is controlled by varying the number of siphons or the elevation of the discharge end of the siphon.

CHAPTER
12
CONVEYANCE
OF IRRIGATION AND
DRAINAGE WATER

Irrigated lands are often located great distances from the sources of their water supplies. Water obtained from natural streams and from surface reservoirs, as a rule, must be conveyed farther than water obtained from underground reservoirs. Main conveyance canals of irrigation projects vary from a few kilometers to 150 or more kilometers in length. Some projects convey water several hundred kilometers from storage reservoirs in the mountains by commingling stored water with water of natural rivers and then again diverting it into large canal systems in the valleys. Days are required on some projects to convey the water from points of diversion to points of use.

The same principles governing flow in irrigation canals also govern flow in drainage ditches. Therefore, the material of this chapter applies to all conveyance channels associated with irrigation projects.

12.1 HYDRAULICS OF FLOW

Irrigation water is conveyed in either open or closed conduits. Hydraulically, the two methods are similar; however, slightly different forms of the equations are used because with pipe flow, pressure head and elevation head differences are usually measured to determine the flow rate while in an open-channel flow the pressure head does not change and the slope of the water surface is the criterion of flow. Derivation of hydraulic equations and details of application are thoroughly covered in hydraulics texts and are not duplicated here. Only those results applicable to irrigation and drainage are summarized. Older empirical equations of flow are bypassed in favor of expressions which have a sounder physical basis and which are being used extensively in irrigation practices.

The most basic equation of water flow is the continuity equation which relates the flow Q past a cross section to the velocity and area A at two different sections:

$$Q = A_1V_1 = A_2V_2 \qquad (12.1)$$

When area is measured in square meters and velocity in meters per second, discharge will be in cubic meters per second.

The second fundamental equation in the hydraulics of irrigation and drainage water conveyance is the Bernoulli equation:

$$\frac{V_1^2}{2g} + \frac{p_1}{w} + y_1 = \frac{V_2^2}{2g} + \frac{p_2}{w} + y_2 + h_L \qquad (12.2)$$

where p = pressure intensity at any point
y = elevation of the point above a common datum
w = weight of unit volume of water
h_L = head loss, energy loss per unit weight of fluid between points 1 and 2
g = acceleration due to gravity

The third basic equation is for head loss h_L as water flows between points 1 and 2 in the system. This is given by the Darcy-Weisbach equation:

$$h_L = f\ \frac{L}{d}\ \frac{V^2}{2g} \qquad (12.3)$$

where f = coefficient of friction loss
L = the length between the two points
d = diameter of the conduit
V = average velocity

The fourth relationship that is also very useful relates the head h to the pressure intensity p and the unit weight of water w:

$$h = \frac{p}{w} \qquad (12.4)$$

Flow through a pipe system in which these four equations apply is shown in Fig. 12.1. Hydraulic terminology commonly used is also shown. At all points within the pipe line, Equations 12.1 and 12.2 can be applied. Equation 12.3 will give the loss of head between any two points of the line. Loss of energy and flow through valves, bends, and transitions of a pipe system can be calculated by these procedures and by utilizing the tables outlined in Chapter 14.

The fifth relationship which has extensive value in open channel flow is the hydraulic radius R. This is defined as:

$$R = \frac{A}{P} \qquad (12.5)$$

where A = cross-sectional area of the flow
P = wetted perimeter of the flow in contact with the channel

Figure 12.1.   Illustrating the flow of water from a large reservoir through a pipe line showing how the energy changes from point to point.

When the hydraulic radius is calculated for a full round pipe, it is found to have the value of $R = d/4$; this value changes somewhat with shape, but for practical purposes in most irrigation problems the value of hydraulics radius can be assumed as $d/4$. Substituting this quantity in Equation 12.3, replacing g with its numerical value and solving for V, the following equation is obtained:

$$V = \frac{8.8}{\sqrt{f}} \sqrt{R \frac{h_L}{L}} = \frac{8.8}{\sqrt{f}} \sqrt{RS} \qquad (12.6)$$

where S is the slope of the water surface or piezometric head line, V is in meters per second and R is in meters.

An equation of similar form was obtained by Chezy, wherein C is the coefficient of roughness in English units.

$$V = \frac{C}{1.8} \sqrt{RS} \qquad (12.7)$$

Another equation was derived by Manning utilizing the Chezy equation but defining differently the coefficient of roughness n.

$$V = \frac{1}{n} R^{2/3} S^{1/2} \qquad (12.8)$$

where n is Horton's values of roughness as tabulated in Table 12.1. The similarity of Equations 12.6, 12.7, and 12.8 readily indicates that the coefficients, f, C, and n are related in the following manner:

$$\frac{8.8}{\sqrt{f}} = \frac{C}{1.8} = \frac{1}{n} R^{1/6} \qquad (12.9)$$

Unfortunately the values of the coefficients of roughness C, f, and n change with flow conditions within the conduit; hence, it is not possible to assign a fixed value for a given pipe and have that value hold for all flow conditions. The variation which occurs and the values which can be used in design are shown in Fig. 12.2 for the normal range of turbulent flow in irrigation channels. The following discussion will deal with the variations in coefficient C, but the comments also apply to f and n which are related to C in the manner shown in Equation 12.9.

The abscissa in Fig. 12.2 is the Reynolds number N_R, a dimensionless quantity where V is average velocity, d is diameter, R is hydraulic radius, and ν is kinematic viscosity. When Reynolds number is computed, care should be taken to ensure that proper dimensions of the variables are used so that all dimensions will cancel. The ratio of e/d is relative roughness of channel and e is absolute roughness. Thus, only in the wholly-rough region is the value of the roughness coefficient a constant with changes in velocity. Furthermore, absolute roughness e of one kind of pipe,

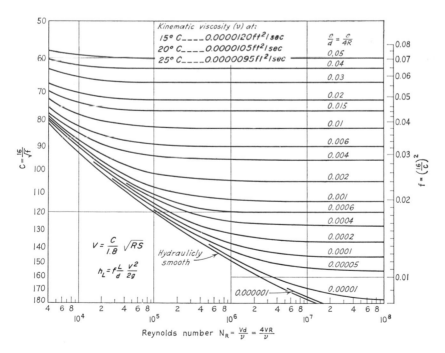

Figure 12.2.   Roughness values for turbulent flow in conveyance structures.

TABLE 12.1   Horton's Values of n. To Be Used with Kutter's and Manning's Formulas

Surface	Best	Good	Fair	Bad
Uncoated cast-iron pipe	0.012	0.013	0.014	0.015
Coated cast-iron pipe	.011	.012[a]	.013[a]	
Commercial wrought-iron pipe, black	.012	.013	.014[a]	.015
Commercial wrought-iron pipe, galvanized	.013	.014	.015	.017
Smooth brass and glass pipe	.009	.010	.011	.013
Smooth lockbar and welded "OD" pipe	.010	.011[a]	.013[a]	
Riveted and spiral steel pipe	.013	.015[a]	.017[a]	
Vitrified sewer pipe	$\begin{cases}.010\\.011\end{cases}$	.013[a]	.015	.017
Common clay drainage tile	.011	.012[a]	.014[a]	.017
Glazed brickwork	.011	.012	.013[a]	.015
Brick in cement mortar; brick sewers	.012	.013	.015[a]	.017
Neat cement surfaces	.010	.011	.012[a]	.013
Cement mortar surfaces	.011	.012	.013[a]	.015
Concrete pipe	.012	.013	.015[a]	.016
Wood stave pipe	.010	.011	.012	.013
Plank Flumes:				
Planed	.010	.012[a]	.013	.014
Unplaned	.011	.013[a]	.014	.015
With battens	.012	.015[a]	.016	
Concrete-lined channels	.012	.014[a]	.016[a]	.018
Cement-rubble surface	.017	.020	.025	.030
Dry-rubble surface	.025	.030	.033	.035
Dressed-ashlar surface	.013	.014	.015	.017
Semicircular metal flumes, smooth	.011	.012	.013	.015
Semicircular metal flumes, corrugated	.0225	.025	.0275	.030

Table continued on next page.

for example cast iron, is reasonably constant; however, relative roughness e/d changes as size of conduit changes. Hence, coefficient of friction varies as flow conditions defined by the Reynolds number vary, and relative roughness varies more than absolute roughness. Values of absolute roughness for different kinds of pipe are tabulated in Table 12.2.

In solving practical flow problems it becomes obvious that the velocity required in the Reynolds number is not known; therefore, it is necessary to estimate the velocity, obtain a solution, and check the assumed value. If the assumed velocity is consid-

TABLE 12.1   Horton's Values of n. To Be Used with Kutter's and Manning's Formulas (continued)

Surface	Best	Good	Fair	Bad
Canals and ditches:				
Earth, straight and uniform	.017	.020	.0225[a]	.025
Rock cuts, smooth and uniform	.025	.030	.033[a]	.035
Rock cuts, jagged and irregular	.035	.040	.045	
Winding sluggish canals	.0225	.025[a]	.0275	.030
Dredged earth channels	.025	.0275[a]	.030	.033
Canals with rough stony beds, weeks on earth banks	.025	.030	.035[a]	.040
Earth bottom, rubble sides	.028	.030[a]	.033[a]	.035
Natural stream channels:				
(1) Clean, straight bank, full stage, no rifts or deep pools	.025	.0275	.030	.033
(2) Same as (1), but some weeds and stones	.030	.033	.035	.040
(3) Winding, some pools and shoals, clean	.033	.035	.040	.045
(4) Same as (3), lower stages, more ineffective slope and sections	.040	.045	.050	.055
(5) Same as (3), some weeds and stones	.035	.040	.045	.050
(6) Same as (4), stony sections	.045	.050	.055	.060
(7) Sluggish river reaches, rather weedy or with very deep pools	.050	.060	.070	.080
(8) Very weedy reaches	.075	.100	.125	.150

[a] Values commonly used in designing.
From King, H. W. *Handbook of Hydraulics*, McGraw-Hill, New York, 1938.

erably in error, the computed velocity should be used and a new velocity calculated. The process should be repeated until satisfactory accuracy is obtained.

The roughness of open channels has not been defined as well as the roughness of pipes, nor can it be, because of the large variation which occurs in nature. This condition presents a challenge to irrigation engineers to develop criteria that will better define the absolute roughness of channels. It is important to remember that under the best conditions, values vary with flow conditions; hence, allowance must be made when designing for extreme conditions.

The Manning coefficient of roughness n has been the object of considerable field study, Table 12.1. The values shown in Table

TABLE 12.2   Value of Roughness e for Various Conduit Materials
(Computed from Albertson, Barton, Simons)

	Relative Roughness (e in cm)
Glass, drawn brass, copper, and lead	Smooth–0.00015
Commercial steel or wrought iron	0.005
Asphalted cast iron	0.012
Galvanized iron	0.015
Cast iron	0.026
Wood stave	0.02–0.09
Cement-lined steel	0.04
Concrete	0.03–0.3
Riveted steel	0.09–0.9
Corrugated metal pipe	3.0–6.0
Large tunnels, concrete or steel lined	0.06–0.12
Blasted-rock tunnels	30–60

12.1 vary only with the conditions of the channel surface, whereas values of C, Fig. 12.2, vary both with relative roughness and with flow conditions. For this reason some error can be expected when n values, Table 12.1 are used. However, the value R to the $1/6$th power in Equation 12.9 does in part make n an absolute roughness, since the ratio $R^{1/6}$ (power) to n is in effect a relative roughness. Nevertheless, the value of n will still vary somewhat because relative roughness is not a simply $1/6$th power function and flow conditions as characterized by the Reynolds number have not been considered. This observation has been borne out in practical construction, particularly where normal values of roughness have been used to calculate the flow occurring in large structures. For normal irrigation work the values of Table 12.1 can be used with reasonable confidence in Equation 12.8.

12.2 EARTH CANALS

The most common type of irrigation conveyance channel is the one excavated in natural material along the line that water must be conveyed. When used without artificial lining of bed or sides, such a channel is called an earth canal. Excessive velocities of water in earth canals cause erosion. Very few natural materials will stand velocities in excess of 1.5 meters per second. The low initial cost constitutes the major advantage of earth canals. The disadvantages are:

(a)   excessive seepage losses.

(b)   low velocities and therefore, relatively large cross-sectional areas.

(c)   danger of breaks due to erosion and burrowing of animals.

(d)   favorable conditions for growth of moss and aquatic weeds which retard the velocity and cause high annual maintenance costs, Fig. 12.3.

The sides of earth canals are usually built as steep as the earth will stand when wet. The slopes of the sides vary from three horizontal and one vertical to one horizontal and one vertical, for very stable materials. The relation of bed width b, to depth of earth canals d, is determined according to the topographic conditions. The bed width may be less than the depth, or it may be ten or more times the depth. The best hydraulic cross section under favorable structural conditions is:

$$b = 2d \tan \frac{\theta}{2} \qquad (12.10)$$

where θ is the angle between the side slope and the horizontal. This relation applies also to lined canals. For rectangular channels $\tan \theta/2 = 1$, and hence, the bed width equals twice the depth for best hydraulic characteristics.

Figure 12.3.   Diesel tractors pulling a large chain weighing 1000 kg to loosen moss and weeds in bottom of a canal. (Courtesy Caterpillar Tractor Co.)

12.3 STABLE CHANNELS

Irrigation channels should be stable—having negligible scour, and negligible deposition of sediments. Scour can be eliminated by low velocities, but Equation 12.1 ($Q = AV$) shows that as velocity decreases the area must increase if a given quantity of water is to be conveyed. Large canals are costly to construct, costly to maintain, and have greater loss of water from seepage and evaporation than small canals.

Another factor to be considered is that as the velocity doubles, the head loss increases by four, Equation 12.3. Since most irrigation canals are constructed at a minimum grade in order to reach as much arid land as possible, it is usually desirable to maintain small velocities. The requirement that the velocity cannot be less than a certain minimum to keep silt in suspension works against low energy loss and promotes scour. Therefore, a compromise is always necessary with the compromise being tailored to fit the conditions that exist at the site.

Recommendations on limiting velocity, together with corresponding values of roughness, n, are shown in Table 12.3. The values shown apply to water depths of 1 meter or less in canals that have been brought to capacity gradually. For depths of water over 1 meter, the velocity should be increased 0.15 meter

TABLE 12.3 Limiting Velocities for Essentially Straight Canals After Aging

Material	Value of Manning's n	Clear Water	Water Transporting Colloidal Silts
		Velocity (m/s)	
Fine sand, colloidal	0.020	0.4	0.8
Sandy loam, noncolloidal	0.020	0.5	0.8
Silt loam, noncolloidal	0.020	0.6	0.9
Alluvial sites, noncolloidal	0.020	0.6	1.1
Ordinary firm loam	0.020	0.8	1.1
Volcanic ash	0.020	0.8	1.1
Stiff clay, very colloidal	0.025	1.1	1.5
Alluvial site, colloidal	0.025	1.1	1.5
Shales and hardpans	0.025	1.8	1.8
Fine gravel	0.020	0.8	1.5
Graded loams to cobbles when noncolloidal	0.030	1.1	1.5
Graded silts to cobbles when colloidal	0.030	1.2	1.7
Coarse gravel, noncolloidal	0.025	1.2	1.8
Cobbles and shingles	0.035	1.5	1.7

per second. For canals with sinuous alignment, a reduction of 25 percent is recommended.

12.4 LINING CANALS

Irrigation canals are lined for purposes of (1) decreasing conveyance-seepage losses, (2) providing safety against breaks, (3) preventing weed growth, (4) retarding moss growth, (5) decreasing erosion from high velocities, (6) cutting down maintenance costs, (7) reducing drainage problems, and (8) increasing the capacity to convey water.

Detailed cost studies are essential to the determination of economic advisability of lining canals. The annual cost of lining is the true cost, not the initial cost. Annual cost includes annual depreciation, maintenance, and interest. All of these factors should be considered when determining economic feasibility.

From the viewpoint of an irrigation project, the most important single factor in a study of advisability of lining is the annual value of water saved by decreasing conveyance losses. In localities where water is very limited, public interests are advanced by lining canals, thus contributing to a more economical use of the available water supply.

Excess canal seepage contributes to waterlogging of farm lands, salt and alkali concentration in the soils, costly road maintenance and drainage activities, ground-water seepage into basements of buildings, and other conditions that concern the public. Although it is difficult, costly, and sometimes quite impractical to measure accurately the degree of contribution to these adverse conditions by any one canal, the public should encourage reduction of canal seepage to protect public interest.

12.5 MATERIALS FOR CANAL LINING

The most used materials for canal lining include: concrete, rock masonry, brick, bentonite-earth mixtures, natural clays of low permeability, and different rubber, plastic, and asphalt compounds. A few canals have been lined under low-cost methods with clay, Fig. 12.4 and with asphalt membrane. For the main canal of the Yakima Project in the state of Washington, Fig. 12.5, concrete prevents both seepage and erosion.

For smaller irrigation canals and laterals, precast concrete slabs made at regular concrete mixing plants, hauled to, and placed in canals, are helpful, Fig. 12.6. Both frequent inspection of the concrete slab joints and careful maintenance are essential

Figure 12.4. A canal with a clay blanket lining. The white sand in which the canal is built has a permeability 1800 times greater than the clay used for lining.

to prevent erosion of soil materials under the slabs with resulting settlement and damage to the lining. Masonry rock and brick work make good canal linings, Fig. 12.7.

Concrete lining has a high initial cost but several advantages. It is durable, maintenance costs are at a minimum, and capacity is increased because of the relatively smooth surface. Placement costs, which are normally fairly high, are reduced when mechanized methods are used such as shown in Fig. 12.8 and

Figure 12.5.   A concrete-lined canal. (Courtesy Bureau of Reclamation.)

Figure 12.6. Placing precast concrete slabs with tongue and groove joints. (Courtesy Bureau of Reclamation.)

Figure 12.7. A serviceable and attractive small canal lined to solve the seepage, erosion, and weed problems. (Courtesy Work Project Administration.)

245

Figure 12.8.   Transit-mix trucks are frequently used for mixing concrete and conveying it to the slipform for placement. This method is particularly favored for reasons of economy when the concrete can be discharged from the mixer directly into the slipform. The tractor and cable keep the form moving forward. (Courtesy Bureau of Reclamation.)

12.9. Pneumatically applied portland cement mortar is used also to line ditches and canals. Compressed air drives a mixture of cement and sand through a hose to a nozzle where water is added. The mixture is blown onto the surface to be covered by air pressures of from 2 to 6 atm. Hoses of 15 to 45 meters in length operate best although they can be used for horizontal distances up to 150 meters and vertically up to 45 meters. For ordinary purposes one part of portland cement is mixed with four parts of sand by volume. The nozzle is held perpendicular to the surface with the nozzle from 1 to 15 meters from the surface. An experienced operator, or nozzle man, is essential.

Earth materials are used for canal linings, especially when lower initial cost is desirable or necessary, and when suitable materials are available. Three methods are employed: (1) placing a blanket of relatively impervious material over or within the permeable canal bed; (2) dispersing clay in the water and having it filter out as the water seeps through the bed: (3) chemically stabilizing the earth material to make it less pervious.

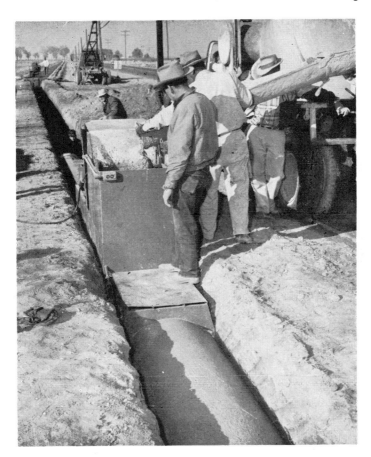

Figure 12.9. Continuous concrete pipe laid in place in a preformed ditch. An inflated rubber tube forms and holds the inside shape until the concrete sets. (Courtesy Fullerform Continuous Pipe Co.)

Dirmeyer[1] outlines specifications for the dispersion methods using bentonite, a montmorillonite clay that swells from 8 to 12 times its original volume upon wetting to seal earth and rock canals. The following guidelines should be used.

1. The Dispersion method shall be used in sandy channels having sufficient flat grade for ponding and having bottom and sides responsive to harrowing so that the bentonite can be made to penetrate the soil rather than to form a temporary surface seal. The ditch section shall be effectively dammed by either waterproofing and existing

[1]R. C. Dirmeyer, Jr., "Evaluation Report on Recent Bentonite Sealing Work in Wyoming Canals," *Wyoming Agr. Ext. Ser.*, March, 1959.

check structure or constructing a temporary earth dam. The bentonite used shall have a grit or sand content of 7 percent or less, and a colloidal mixture shall be ponded above the high water line at the upper end of the pond and be kept ponded for at least 48 hours. After the maximum ponding depth is attained, the ditch sides and then the bottom shall be stirred by a harrow, disk, or similar implement. This shall be done at least twice each day for 2 days and in a manner producing a minimum of bank sloughing.

Known extreme high loss areas shall be blanketed with a tamped soil-bentonite mixture prior to ponding. Additional bentonite shall be added whenever an unexpected insufficiency occurs because of bentonite flocculation, or excessive loss of mixture during filling.

2. The Multiple Dam Method shall be used in gravelly or rocky channels having steep grade and ditch bottom and sides unsuitable for harrow operation. The ditch section shall be divided by an adequate number and spacing of dams built of a mixture of granular high-swell bentonite (same bentonite quality as above) and some supplemental bridging agent such as a local low-grade bentonite or wet sawdust. Composition of this mixture material shall be acceptable to the responsible technician. Water shall be released from the upstream end. As the first dam is over-topped, equipment such as a dragline or back-hoe shall be used to help break up the dam and obtain a lumpy mixture at maximum speed. The same process shall be repeated for the remaining dams.

In case of extremely rocky and open materials, additional quantities of the sedimenting mixture shall be spread or water jet sluiced on the ditch bank areas and whenever deemed necessary by the technicians.

The most commonly used method of stabilizing soils is to mix about 10 percent of portland cement with a moist soil and compact the mixture. Thorough mixing, often difficult to secure in practice, is essential for a satisfactory, relatively durable lining.

Asphalt, plastic, and rubber compounds are becoming more widely used for linings in canals and ponds. Thin films are used where only an impermeable layer is desired. Thin layers are often covered with earth for protection. Thicker layers and built-up sections are used when resistance to abrasion and mechanical damage, as well as durability is needed. These products are being increasingly used when lower initial cost is required. Their feasibility will continue to increase as better products and procedures are developed.

12.6 SEEPAGE FROM CANALS

Seepage of water from irrigation canals is a serious problem. Not only is water lost, but also drainage problems are often aggravated on adjacent or lower lands. Occasionally water that seeps out of a canal re-enters the river in the valley where it can be rediverted, or enters an aquifer where it can be reused. It is a more serious economic loss when the water seepage losses are not recoverable. However, serious legal problems may occur as a result of seepage water which is subsequently used. Users of a return flow to the river may have established a water right so that water saved by lining a higher canal will still have to be delivered to lower users. Also, economic and legal liabilities may result from water seepage from upper canals causing drainage problems on lower lying lands. The economic and legal problems associated with seepage from canals are sometimes very complex.

Several methods used to measure seepage from canals are inflow-outflow, ponding, seepage meters, wells, laboratory tests of permeability of soils, and special methods, including electrical resistance and tracing of natural and radioactive salts. The method best suited to a canal will depend on the depth and velocity of flow, ability to drain the canal, the material in the bed, and rate of seepage. The three most commonly used methods for field measurement are discussed briefly.

The inflow-outflow method consists of measuring the flow into and the flow from a selected section of canal. The accuracy of this method increases with the difference between the quantity of inflow and outflow rates. Water level should be held constant during the measurement and allowance should be made for rainfall and evaporation.

The ponding method consists of constructing dikes in the canal, filling the intervening section with water, and measuring the drop in water surface per hour. This is the most accurate method, provided that leaks through the dikes are either prevented or measured, and provided allowance is made for rainfall and evaporation. Measurements of water level should be made at both ends of the pond to eliminate the effect of wind on the water level. The principal disadvantages to this method are that tests can be made only when the canal is not in use and that the expense of constructing dikes and supplying and regulating the water is considerable.

Seepage meters are used to obtain measurements of seepage from relatively small areas of canal surface. Measurements can be taken without disturbing the flow in the canal unless velocity is excessive. Seepage meters generally consist of a cylinder with a dome or cone on the top to allow for trapped air to be removed from the system through a valve attached to the top of the cone. The cylinder is pressed gently into the soil and the unit is filled with water. A Marriotte siphon can also be used to obtain a constant head.

Extensive studies indicate that several factors contribute to error in the seepage meter, so that these meters give only an indication of the order of magnitude of seepage rather than absolute rates of seepage. The surface is also disturbed by installing the meter, and the area under test is small and may not be representative of the canal section. To obtain results indicative of actual seepage, several readings are necessary with representative readings taken on the sides as well as on the bottom of the canal.

12.7 KEEPING CANALS CLEAN

One of the vital maintenance problems in conveyance of irrigation water is cleaning canals. Growth of weeds and willows on canal banks, and moss and other aquatic plants in canals, greatly retards water velocities decreasing canal capacities. Silt and clay sedimentation in canals also restricts water flow. Dense weed growth prevents proper inspection of the canal bank and provides winter cover that allows burrowing animals to ruin the canal banks.

Weed top growth falls into the water and catches floating debris, retards flow, and causes failure of banks. Large perennials such as willows, tamarisk, and cane make it almost impossible to clean a canal and remove the smaller weeds. In addition to the operational problem, many land weeds developing seed along a canal bank are a source of weed infestation to crop land.

Water weeds such as tules, cattails, pondweeds, coontail, and chara also reduce the flow. These weeds, like those on the banks, often clog measuring flumes, weirs, spillways, and other parts of the irrigation system, causing delays and additional costs in cleaning. Weed growths cause sand and silt bars to build up in the channel, which retard the velocity of flow, increase the seepage loss, and at times cause overflows of the bank.

Hand-labor methods of canal cleaning are being replaced by the use of bulldozers, dragline excavators, and tractor-drawn chains, Fig. 12.3. The most common methods of controlling weeds on canal banks are: pasturing, mowing, burning, and the application of chemical weed killers. Methods for controlling water-weed growths may be classed as: mechanical, drying, shading, and chemical.

Chemicals, primarily aromatic solvents, have been very effective in killing weeds in canals and ditches. When applied in recommended quantities, no injury to crops or to livestock has been observed, but these chemicals are deadly to aquatic animal life.

Aromatic solvents are most economical in small channels, while mechanical methods of weed control are more practical and more economical on larger canals.

Weeds along rivers, canals, and ditches have been estimated to use, and thereby largely waste, 25 percent of the available irrigation water in the western United States. Water losses by evaporation and consumptive use are increased by weeds and growth along irrigation water courses.

CHAPTER
13

WELLS FOR
IRRIGATION WATER

Throughout man's history wells have been an integral part of his life and activity, supplying clean pure water where surface supplies were not adequate. The Bible frequently refers to wells, wherein they are a source of life-giving water and a symbol of security and well-being. Joseph's well in Cairo, Egypt, constructed some 3700 years ago, is still in operation. The ancient Chinese drilled wells as deep as 1100 meters with primitive churn-drill methods. Some were cased with bamboo. The ancient ganats of Persia (Iran) have been and still are the basis of existence in many parts of that land. Milligan[1] clearly and interestingly states the historical significance of wells as follows:

The well was and still is, in many areas, the center of religious social activity. These areas include Southern Europe, the Middle East, the Far East, Latin America, and many islands of the sea. The pagans worshipped their wells; the Greeks consecrated them to their Gods; the Moslems constructed mosques over them; and the Hindus constructed temples near important wells.

The ancient ganats of Persia were shafts started from the surface and dug slightly upgrade toward the ground-water supply, Fig. 13.1. The essentially horizontal shaft served as a conveyance channel and continued sometimes for 30 to 50 kilometers to the source of the ground-water supply. Approximately every 50 to 300 meters, depending upon the depth, a vertical shaft was dug to the tunnel through which earth was removed and air, food, and general access provided to the diggers. The skill developed by diggers of ganats is still passed from father to son and the accuracy attained by these artisans is truly pnenomenal.

The Maui tunnels were developed on the island of Maui in the Hawaiian Islands, Fig. 13.2. These wells are vertical shafts with horizontal tunnels or galleries extending in one or two directions from the central shaft. Frequently these tunnels intercept porous

[1]Cleve H. Milligan, "Pumping Ground Water for Irrigation and Drainage," *Trans. Am. Soc. Civil Eng.*, Paper No. 2857, published as Proceedings, February 1955.

Shaft constructed to remove soil,
for light and ventilation, and
for maintenance access

Depth to
150 m

Pile of soil removed from hole,
protects hole from flood flows

Tunnel

Ground
surface

Tunnel
intercepts
water table

Length - from 1 to 50 km.

Water table

Figure 13.1.   Ganats constructed in Persia (Iran) to intercept water source and convey water to agricultural lands.

lava formations and tubes which result in extremely copious flows of water.

Dug wells have been the principal source of water until recent times, when modern well-drilling methods have been developed and used extensively. Dug wells are usually shallow wells, gen-

Figure 13.2.   Tunnels constructed in lava formation to collect water for irrigation. (Courtesy of Hawaiian Commercial and Sugar Co.)

erally less than 20 meters in depth and more than 1 meter in diameter. In the past they were usually excavated by hand, and even today in many areas this is the principal method of construction.

The Ranney well is a modification of the dug well, wherein a large central shaft is constructed to a depth below the water table. Wells are then drilled horizontally in a radial direction from the bottom of the shaft. Large quantities of flow can usually be obtained from a well of this type.

Drilling methods developed by the French in the early 1800s are still used extensively. The most common well is the vertical cased well. It varies in size from the small-diameter well of shallow depth with sufficient capacity to supply domestic needs for a family and water for livestock, to the large-diameter well of great capacity for agricultural, municipal, and industrial purposes. Some vertical water wells extend 500 meters into the earth. Because of the great variety of types, sizes, capacities, and drilling conditions, many different methods are used to drill vertical wells.

The importance of wells and ground-water development is indicated by the size of the well-drilling industry. It is estimated that there are more than 10,000 well-drilling contractors in the United States, operating approximately 22,000 drilling machines. The number of wells drilled per year by these contractors is estimated to be about 750,000. The number of wells in operation in 1977 has been estimated to be 18,000,000. Approximately 25 percent of the total water supplies of the United States are obtained from ground water.

13.1 DRILLING METHODS AND EQUIPMENT

Excavation by hand is the oldest method used for construction of wells. When mechanical equipment is not available, the material is removed from the hole by hand and bucket. When power is available, the material is removed by clamshell or orange peel buckets attached to excavating equipment. Lining or cribbing is lowered as the well deepens to prevent cave-ins, particularly in unconsolidated sedimentary materials. Occasionally large shallow open cuts through water-bearing materials serve satisfactorily as collecting pits for ground water.

Cable Tool Method

The cable tool method of drilling, often referred to as the standard or the percussion method is the oldest and most versatile drilling procedure. The first recorded use of percussion boring tools was in China about 1600 B.C. for drilling brine wells a few hundred feet in depth. By 1500 A.D. holes were being drilled to a depth of 600 meters.

A cutting tool is suspended from a cable, and vertical oscillation of the tool cuts and loosens material at the bottom of the hole. A special tool, sometimes a modification of a cutting tool, is used to remove the material from the hole. Generally, water is required for drilling, and the holes are usually less than 40 centimeters in diameter. A typical cable tool outfit is shown in Fig. 13.3.[2]

The material at various depths in the hole can be determined quite accurately by this method. Noting the material and the depth from which it comes is referred to as logging the well.

[2]David Keith Todd, *Ground Water Hydrology*, Wiley, New York, 1959.

Figure 13.3.   Drilling a well using the cable tool method. (After Todd.)

Such information is required often by the governmental agency responsible for ground-water development and is very helpful for interpreting the performance of the well and anticipating what can be expected at other locations in the vicinity.

Difficulty is experienced when drilling through unconsolidated sand, especially quicksand, with the cable tool because of material caving in around the bit. The variety of tools used by the cable tool method can be grouped into five general classes, Fig. 13.4. The cutting tool or bit cuts, softens, and grinds the materials so that the second type of tool, the bailer, can pick up this material and remove it from the hole. Three auxiliary tools are used to facilitate the operation of the cutting tools and the bailer: the rope socket for attaching the cable to the other tools, the jars which allow vertical movement in order to loosen tools should then be stuck in the hole, and the drill stem to add weight and length to the drill so that it will cut rapidly and vertically.

Hydraulic Rotary Method

A bit is rotated rapidly on the bottom of a string of drill pipe. A mud of clay and water, the consistency of which is varied with

Rope   Jars   Drill stem   Drilling bit   Bailer
socket

Figure 13.4.   Basic well-drilling tools for the cable tool method. (After Todd.)

drilling conditions, is forced through the drill pipe and carries the cuttings around the outside of the pipe to the ground surface. With the proper consistency of rotary mud, and correct pressure, the sides of the hole do not slip in during drilling and no casing is used during the operation. The fluid circulating through and around the drill cools the bit.

The rotary method permits the drilling of a test hole without having to invest in casing. Hence, exploratory work is generally cheaper when done by this method. If favorable conditions are found in the hole, casing is lowered and perforated sections inserted, or perforations are made at the depths of the more favorable water-bearing aquifers. Clay which is deposited on the walls of the hole from the rotary mud is removed in those sections opposite the perforations by washing and development procedures. If subsurface conditions are found to be very favorable, the well can be reamed to a large size and a large casing inserted. Because of ease and rapidity of drilling, rotary holes are usually considerably larger than the casing with the space between being backfilled with graded gravel.

One of the principal disadvantages of the rotary method is that it is sometimes difficult to remove all of the mud from the water-bearing aquifer. If the consistency is not ideal, the mud may penetrate the aquifer for some distance and then be difficult to remove by normal development methods. Also, accurate well-logs are more difficult to obtain by the rotary method than by other methods of drilling. Since the mud mixes with the drill cuttings, the amount of fine material is difficult to ascertain in some circumstances. An experienced operator can minimize both problems.

Reverse Rotary Method

When drilling mud is forced down the outside of the drill pipe and comes up inside the pipe carrying the drillings with it, the method is called the reverse rotary method. One of its particular advantages is that the velocity of the upward flow is much greater for a given discharge and thereby can carry larger size sediments and small rocks to the surface. Reverse circulation is the cheapest method of drilling large diameter holes in soft, unconsolidated formations. Most wells drilled by this method have a diameter of 60 centimeters or larger. Sand, silt, or soft clay formations and the absence of clay or boulders are conditions favorable to reverse rotary drilling. The drill pipe must be pulled periodically and those cobbles and boulders larger than the hole in the drill pipe or the openings in the drill bit must be removed by an "orange peel" bucket.

Care must be taken not to seal with mud the low pressure zones which may otherwise be excellent aquifers. Procedures for casing and developing wells drilled by the reverse rotary method are the same as those used in the hydraulic rotary method.

13.2 DRILLING IMPROVEMENTS

Development of more efficient and effective drilling equipment consists primarily of improvements in the drilling head tool. Diamond studs and percussion hammers are being used in cable tool drills with increased drilling rates. Turbine drills have been developed powered by rotary drilling mud and others are powered by air. The turbine with vertical axes is attached to the lower end of the stationary drill stem. Drilling rate and efficiency are higher than those obtained with a conventional rotary rig.

A bottom-hole rotation device called the "electrodrill" is being developed. One down-hole-fluid motor operates using drilling mud and another has two air motors connected by a gear reduction to a rotary bit for use with air as the drilling fluid.

13.3 WELL CASING PERFORATIONS AND SCREENS

A well can be left uncased only in consolidated formations. A casing is necessary when the formation is unconsolidated. The casing must have suitable openings to permit the water to enter freely but restrict the entry of surrounding material. This can be accomplished by using a perforated well casing, perforated casing and a gravel pack, a well screen, or a well screen and a gravel pack.

Perforated Well Casing

The well casing can be perforated while in the well by a special tool which simply tears a series of holes in the section of casing that penetrates the water bearing strata. If the depth to the water-bearing aquifer is known before casing is placed in the well, perforations can be cut at the factory or at the drilling site with a cutting torch.

Perforated Casings and Gravel Packs

The slots in a perforated casing may be as large as 1 centimeter wide and can only be used where the water bearing strata consists of coarse gravels. When large quantities of fine sand exist in the aquifer, a gravel envelope is often used around the outside of the casing to screen out the sand and permit only water to enter the well. To use a gravel pack around the pipe, the diameter of the well hole must be larger than the diameter of the well casing by twice the desired thickness of the pack. For example, if the casing diameter is 25 centimeters and the gravel-pack thickness needs to be 10 centimeters, then the hole diameter must be 45 centimeters. Seldom is it necessary to place more than a 15-centimeter thickness of gravel around the casing.

For deep wells the gravel can be placed best if dropped to the desired depth through a 5- or 7.5-centimeter pipe outside the casing. This practice will prevent the gravel bridging at some intermediate depth.

When local materials permit, the gravel should generally be graded from sand up to ¼-inch gravel. Additional gravel should be placed as the pack settles during development and subsequent operation. The criterion mentioned above for the relationship of extreme size to surrounding material will apply to perforated pipe and the surrounding gravel pack. Any gravel pack extending to the surface will permit contamination of the subsurface waters by the generally less-pure surface water. Hence, if the gravel-packed well is to be used for domestic supply, care must be taken to prevent contamination.

Well Screens

When considerable fine sand is in the aquifer, a screen may be advisable in place of perforations. A well screen is made by wrapping a specially made wire around a metal, cylindrical frame of vertical bars. The spacing between consecutive wire wraps controls the size of particles which can enter the well. The slot opening should pass about 70 percent of the surrounding material. Excessive entry of sand is extremely dangerous, because the earth above the cavity caused by the removal of the sand might cave and either collapse the casing or seal off the screen. It may also cause subsidence of the ground above.

The slot opening in the screen should be such that the velocity of water passing through the slot should not exceed 6 centimeters per second. A low velocity ensures that the energy loss through the screen is negligible. High velocities cause low pressures, as water passes through the screen and produces increased incrustations from the release of calcium carbonate and other dissolved mineral salts. Excessive incrustations and corrosion will also result on any part of the well screen or perforations above the pumping level of water in the well.

Well Screens and Gravel Packs

A gravel pack may be used in conjunction with a well screen when aquifer conditions would require slot openings in the screen so small that high velocities would occur, or insufficient structural strength would result. Under such conditions, the gravel pack would be designed and placed using the same design criteria described previously for a perforated well casing.

13.4 DEVELOPMENT OF WELLS

Developing the irrigation well is almost as important as drilling it. The primary purpose of developing a well is to increase the water yield. This is accomplished by increasing the permeability of the formation through which water moves toward the well. The secondary but nevertheless very important reason for developing a well is to determine the water supply available and the needed characteristics of pump and power unit to be installed.

The method used to increase the yield of a well depends upon the formation in which the well is drilled and upon the equipment available. If the water-bearing material is sand and gravel, marked improvement is obtained by removing the sand, silt, and clay from the water-bearing formation in the area immediately around the perforated portion of the casing.

When the well penetrates a consolidated sedimentary formation, it may be necessary to dissolve part of the limestone cementing agent to create larger openings adjacent to the well. When a well penetrates a rock formation, it is often desirable to fracture the rock to create additional crevices through which water can flow into the well.

In all cases it is essential to remove the material which moves into the well and settles to the bottom during the development operation. This can be done with a boiler or a sand pump. As explained earlier, care should be taken in the development of a well not to remove excessive quantities of fine materials in order to prevent the collapse of overbearing layers which might destroy the well.

Development by Pumping

Developing a well by pumping requires a variable speed pump of large capacity. A new pump should not be used because of the damage and lowering of efficiency caused by the wearing action of sand moving through the pump. Pumping should be started slowly. Since a constant speed will cause bridging of sand particles and a marked lowering in the capacity of the well, the speed should be increased in steps and held constant between steps until no further sand is removed. Pumping should be continued until the maximum discharge is reached and no further sand is withdrawn. The pump is then shut off and the water level allowed to return to normal position. This cycle is repeated until no further sand is removed by this method.

A process known as "raw hiding" is often used, wherein the pump is started and stopped very quickly. This allows for a decrease in pressure followed by a back-surge on the well which gives a flushing action. No foot valve is installed during this program so that the flow can move both ways in the well.

Development by Surging

Surging is one of the commonest and most effective methods of developing a well in sand and gravel formations. A plunger is moved up and down inside the well opposite the perforated casing forcing water to move alternately into and out of the well. This surging action flushes out smaller particles and deposits them in the well where they can be removed by bailing. Surging is started slowly at first and increased in speed as development proceeds. Solid surging blocks are most commonly used. However, on wells of low yields, a valve-type block should be used. On the downward stroke, water passes through the valve and also is forced into the aquifer. On the upward stroke, water is drawn rapidly into the well.

Development by Compressed Air

Developing with air is best suited to small wells. The depth of water in the well should be at least two-thirds the depth of the well. Air pressure of 300 to 450 kilograms per square centimeter are necessary. The most effective development with air is a combination of surging and pumping. For irrigation wells a 5 centimeter diameter air pipe is normally lowered inside a 15- to 20-centimeter discharge pipe, the end of which is near the bottom of the well. With the air pipe extending below the end of the discharge pipe, a large quantity of air is suddenly released causing a surge of water outward into the acquifer. By raising the air pipe within the discharge pipe an air jet pumping action is created which causes water to flow into the well. Hence, raising and lowering the air line with respect to the end of the discharge pipe, causes a reversal in the direction of flow which dislodges sand, silt, and clay and carries it into the well. When the flow of sand stops, the discharge pipe is raised and the process repeated until the entire perforated portion of the well has been subjected to the surging and pumping action.

"Backwashing" is also used to develop a well. In this process air is forced into the capped well until the water level is lowered to the top of the perforations. Then air is suddenly released causing a reversal of flow within the well. Surging action removes fine material and improves yield.

Other Development Methods

Yield in wells is also improved by using dry ice, chemicals, and explosives with varying degrees of success. Greater skill and experience is usually required when these methods are used than is required in the procedures described previously. Dry ice (solid carbon dioxide) is also used to create pressure surging in wells and operates much the same as backwashing. Muriatic acid (hydrochloric—usually a 15 percent solution) is sometimes added to loosen the sand, when cemented with lime, and to widen interstices in rock. Other chemicals are also used as dispersing and cleansing agents to remove clay and silt and for the purpose of removing lime, iron, and other incrustations. Explosives are used to shatter rock formations and remove incrustations. Special low-power vibratory explosives have been developed which cause sufficient vibrations to shake loose incrustations in well casings.

13.5 SAND SEPARATORS

Because sand is detrimental to pumps and piping systems, considerable effort has been applied to removing sand from irrigation water. Sand is erosive to pumps and sprinkler equipment. Accumulations of sand in sprinkler pipe and trickle irrigation systems can make these systems inoperative.

Settling basins are very effective, but often impractical. A device that creates a vortex is used for removing sand from water in a closed piping system under pressure. High tangential velocities produce high centrifugal forces causing solid particles to move to the outside of the container where they are removed by flushing.

Commercial separators are now available with capacities up to 400 liters per second and pressure loss of only 0.7 kilograms per square centimeter.

13.6 CONTRACTS FOR DRILLING

A definite agreement between the owner and driller is essential. A written contract minimizes misunderstanding, but a rigid contract will increase cost. The best assurance of obtaining a good job is to employ a reliable driller. Cut-rate operations usually result in the owner's paying excessively for a job poorly done. The following items should be clearly understood before drilling is commenced.

1. Starting and completion dates.

2. Diameter and thickness of casing.

3. Vertical alignment, essential for installation of a deep-well pump.

4. Well record, showing an accurate log of material encountered in drilling, water-bearing formations, perforations, static water level, draw-down, discharge, and development work.

5. Perforating or screening procedure and cost.

6. Procedure for developing the well and cost.

7. Test discharge and draw-down procedure and cost.

8. Price for drilling, including time for moving and setting up equipment.

9. Price if a change of diameter is made.

10. Method and time of payment.

13.7 HYDRAULICS OF WELLS

The general behavior of a well and the hydraulic formulas which apply to it depend upon the formation in which the well is drilled. Figure 13.5 illustrates four wells and a spring in a typical valley. The stream entering the valley from the canyon on the left passes over a gravel fan where considerable seepage occurs. Well A drilled in this zone is not flowing, with the water level in the well at the level of the water table adjacent to the well. Well B flows because the top of the well is below the piezometric surface of the confined aquifer.[3]

[3]The piezometric surface is the height water will stand in a piezometer or pipe open at the end which extends into the aquifer. The height water will rise in the pipe above the bottom of the pipe is equal to the pressure divided by the unit weight of the water, $h = p/w$.

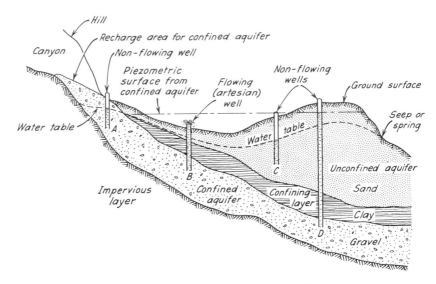

Figure 13.5. Wells in typical valley.

Well C is similar to well A, and both are referred to as un-confined wells, often called gravity wells or water-table wells. Well D is similar to well B, except that the well D does not flow without a pump, since the top of the well is above the piezometric surface in the aquifer. Wells B and D are called confined wells, and are also referred to as artesian wells and as pressure wells. Well B may also be called a flowing well.

Water Yield of Wells

The stream of water obtained for irrigation from a well with a pumping plant is determined by one or both of two major factors: (1) the capacity of the pump and the power of the motor or en-gine, and (2) the capacity of the well, which depends on the draw-down of the water surface or pressures, the depth and ef-fective diameter of the well, and the permeability of the water-bearing material. Pump capacities and energy requirements to lift given quantities of water through specified heights are well understood and may be predicted with fair precision. It is far more difficult to predict the energy requirement to move specified quantities of water through water-bearing sands or gravels into a well, because of the uncertain permeability of the sands and gravels through which the water flows.

Confined Wells

Radial horizontal flow in a confined layer of thickness t to a fully penetrating well is shown in profile in Fig. 13.6. The flow Q

toward the well can be represented as $Q = AV$. The area A of a cylinder of radius r and height t is $2\pi rt$. The velocity V through the cylinder is given by the Darcy Law:

$$V = k \frac{dh}{dr} \tag{13.1}$$

where k is permeability and dh divided by dr is slope of the draw-down curve at radius r. Therefore, the quantity of radial flow toward a well is:

$$Q = AV = 2\pi rtk \frac{dh}{dr} \tag{13.2}$$

Separating the variables and integrating this differential equation between limits of h_e and h_w in the vertical plane and r_e and r_w in the horizontal plane, the result is:

$$Q = \frac{2\pi kt(h_e - h_w)}{2.3 \log_{10}(r_e/r_w)} \tag{13.3}$$

The draw-down $(h_e - h_w)$ substantially is directly proportional to the discharge, whereas the discharge varies as the logarithm of the well radius. Hence, when the well radius is doubled, the discharge increases only 10 percent, and increasing the radius four times will increase the discharge only about 20 percent.

Unconfined Wells

A confined well becomes an unconfined well when the flow is not restricted by the impervious layer above the flow, Fig. 13.7. In unconfined wells the surface of the flow is the water table, which

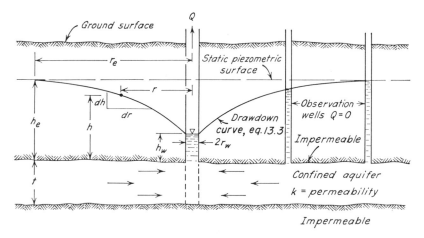

Figure 13.6.   Radial flow in a confined aquifer to a fully penetrating well.

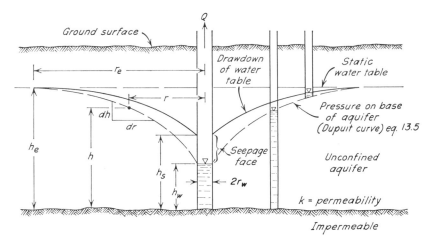

Figure 13.7.   Radial flow in an unconfined aquifer to a fully penetrating well.

is the plane of atmospheric pressure below which the pores of the aquifer are essentially saturated.

The hydraulics of the unconfined well can be developed in the same manner as that of the confined well, except that the constant thickness of the aquifer t is replaced by a variable thickness h. Hence, Equation 13.2 becomes:

$$Q = AV = 2\pi rhk \, \frac{dh}{dr} \tag{13.4}$$

and integrating between the same limits, the classical Dupuit equation is obtained:

$$Q = \frac{\pi k(h_e^2 - h_w^2)}{2.3 \log_{10} (r_e/r_w)} \tag{13.5}$$

The discharge Q now varies as the difference of the squares of h_e and h_w, whereas in a confined well discharge it varied linearly with h, provided that in each case there is no interference from adjacent wells. The same relationship exists between discharge Q and the well radius as exists in confined wells.

Limitations of Equations

Both Equations 13.3 and 13.5 are developed assuming radial horizontal flow through uniform material perpendicular to vertical cylindrical surfaces. Also, steady flow (no variation with time) is assumed. By using the Darcy equation, $V = k(dh/dr)$, the flow is assumed to be laminar. In the case of confined flow the further assumption is made that the thickness t of the aquifer is constant.

Seldom, if ever, in field conditions are these assumptions met fully. However, they are reasonably close in most cases. Being aware of the assumptions and knowing the effect of the error involved, allowance can generally be made so that results from theoretical equations are nevertheless very useful.

Unconfined well flow near the well is not horizontal as assumed. Hence, the nearer the well the greater the error that can be expected. Equation 13.5 shows that the depth of water outside the well approaches the depth within the well h_w as the radius approaches the well radius r_w. However, in actuality a seepage face h_s always occurs above the water surface in an unconfined well. Hence, the correct boundary condition is given by:

$$Q = \frac{\pi k(h_e^2 - h_s^2)}{2.3 \ (\log_{10} r_e/r_w)} \tag{13.6}$$

In Equations 13.3, 13.5, and 13.6, h_e and r_e can be replaced by h and r at any intermediate point. Since h_s generally is not known and there is no mathematical expression relating the commonly measured h_w to h_s, a functional relationship has been developed as shown in Fig. 13.8 A dimensionless discharge number $\dfrac{Q}{kr_w^2}$ characterizes the draw-down of the water table around the well—the greater the number, the greater the draw-down. For an unconfined pumped well, the water table adjacent to the well may be considerably higher than the water surface inside the well. When wells are being used to lower the water table, the height of the seepage face h_s becomes very important.

Unsteady Flow from Wells

Flow conditions in a well are seldom steady. Either the discharge or the depth of water is usually changing with time. Nevertheless, Equations 13.3, 13.5, and 13.6 can be applied to unsteady flow because the velocity of flow through the aquifer toward the well is so small that kinetic energy is negligible.

Difficulty is experienced in selecting a value of the radius of influence r_e, since the draw-down curve approaches asymptotically the original water-table condition. However, a considerable error in r_e will not seriously affect the calculated draw-down or discharge, since the logarithm of the radius is used in the formulas. A difference of less than 10 percent results when a radius of 150 meters is used in lieu of 300 and 600 meters causes another 10 percent change in Q or h. A reasonable value of r_e for a short period of pumping or for a tight aquifer is 150 meters. For long periods of pumping or very permeable aquifers a value of 600 meters may be used as a reasonable radius of influence.

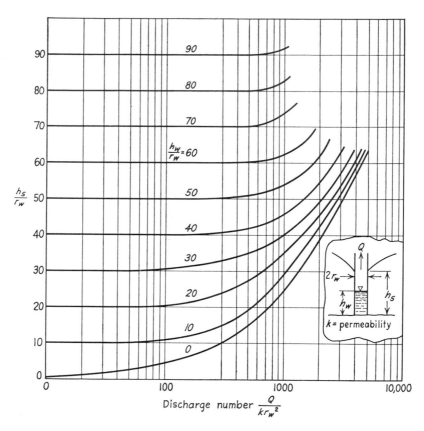

Figure 13.8.   Seepage face and depth of water in the well related to discharge number for an unconfined radial well.

When the formulas are used, the correct mathematical definition for r_e is the intercept of the draw-down curve with the unaffected water table. For a confined well, plot h against $\log r$ and extrapolate the straight line resulting over the major portion of the radius to the water-table line. For the unconfined well, plot h^2 against the logarithm of r and similarly extrapolate the resulting straight line to the original water-table line. The intercept of the draw-down line and water-table line is the radius of the influence r_e.

Conditions at an Unconfined Well

Three dimensionless parameters (Q/kr_w^2, h_s/r_w, h_w/r_w) have been found useful in describing flow conditions to an unconfined well. The discharge number, Q/kr_w^2, is indicative of the cone of influence, with large numbers indicating steep cones and small numbers indicating shallow cones of depression of the water table. The parameters h_s/r_w and h_w/r_w index the geometry adjacent to the well. Figure 13.8 relates the seepage face and the depth of the water in the well to the discharge number. These parameters avoid the necessity to refer to the radius of influence and permit calculating the height of the seepage face, $h_s - h_w$.

13.8 DRAW-DOWN DISCHARGE RELATIONS

In pumping from thick unconfined water-bearing formations, or from confined ground-water or artesian formations, the discharge is directly proportional to the draw-down as shown in Fig. 13.9, provided that there is no interference from the pumping of other wells in close proximity. Shallow unconfined water-bearing formations sustain a gradual decrease in well discharge per meter of draw-down as the draw-down increases. This relation is shown in the upper curve of Fig. 13.9. In general, water flows radially through rather homogeneous water-bearing sands and gravels toward the well through a series of concentric cylindrical surfaces having the well as a vertical axis. Therefore, for constant yield as the water approaches the well, its velocity must increase because the cross-sectional area through which it flows is continuously decreasing. Consequently the hydraulic slope (driving force per unit weight) must increase as the water approaches the well, and the draw-down curve becomes steeper as the well is approached. Where the capacity of the pump exceeds

Figure 13.9. Typical discharge draw-down relation of a well in a thin water-bearing formation and one in a thick water-bearing formation or an artesian stratum.

Figure 13.10.   A pump installation where the water is pumped 50 cm higher than necessary. The stilling basin with three baffles is very effective. (Courtesy Soil Conservation Service, U.S.D.A.)

the capacity of the well, the draw-down is excessive in the immediate vicinity of the well. It is therefore desirable, if practical, to provide a large "effective" diameter well to avoid excessive draw-down (and excessive power requirements) in order to drive the water into the well.

13.9 POLLUTION

Prevention of pollution of water supplies is becoming increasingly important. Normal types of water pollution generally are not detrimental to irrigated crop production. However, well water used for irrigation is also consumed occasionally or regularly by people. The general assumption is that well water is good water to drink, hence, care needs to be taken to keep the water suitable for human consumption.

Chemical and biological contaminants can enter a well that is being drilled, or during operation of the well, or following abandonment of the well. Contaminants can enter through the pump, under the pump, round the casing, and through the gravel pack. Contaminated water can also move from one aquifer to another through or around the well and contaminants might also seep through the soil or trickle down a fractured zone into the water-bearing aquifer.

Frequent deficiencies in well construction resulting in bacterial contaminations of the well water include: (1) insufficient and substandard well casing; (2) inadequate seal between the well casing and the bore holes; (3) poor welding of casing joints; (4) lack of sanitary covers; and (5) well pits to protect from freezing.

With care and understanding, well water can be protected against chemical and biological contamination that would make it unfit for drinking.

PUMPING WATER FOR IRRIGATION AND DRAINAGE

Large areas of land in arid regions are so situated that available water cannot be brought economically to them by gravity flow in canals and pipes. In many areas, surface or underground water is pumped to the land to be irrigated. Pumping is also important for irrigation by sprinkling.

Pumping for drainage may be as important as pumping for irrigation. Drainage waters must frequently be lifted to a higher elevation where they can be used for irrigation. Also, pumps are frequently required to lift drainage water into a higher channel where it can flow out of the area by gravity.

14.1 POWER REQUIREMENTS FOR PUMPING

Work is defined as force times distance. Power is defined as work per unit of time or the rate of doing work. Work is required to lift water out of a well and the amount of water delivered in a unit of time can be related to power and is referred to in units of horsepower by the following formula:

$$HP = \frac{Qh}{273} \tag{14.1}$$

where HP = horsepower
Q = discharge in m³/hr
h = vertical lift, in m

or

$$HP = \frac{Qh}{76} \tag{14.2}$$

where Q = discharge, in l/s
h = vertical lift, in m

14.2 EFFICIENCY

Efficiency is output divided by input. Thus the efficiency of a motor would be the energy output of the motor divided by the input energy to the motor. Input power to a pump is often called brake horsepower and the output power is called water horse-power. Hence, for pumps efficiency is:

$$E_p = \frac{\text{output}}{\text{input}} = \frac{\text{WHP}}{\text{BHP}} \qquad (14.3)$$

Thus combining Equations 14.2 and 14.3, the output can be defined as:

$$\text{WHP} = \text{BHP} \times \text{Efficiency} = \frac{QhE_p}{76} \qquad (14.4)$$

Table 14.1 shows the horsepower required for various discharges and vertical lifts assuming an efficiency of 50 percent for the pumping plant. Special care should be taken to keep the efficiency of the pump and motor high through proper maintenance and timely adustments. Occasional field tests to determine efficiency are advisable. An over-all efficiency of a new electric motor and water pump should be about 65 percent. The motor should be at least 95 percent efficient and the pump about 75 percent efficient. Reconditioned pump and motor units should operate at 60 percent efficiency or better. If the efficiency is as low as 40 percent, the unit should be repaired and/or adjusted. The life of a pump is generally from 20 to 25 years.

TABLE 14.1   Horsepower Required to Lift Different Quantities of Water to Elevations of 10 to 90 m

(Efficiency of pumping plant 50% of theoretical.
Use for estimating only.)

Discharge		Horsepower Required for Elevations of								
m³/hr	l/s	10 m	20 m	30 m	40 m	50 m	60 m	70 m	80 m	90 m
10	2.78	0.7	1.5	2.2	2.9	3.7	4.4	5.1	5.9	6.6
20	5.56	1.5	2.9	4.4	5.9	7.3	8.8	10.3	11.7	13.2
30	8.34	2.2	4.4	6.6	8.8	11.0	13.2	15.4	17.6	19.8
40	11.1	2.9	5.9	8.8	11.7	14.7	17.6	20.5	23.4	26.4
50	13.9	3.7	7.3	11.0	14.7	18.3	22.0	25.6	29.3	33.0
60	16.7	4.4	8.8	13.2	17.6	22.0	26.4	30.8	35.2	39.6
70	19.5	5.1	10.3	15.4	20.5	25.6	30.8	35.9	41.0	46.2
80	22.2	5.9	11.7	17.6	23.4	29.3	35.2	41.0	46.9	52.7
90	25.0	6.6	13.2	19.8	26.4	33.0	39.6	46.2	52.4	59.3
100	27.8	7.3	14.7	22.0	29.3	36.6	44.0	51.3	58.6	65.9
125	34.8	9.2	18.3	27.5	36.6	45.8	54.9	64.1	73.3	82.4
150	41.7	11.0	22.0	33.0	44.0	54.9	65.9	76.9	87.9	98.9
175	48.6	12.8	25.6	38.5	51.3	64.1	76.9	89.7	102.6	115.4
200	55.6	14.7	29.3	44.0	58.6	73.3	87.9	102.6	117.2	131.9
250	69.5	18.3	36.6	54.9	73.3	91.6	109.9	128.2	146.5	164.8
300	83.4	22.0	44.0	65.9	87.9	109.9	139.9	153.8	175.8	197.8

Low efficiency means considerably more power cost, a system not operating at design flow or pressure, and more time required to apply the needed irrigation water. Note that increasing efficiency from 35 to 65 percent reduces by 46 percent both power cost and time required to apply a given quantity of water.

In areas where pumping equipment is not readily available, the purchaser will need to guard against being sold a unit with discharge and pressure characteristics that do not fit his specific requirements. Efficiencies of most pumps are considerably lower when not operating at design capacities.

14.2 PUMPING LIFTS

The vertical distance water is lifted for irrigation purposes varies widely. In irrigation practice the maximum economic height of lift is determined by cost limitations, not by mechanical or power limitations. The discussion in the preceding section, and also Table 14.1, indicates that for any given size of irrigation stream the power requirement is proportional to the lift. The difference in elevation of the water surface in a pond, lake, or river from which pumped water is taken, and the water surface of the discharge canal into which water flows from a submerged discharge pipe is known as the "static head."

In pumping from ground-water sources the static head is the difference in elevation between the water surface in the well and the water surface of the discharge canal. While the well is being pumped the water level lowers in the well. The pumping lift from a well then becomes the difference in elevation between the lowered water surface in the well when pumping and the water surface of the discharge canal plus the energy lost within the pump and pipe while pumping. It is nearly always desirable to avoid excessive draw-down in the well in order to reduce excessive power requirements. Poorly designed pumps, wear, and plugging of well perforations are the main reasons why pumping-plant efficiencies range from about 75 percent under very favorable conditions down to 20 percent or less under unfavorable conditions.

Because of the large number of variable factors influencing profits from pumping water for irrigation, it is impracticable to set specific limits of profitable pumping lifts that will apply in different localities for any length of time. Those who contemplate irrigation pumping should keep in mind the fact that the cost of pumped water is roughly proportional to the height of lift. Proposals to pump water through high lifts should be carefully considered before investments are made; on the other hand, good dependable water supplies that may be made available for irrigation by pumping only a few meters should not be overlooked.

14.3 PRIMITIVE IRRIGATION PUMPING

Pumping water for irrigation has been practiced for centuries in Egypt, India, and other older countries where irrigation is essential to agriculture. One of the devices used in early Egypt and India, known as the shaduf, is illustrated in Fig. 14.1. This device makes use of the principle of the lever with a suspended fulcrum and a counterweight. The bucket, suspended from the long end of the pole, is sometimes made of leather, stiffened near the top with a wooden hoop. The operator throws his weight on the

Figure 14.1. The shaduf, used in Egypt and India to lift water. (Courtesy U.S.D.A.)

sweep, the bucket fills, and the counterweight raises it to the next higher channel into which the water is poured. A single shaduf is operated by one man lifting water less than 2 meters, but the devices are sometimes installed in series of three or four, thus raising the water 6 meters or more. With the shaduf one man can raise approximately 1 liter per second, about 2 meters.

14.4 MODERN IRRIGATION PUMPING

In contrast to the primitive methods of pumping water for irrigation, pumping machinery of high efficiency is used on many irrigated farms. Substantial advancement has been made in the design and operation of pumps. Costs are greatly reduced by obtaining the necessary energy for pumping from fuels, rather than using the energy of men or animals.

For example, assume that for irrigation pumping, 1 kilowatthour of electric energy may be purchased at a cost of $0.02. A strong healthy man, in an hour, can generate less than ⅛ kilowatthour. At the rate of $2.00 per hour for man-labor, the cost of a man working 8 hours to generate 1 kilowatthour would be $16.00 as compared to $0.02 for electricity. Hence, manpower would cost about 800 times as much as electrical energy to do the same pumping. If the man is only paid $2.00 per day, manpower would cost 100 times as much as electrical energy.

Modern irrigation pumping methods are based on years of painstaking laboratory research, together with careful study of field pumping conditions by competent engineers. Out of these investigations there have come into use pumps of many different classes and types, each suited to different conditions of operation.

14.5 PUMP CHARACTERISTICS

In order to use modern pumps most profitably to obtain irrigation water, it is essential to select pumps well adapted to the particular conditions of operation and to obtain a relatively high efficiency. If the quantity of water pumped is appreciably less than the quantity for which the pump is designed, and the head is excessive, a low efficiency results. Likewise, a pump may deliver more water than it is designed to deliver at a head lower than normal and cause the efficiency to be low. The interrelations between speed, head, discharge, and horsepower of a pump are usually represented by curves which are designated "characteristic" curves, Fig. 14.2. Knowledge of pump characteristics enables one to select a pump which fits operating condi-

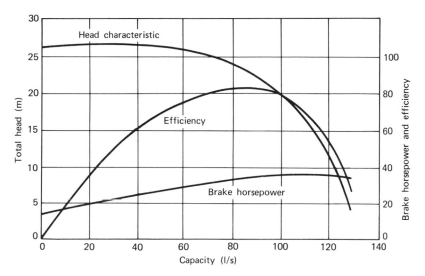

Figure 14.2. Characteristic pump performance curves for a particular pump.

tions and thus attain a relatively high efficiency with low operating cost.

The head-capacity curve shows how much water a given pump will deliver with a given head. As the discharge increases, the head decreases. The resulting efficiency is observed in Fig. 14.2 to increase from 0 when the discharge is 0 to a maximum of 82 percent when the discharge is 86 liters per second and the head is 23 meters (and then decreases to 0 at 0 head). The brake horsepower curve for a centrifugal pump usually increases over most of the range as the discharge increases, reaching a peak at a somewhat higher rate of discharge than that which produces maximum efficiency. The curves shown in Fig. 14.2 vary with the speed of the pump. Therefore speed must be considered when selecting a pump to secure maximum efficiency. Each of the curves shown in Fig. 14.2 varies with the type of pump.

14.6 TYPES OF PUMPS

Pumps for irrigation purposes are of many different makes. In general they range from pumps with small discharge and high heads to large discharges with low heads. Small-discharge and high-head pumps are centrifugal pumps used for sprinkler irrigation and where water is lifted considerable distances. Pumps with large discharge and low heads are used in drainage and where large quantities of water are lifted only a few meters. Variation in pump characteristics are shown in Fig. 14.3.

Specific speed $N_s = \dfrac{rpm\sqrt{gpm}}{H^{3/4}}$		Cross section	Type of pump	Head–discharge characteristics
(a)	500		Centrifugal (Radial flow)	High head Small discharge
(b)	1000			
(c)	2000		Francis	Intermediate head and discharge
(d)	3000			
(e)	5000		Mixed flow	
(f)	10000		Propeller flow (axial flow)	Low head Large discharge

Figure 14.3. Variation in impeller design, specific speed, and head-discharge characteristics of pumps.

Pump Characteristics

The characteristics of impeller-type pumps can be summarized by illustrating the relationship between discharge Q, speed N, diameter D, width of impeller W, head H, and power P with subscript zero referring to original conditions.

Centrifugal Pumps

Water is drawn into a centrifugal pump axially and leaves radially as shown in Fig. 14.3 (a). Centrifugal pumps are built on both horizontal and vertical shafts. Horizontal-shaft centrifugal pumps have the advantages of being efficient, simply constructed, relatively free of trouble, low cost, easy to install, and capable of high speeds; therefore, they are usually connected directly to electric motors. One of their principal disadvantages is their limited suction lift and susceptibility to losing prime.

The practical suction lift will decrease with elevation, being not over 6 meters at sea level and generally less than 4.5 meters at a 1500 meter elevation. If pumping seals are in excellent condition the lift can be greater, but it is not wise to design for higher lifts unless the pump can be maintained in very good condition. Because of the priming problem and limited lift, it is usually advantageous to set pumps as near the water surface as is convenient, yet protect them from submergence during high water. It is especially important to avoid submerging electric motors.

Before a horizontal-shaft centrifugal pump can be started, it is necessary to fill the suction pipe and pump case with water and thus expel all air. This operation of filling suction and pump case is designated "priming the pump." Priming is essential on a horizontal-shaft centrifugal pump because of the nonpositive action of the impeller. These pumps must be primed before they will lift water from the source of supply. Priming can be accomplished by adding water to the intake line in the pump from an outside source, such as a storage tank. In this case a foot valve on the end of the intake pipe is essential to hold the water in the intake line while it is being filled. A good foot valve will also hold water in the line so that priming will not be necessary each time the motor is started. Priming can also be accomplished by sucking out the air with various devices, thereby lowering the pressure in the line causing the water to rise from the source to the pump by atmospheric pressure. Hand primers, exhaust primers, manifold primers, and dry vacuum pumps are used to remove air when priming a pump.

Deep-Well Turbine Pumps

When the impeller is suspended vertically on the drive shaft within a long discharge pipe, the pump is termed a deep-well turbine unit. The impeller may be a centrifugal unit, an axial-flow unit, or any design between these extremes, depending upon the desired head-discharge relationship. The bowl of the pump houses the impeller and guide vanes. When several bowls are connected in series to obtain the desired total head, the pump is referred to as a multiple-stage pump. These bowl assemblies are nearly all located beneath the water surface.

High efficiency is dependent upon a close fit of the impeller against the pump housing. Adjustments in position are made by raising or lowering the impeller shaft. Many deep-well pump heads are equipped with a clutch or ratchet to prevent the pump from turning in the wrong direction because of the backflow of water when the power is shut off, which may unscrew the fittings.

Deep-well turbine pumps are used for irrigation when the water surface is below the practical lift of centrifugal pumps. Successful installations have been made when water is as much as 300 meters below the ground surface.

The pump is driven by an electric motor or other source of power located at the ground surface and connected by a long vertical shaft held in position by bearings built in the discharge pipe or column. Being submerged, the deep-well pumps have the advantage of requiring no priming and of meeting rather wide fluctuations of water surface without having to reset the pumps. They have the disadvantage of the operating parts being inaccessible and consequently difficult to inspect.

Low efficiency is more common in vertical-shaft pumps than in horizontal-shaft pumps. This is because deep-well pumps are frequently permitted to continue running after bearings are worn badly, and sometimes until the pumps fail to operate, before repairs are provided. Figure 14.4 shows a typical section of a deep-well turbine pump.

Submersible Turbine Pumps

A deep-well turbine pump close-coupled to a small diameter submersible electrical motor is termed a submersible pump. The characteristics of the pump unit are similar to the conventional deep-well submersible pump having the same flexibilities and design possibilities. Efficiency is increased by direct coupling and effective cooling resulting from complete immersion which permits a reduction in amount of iron and copper in the core. Submersible pumps have been used in wells over 4000 meters deep. Units with more than 250 stages have been used. Submersible motors as large as 250 hp are in use in 20-centimeter casings, and larger motors are available for larger casing sizes. Pumps are available for 10-centimeter and larger well casings.

The principal advantage of submersible pumps is that they can be used in very deep wells where long shafts would not be practical. These pumps are also excellent in crooked wells where the ordinary deep-well turbine might not operate. Bending in the pump column of a conventional turbine will cause the shaft to crystallize and rupture. Submersible pumps are also useful where the installation may be flooded or where an above-ground pump house would be inconvenient, unsightly, or hazardous.

Propeller and Mixed-Flow Pumps

Propeller (axial-flow) and mixed-flow pumps are used for low-head high-discharge operation. By varying the vane pitch and curvature of the impeller, a wide range of head-capacity requirements can be met without changing the impeller diameter. Multiple stages may be used to obtain a higher head.

The impeller of the pump should have sufficient submergence to minimize cavitation. The impeller must be deep enough below the water surface to minimize the formation of vapor on the

Construction
Berkeley vertical turbine pumps
Enclosed column   Open column

Driver

Oil reservoir

Rotary seal optional

Stuffing box assembly

Automatic sight feed oiler and solenoid valve

Head shaft bearing

Discharge head

Shaft bearing and inner column coupling

Tension nut assembly

Discharge head

Shaft bearing spider

Shaft sleeve coupling

Outer column coupling

Outer column coupling

Inner column

Shaft bearing

Shaft coupling

Outer column

Shaft

Outer column

Shaft

Inner column seal

Top case bearing

Top case

Weep holes

Top case

Top case bearing

Bowl shaft

Bowl

Bowl

Choice of impellers
Semi-open impellers   Enclosed impellers

Taper sleeve

Taper sleeve

Enclosed impeller

Open impeller

Bottom case

Bottom case

Bottom case bearing

Bottom case bearing

Figure 14.4. Vertical turbine pump showing main features. (Courtesy Berkely Pump Co.)

blades at the point of maximum local velocity. The formation and rapid collapse of vapor bubbles occurring several times a second causes a fluctuating local pressure which results in metal fatigue with serious pitting of the blade surface and reduced efficiency. Submergence depths, as well as clearance depths from the bottom of the pump and between adjacent units, as recommended by the manufacturers, should not be ignored.

A tendency of an axial-flow pump to overload makes it inadvisable to use this type of pump when it is necessary to throttle the discharge to secure reduced delivery. Also, the overloading tendency at increased heads makes it important to select a motor which will provide ample power under all pumping conditions.

Propeller pumps have a low initial cost. Each stage is generally limited to less than 3 meters of lift. Additional stages may be added to increase the head to 10 meters or more. For higher heads a mixed-flow pump should be used, since single stages could economically lift up to 8 meters and multiple stages can be used for up to 40 meters.

14.7 FRICTION LOSSES IN PUMP SYSTEMS

As water is lifted by a pump from one elevation to another, several factors must be considered when determining horsepower required. The head loss due to friction in each component of the pumping and pipe system must be added to the total vertical height water is lifted to the point of delivery. These sources of loss include the foot valve, intake pipe, pump, discharge pipe, and all intermediate valves, elbows, and other fittings. Figure 14.5 lists the loss of head due to friction in elbows, valves, and other fittings in terms of the equivalent length of pipe. The combined equivalent pipe losses due to fittings can then be added to the losses in the straight section of the pipe to obtain the total equivalent length of pipe. With this total loss of head, Table 14.2 then gives the total head required in the entire system due to change of elevation and fittings. Table 14.2 is for pipe 15 years old. Since the system will be operating for many years, it is wise to figure the power requirement on the average, or longer, life to ensure reasonably adequate power during the entire period of operation of the irrigation system.

The following example will illustrate the use of Fig. 14.5 and Table 14.2 for estimating total horsepower requirements. Assume for simplicity, that the entire pumping system is composed of 15.2 centimeter-diameter pipe and fittings; assume further that the discharge is 25 liters per second with a static life of 30 meters and a pump efficiency of 70 percent.

1. Foot valve, assuming equivalent to an open globe valve   52 m
2. Gate valve, fully open   1 m
3. Three 90° medium sweep elbows   13 m
4. Length of 15-centimeter pipe in intake and discharge lines   40 m
   Total equivalent length of pipe   106 m

For sudden enlargements and sudden contractions the equivalent length is in feet of pipe of the smaller diameter, d. The dashed line shows the determination of the equivalent length of a 6-in. standard elbow.

Figure 14.5. Minor friction losses caused by elbows, valves, and other fittings. (Courtesy Oregon Agr. Exp. Sta.)

TABLE 14.2   Loss of Head, in m, Due to Friction per 1000 m of 15-Year-Old Ordinary Steel Pipe[a]

$$V = CR^{.63}S^{.54} \text{ where } c = 100$$

Rated Flow (m³/hr)	l/s	¾ in. (1.90 cm)	1 in. (2.54 cm)	2 in. (5.08 cm)	2½ in. (6.35 cm)	3 in. (7.62 cm)	4 in. (10.61 cm)	5 in. (12.70 cm)	6 in. (15.24 cm)	8 in. (20.32 cm)	10 in. (25.40 cm)
						Nominal Diameter of Pipe					
0.25	0.069	10	2.4								
0.5	0.139	36	9								
0.75	0.208	82	18								
1.00	0.278	145	33	1.2	0.4						
1.5	0.417	300	68	2.5	0.8						
2	0.556		110	4.0	1.5	0.6	0.1				
3	0.834			8.9	3.0	1.2	0.3				
4	1.11			15	5.0	2.0	0.5	0.2			
5	1.39			22	7.5	3.0	0.7	0.3			
6	1.67			31	10	4.2	1.0	0.4			
8	1.95			54	18	7.3	1.8	0.6	0.3		
10	2.78			80	27	11	2.7	0.9	0.4		
12	3.34			110	32	15	3.8	1.1	0.5		
14	3.89			150	38	20	5.0	1.3	0.7	0.2	
16	4.45			190	65	26	6.5	2.2	0.9	0.2	
18	5.00			240	80	32	8.0	2.7	1.1	0.3	
20	5.56			265	95	38	9.0	3.2	1.3	0.3	
25	6.95			430	150	60	14	5.0	2.0	0.5	0.2
30	8.34					80	20	7.0	2.8	0.7	0.2
35	9.73					108	27	9.5	3.7	1.0	0.3
40	11.1						34	12	4.8	1.2	0.4
45	12.5						43	14	5.8	1.5	0.5
50	13.9						50	18	7.2	1.8	0.6
55	15.3						60	22	8.8	2.4	0.8
60	16.7						70	25	10	2.6	0.8
65	18.1						84	30	12	3.0	1.0
70	19.5						98	34	13	3.4	1.2
80	22.2							42	17	4.3	1.5
90	25.0							54	22	5.5	1.9
100	27.8							64	26	6.6	2.3

Table continues on next page

Loss of head in meters (25 liters per second through 15-centimeter pipe results in 22 meters loss per 1000 meters given in Table 14.2), 22 × 0.106 = 2.3 m.

Total dynamic head = static lift of 30 m + friction loss of 2.3 m = 32.3 m.

Use 33 meters to allow for losses in entrance and exit.

Required power output at a pump which is also referred to as the water horsepower is found by Equation 14.2 to be:

$$\text{WHP} = \frac{Qh}{76} = \frac{25 \times 33}{76} = 10.9$$

The required power input to the pump, referred to as brake

TABLE 14.2   Loss of Head, in m, Due to Friction per 100 m of 15-Year-Old Ordinary Steel Pipe (continued)

Rated Flow (m³/hr)	l/s	¾ in. (1.90 cm)	1 in. (2.54 cm)	2 in. (5.08 cm)	2½ in. (6.35 cm)	3 in. (7.62 cm)	4 in. (10.61 cm)	5 in. (12.70 cm)	6 in. (15.24 cm)	8 in. (20.32 cm)	10 in. (25.40 cm)
						Nominal Diameter of Pipe					
110	30.6							76	30	8.0	2.7
120	33.4							90	36	9.4	3.0
130	36.2							103	42	11	3.7
140	38.9							120	49	12	4.2
150	41.7							135	55	14	4.8
160	44.5								64	16	5.4
170	47.2								70	18	6.0
180	50.0								80	20	6.6
190	52.8								88	22	7.5
200	55.6								95	24	8.0
225	62.5									30	10
250	69.5									34	12
275	76.5									43	15
300	83.4									50	17
350	97.3									68	22
400	101									88	29
450	125									110	36
500	139									130	44
600	167										60
700	195										80
800	222										103
900	250										130

[a] Pipe coefficients: The values of friction given in this table are for commercial wrought iron or cast iron pipe of 15 years' service when handling soft clear water. In order to be able to use that table for other classes of pipe, the values taken from the table should be multiplied by a coefficient, selected from the list below, to correspond to the required condition:

New smooth brass and steel pipe (Durand) coefficient	=0.6
New smooth iron pipe (Williams & Hazen C = 120)	=0.7
15-year-old ordinary pipe (Williams & Hazen C = 100)	=1.0
25-year-old ordinary pipe (Williams & Hazen C = 90)	=1.2
Portable aluminum with couplers (Williams & Hazen C = 120)	=0.7

Source: Based on original data by Williams and Hazen. Courtesy Goulds Pumps, Inc. and published in Bulletin 481, Oregon Agricultural Experiment Station.

horsepower, is the water horsepower divided by the efficiency of the pump, Equation 14.3, assumed to be 70 percent:

$$BHP = \frac{WHP}{E_p} = \frac{10.9}{0.7} = 15.6$$

14.8 SELECTION OF POWER PLANT

The power plant must be capable of delivering the required power under varying conditions. The variations of operating conditions which must be accommodated by the power unit will influence the choice of unit. Features to be considered in selecting a power unit include:

1. Brake horsepower required.

2. Initial cost.

3. Availability and cost of energy or fuel.

4. Depreciation.

5. Dependability of unit.

6. Portability required.

7. Maintenance and convenience of operation.

8. Labor availability and quality.

In all cases it is good insurance to obtain a power plant from a reputable dealer. Occasionally old power units are used to reduce the initial cost. Frequently, the very low efficiencies, which result in very high fuel costs, more than offset low initial cost. Make allowance in the design for local conditions that will reduce the horsepower rating given by the manufacturer. The operator should carefully follow recommendations of the manufacturer.

14.9 INTERNAL COMBUSTION ENGINES

Two general types of internal combustion engines are used: (1) spark ignition and (2) compression ignition, better known as diesel. Ignition systems generally use gasoline, LP gas, and natural gas fuels while the diesel units use the well-known diesel fuel. Since the initial cost of a diesel engine is much higher than for spark ignition, it is necessary that diesel units be used many more hours per season than are normally required for spark ignition units to be economically feasible. Generally, spark ignition power units or single-phase electrical units are used for horsepower up to 7½. Spark ignition units and three-phase electricity are practical between 7½ and 40 hp; and above 40 hp, diesel units generally become more practical. Natural gas is competitive with diesel fuel in many areas for irrigation pumping.

14.10 ELECTRIC MOTORS

Electric motors which are properly selected, installed, and mounted may give many years of trouble-free power. Advantages of electrical motors are relatively long life, dependability, ease of operation, and low maintenance cost. In some areas frequent power interruptions are a major disadvantage associated with electrical motors. The necessity to construct power lines to each power site and the cost of transformers at pumping locations are disadvantages. Lightning striking an electrical installation can cause serious loss of time and money. Variation in line voltage will cause motors to burn out.

Single-phase motors are not usually feasible for loads of more than 7½ hp. In some areas three-phase power is not readily available. The best electric motor for irrigation pumping is the squirrel-cage induction motor operating on 60 cycle, 220 to 440 volt, three-phase energy. The most common speed is 1760 rpm. Direct connection between motor and pump should be used whenever possible for minimum cost and maximum efficiency.

14.11 ELECTRIC SERVICE SCHEDULES AND COSTS

Power companies usually base charges for electric service in part on the maximum demand of the consumer, regardless of the energy consumed, and in part on the energy actually used. Power companies encourage consumers to avoid high demands for electricity over short periods of time and to strive to use power as many hours per day and days per month as conditions may justify. Consideration of a typical power-rate schedule will clarify the low energy costs of low demand and continuous use, as compared to high energy costs resulting from high demand and short-time use.

A sample irrigation service schedule provides a monthly power charge of $2.90 per kilowatt for the first 100 kilowatt of demand and $1.90 per kilowatt for all additional kilowatts of demand. The monthly charge for kilowatt-hours used by the motor is then added to the monthly demand charge, on the following step rate:

2.8¢ per kWh first 100 kWh per kW of demand

1.7¢ per kWh next 5000 kWh

1.1¢ per kWh next 20,000 kWh

0.8¢ per kWh all additional kWh

Most irrigation installations are of a permanent nature. Usually, the irrigator signs a 10-year contract for service which entitles him to a 5 percent term discount on his monthly power bill. Also, when the irrigator makes his original permanent installation, he includes the transformers with his pump and motor installation, which further entitles him to a discount on his monthly power bill, as follows:

40¢ per kW for first 100 kW of demand, and

20¢ per kW for all additional kW of demand

Example

For a 200-hp motor, if it runs 40 hours (using a measured demand of 160 kilowatts), the monthly charge would be:

Demand:
100 kW at $2.90 per kW = $290.00
60 kW at $1.90 per kW = 114.00
Energy:
40 hr × kW at 2.8¢ per kWh = 179.20
Gross charge $583.20
Less term discount, 5% 29.16
Less voltage discount
40¢ for first 100 kW 40.00
20¢ for next 60 kW 12.00
Net monthly charge $502.04
Average cost per kWh = $502.04/(40 × 160) = 7.8¢

If the irrigator runs the same 200-hp motor with a measured demand of 160 kW continuously for 30 days, 24 hours each day, then 720-hour use equals 115,200 kWh, and his monthly charge would be:

Demand:
100 kW at $2.90 per kW = $ 290.00
60 kW at $1.90 per kW = 114.00
Energy:
16,000 kWh at 2.8¢ per kWh = 448.00
5,000 kWh at 1.7¢ per kWh = 85.00
20,000 kWh at 1.1¢ per kWh = 220.00
74,200 kWh at 0.8¢ per kWh = 593.60
Gross charge $1750.60
Less term discount, 5% 87.53
Less voltage discount
40¢ for first 100 kW 40.00
20¢ for next 60 kW 12.00
Net monthly charge $1611.07
Average cost per kWh = $1611.07/115,200 = 1.4¢

Thus by operating his 200-hp motor 720 hours per month, instead of only 40 hours, the irrigator reduces the unit cost of electricity from cents to 1.4 cent per kilowatt/hour, or a reduction of 82 percent.

A farmer cannot as a rule use irrigation water continuously to advantage. The advantages in the use of large streams obtained by large motors and pumps partly compensate the irrigator for higher costs for electricity. On the other hand, it is frequently advantageous, where electricity is used for pumping, to provide small reservoirs in which to store the water during the night, thus making it possible to irrigate with a stream approximately twice the size of the pump discharge.

14.12 IRRIGATION PUMPING COSTS

In order to estimate costs of irrigation water obtained by pumping, and to compare these costs with costs of water from gravity systems, it is customary to compute all pumping costs in terms of the volume of water delivered annually to the irrigated farm. The factors that determine the annual costs of pumped water are:

a. Interest on capital investment, i.e., on first cost.

b. Taxes.

c. Depreciation on pumping machinery and on housing.

d. Fuel or power and lubricating oils.

e. Attendance.

A well is 36 centimeters in diameter and 43 meters deep, the depth of the water table being 17 meters. While pumping 70 liters per second, the lift is 23 meters. A turbine pump is driven by a 45-hp, high-speed diesel engine. The costs during the irrigation season for pumping 65 hectare-meters during the season are presented in Table 14.3.

It will be noted from Table 14.3 that operating costs only make up 22 percent of the total cost of $314 per hectare-meter of water.

14.13 EFFICIENCY OF PUMPING PLANTS

Irrigation pumping plants invariably operate at considerably less than peak efficiency. Several reasons for low efficiency exist, such as the pipe being too small or having many bends. The most common error is discharging water considerably above the necessary level. The pump may also be worn, and frequently the bowls are not set properly in turbine-type pumps. The drive or coupling between pump and motor may not be an efficient unit. Too frequently a pump or motor is not the best pump or motor for the job to be done. Correct matching of pump, motor, and drive is important. For a new electric motor and pump, an

TABLE 14.3   Cost of Operating a Modern Pumping Plant

Cost of Plant:		
Well	$3920.00	
Pump	4910.00	
Engine (installed)	4400.00	
Shelter	450.00	
		$13480.00
Annual Fixed Costs:		
Interest on $13480.00 @ 9%	1213.20	
Taxes (estimated)	322.00	
Depreciation on pump @ 7%	343.70	
Depreciation on engine @ 7%	308.00	
Depreciation on well and shelter @ 4%	174.80	
		$ 2361.70
Annual Operating Costs: (2580-hr operation/year)		
Fuel 27,000 liters @ 14¢/liter	3780.00	
Lubrication @ $6.70/100 hr	172.86	
Sales tax on above @ 4%	158.11	
Repairs @ $15.70/100 hr	405.00	
Attendance @ $8.30/100 hr	214.14	
		$ 4572.06
		$20413.76
Total cost per ha-m	$314	
Operating cost per ha-m	$ 70	

over-all efficiency of about 65 percent should be expected. When efficiency drops to 35 to 40 percent, the pump should be repaired or replaced.

Efficiency will also be reduced by elevation, heat, accessories, and continuous operation. Hence, industrial-type engines used for irrigation are derated from the maximum rating given by the manufacturer in order to determine safe continuous operating horsepower available for pumping. Following are some of the factors which must be considered:

	Decrease in efficiency(%)
1. For each 300 m above sea level	3
2. For each 3° operating air temperature above 18°C	1
3. For accessories, using heat exchangers	5
4. Radiator, fan	5
5. For continuous load operation	20
6. Drive losses	0–15

Hence, if these losses total 40 percent and the water horsepower energy imparted to the water including pump and conveyance

TABLE 14.4   Performance Standards for New Deep-Well Pumping Plants

Energy Source	Rated Load for Representative Power Units (hp-hr/l)	Performance Standard (WHP-hr/l)[a]
Diesel	3.90[b]	2.92
Gasoline	2.98[b]	2.24
Tractor fuel	2.92[b]	2.20
Propane	2.36[b]	1.76
Natural gas	28.92[c]/1000 liters	21.68/1000 liters
Electric	88% efficient	0.885[d]/kWh

[a]Based on pump efficiency of 75%.
[b]Taken from Test D of Nebraska Tractor Test Reports. Drive loss for flat belt is included.
[c]Manufacturer's data corrected for 5% drive loss.
[d]Not corrected for drive loss. Assume a direct connection.
Source: Paul E. Schleusener, J. J. Sulck, and J. F. Schrunk, "How Efficient is Your Pumping Plant?" Exp. Sta. Qrtrly., Lincoln, Nebraska, Spring, 1955. Courtesy Nebraska Experiment Station.

losses are 30 hp, a unit rated at $50 = 30/(1.0 - 0.4)$ brake horsepower would have to be selected.

A good measure of the efficiency of a pumping-plant is the fuel consumed by the plant. Standards shown in Tables 14.4 and 14.5 have been developed by the Nebraska Experiment Station.[1] Standards of common fuels and electricity are shown in Table 14.4. Fuels listed are mixtures of hydrocarbons which may vary in composition. However, this tabulation is a good relative guide to what can be expected from an efficient pumping plant. Table 14.5 lists fuel requirements for various pumping conditions which can readily be compared to fuel consumed by a given pumping plant. Table 14.6 converts any discharge pressure to an equivalent lift which must be added to the distance the water must be lifted while pumping in order to obtain the correct fuel requirement from Table 14.5

By installing high-compression pistons or cylinder heads, a portion of the loss due to altitude can be regained. If air temperatures over 40°C are encountered, provision should be made for cooling the intake manifold. Fuel consumption per brake horsepower-hour increases if the engine is not fully loaded. If the engine is loaded to a small portion of the rated horsepower for

[1]Paul E. Schleusener, J. J. Sulek, and J. F. Schrunk, "How Efficient is Your Pumping Plant?" Exp. Sta. Qtrly., Lincoln, Nebraska, Spring 1955.

TABLE 14.5   Maximum Fuel Requirements for a Good Pumping Plant[a]

Flow of Water (m³/hr)	Lift (m)[b]	Water Horse-power	Propane	Diesel	Gasoline or Tractor Fuel	Natural Gas	Elec-tricity[c]
					Fuel Required (l/hr)		
	20	7.5	4.2	2.7	3.5	350	8.5
100	50	18.5	10.5	6.2	8.5	860	21
	70	26.0	14.7	9.0	11.7	1200	29
	20	11	6.2	3.7	5.2	510	12.5
150	50	28	15.7	9.5	13.0	1290	32
	70	39	22.0	13.5	18.2	1800	44
	20	15	8.5	5.2	6.7	690	17
200	50	37	21.0	12.5	16.5	1710	42
	70	52	29.5	17.7	23.5	2400	59
	70	19	10.7	6.5	8.5	880	22
250	50	46.5	26.5	16.0	21.0	2150	53
	70	65	36.7	22.2	20.2	3000	73

[a] Based on standards in Table 14.4.
[b] If you have a pressure at the pump discharge, add the number of meters shown in Table 14.6 to the distance the water must be lifted while pumping.
[c] Kilowatthours per hour.
Source: Paul E. Schleusener, J. J. Sulck, and J. F. Schrunk, "How Efficient is Your Pumping Plant?" *Exp. Sta. Qtrly.,* Lincoln, Nebraska, Spring, 1955. Courtesy Nebraska Experiment Station.

TABLE 14.6   Relationship Between Discharge Pressure and Lift

Discharge Pressure (atm)	Equivalent "Lift" (m)
1	10.3
2	20.7
3	31.0
4	41.4
5	51.7
6	62.0
7	72.4

Source: Paul E. Schleusener, J. J. Sulck, and J. F. Schrunk, "How Efficient is Your Pumping Plant?" *Exp. Sta. Qtrly.,* Lincoln, Nebraska, Spring, 1955. Courtesy Nebraska Experiment Station.

the speed, the fuel consumption will be considerably increased. Under these conditions it will be more economical to reduce the engine speed until the proper horsepower is developed. By changing the pulley drive the proper pump speed can be obtained.

CHAPTER
15
DRAINAGE OF IRRIGATED LANDS

Adequate drainage of crop-producing lands requires a general lowering of shallow water tables. Some students, especially in humid regions, find it difficult to think of drainage as being essential in arid regions where irrigation is required for crop production. Experience has demonstrated fully the need for drainage of irrigated lands. In some valleys the higher lands never require drainage, but the need for drainage of the lower valley lands is frequently a result of the irrigation of the higher lands. From 20 to 30 percent of the irrigated lands in arid regions need drainage to perpetuate their productivity. The reclamation of saline and alkali soils has many important phases, but adequate lowering of the water table by drainage is a first and basic necessity.

Irrigation and drainage in arid regions are complementary practices, the necessity for drainage being increased by low efficiencies in the conveyance and application of irrigation water.

In humid regions drainage is of even greater necessity than in arid regions. Excess rainfall produces bogs in low-lying flat areas. In arid regions drainage usually follows irrigation. In humid regions drainage frequently must precede agricultural development and often must precede human habitation. And yet frequently the drained lands are the most productive agricultural lands. The challenge is to correctly evaluate the potential usefulness of the land, to design and install the most economical drainage on those lands justifying drainage, and to maintain a workable drainage system so that maximum and economic benefits will result.

15.1 BENEFITS OF DRAINAGE

Adequate drainage improves soil structure and increases and perpetuates the productivity of soils. Drainage is the first essential in reclamation of waterlogged saline and alkali soils. Even if only those lands which have been farmed are considered, drainage benefits irrigation agriculture and the public in many ways.

For example, adequate drainage (1) facilitates early plowing and planting, (2) lengthens the crop-growing season, (3) provides more available soil moisture and plant food by increasing the depth of root-zone soil, (4) helps in soil ventilation, (5) decreases soil erosion and gullying, by increasing water infiltration into soils, (6) favors growth of soil bacteria, (7) leaches excess salts from soil, and (8) assures higher soil temperatures.

Drainage also improves sanitary and health conditions and makes rural life more attractive.

15.2 SOURCES OF EXCESS WATER

The major sources of excess water that make drainage necessary on parts of the irrigated lands are seepage losses from reservoirs or canals and deep percolation losses from irrigated lands. Efficient water application on the higher irrigated lands reduces the need for drainage of the lower lands. Flooding of low lands due to overflow of rivers and natural drainage channels during periods of maximum streamflow constitute important sources of excess water in certain low-valley areas. The flow of ground water toward waterlogged lands in arid regions may be in any direction. In some areas flow is largely downward through highly permeable surface soils to relatively impermeable subsoils.

In other areas unconfined or free ground water may flow under small hydraulic slopes. In still other areas the major source of excess water may be upward flow from an artesian aquifer. Two or more of these possible sources of excess water contribute to the maintenance of shallow water tables in some soils. Thorough ground-water investigations and subsoil studies are essential to intelligent design of drainage systems. Small-diameter pipes used as piezometers, and auger holes, described in Chapter 5, can be used to determine the depth to the water table and the direction of movement of the ground water. Permeability measurements are very helpful because of the great range in soil permeability. The soil permeability is the dominant variable influencing the feasibility and the cost of drainage.

Gravelly and sandy soils, under natural conditions, are from 25,000 to 50,000 times as permeable as clay soils; and in some drainage studies permeability ratios of 100,000 to 1 have been measured. Subsoil formations and permeabilities thus influence sources of excess water in soils.

However, care should be exercised in interpreting the capacity of soils to transport water. Clays may be well aggregated and may be quite permeable. Within the basin of the Great Salt Lake, the very fine-textured clays often transmit large quantities of water readily through cracks and fissures, apparently formed shortly after the clay was deposited.

15.3 CONTROL OF WATER SOURCES

In some areas, substantial progress toward a general lowering of the water table, and the solution of drainage problems, can be made by control of excess water sources. For example, as stated in Chapter 12, nearly 40 percent of the water taken into canals is lost during conveyance. This seepage water reaches the water table and causes it to rise. Lining canals to reduce these losses should be encouraged.

Control of the source of excess water by percolation from the irrigated land also is a difficult and perplexing task. In a few areas, there is no excess salinity or alkali in the soil, and lowering of the water table is all that is necessary to solve the drainage problem. Lining of the irrigation canals to prevent seepage losses and more efficient application of irrigation water to reduce or eliminate deep-percolation water losses may result in a satisfactory lowering of the water table in these areas, thus removing the need for drainage. In other areas control of excess-water sources may be both impractical and prohibitive in cost, thus making drainage essential. In most arid regions waterlogged soils, leaching of soluble salts from water-logged soils is essential to crop production, and adequate lowering of the water table by artificial drainage must precede leaching.

15.4 REQUIRED WATER-TABLE DEPTHS

Adequate crop production and perpetuation of soil ferility in irrigated areas generally require water-table depths of 1.5 meters or more. In many cases, even where drainage systems have been installed, the water table during part of the year may be less than 1 meter. A summary of required water-table depths approved by irrigation authorities and also by financial institutions that are interested in long-time loans for improving of irrigated land follows.

Classification	Range in Water-Table Depth Range in Water-Table Depths
Good	Static water table below 1.2 m may rise to 1.8 m for a period of about 30 days of the year.
Fair	Water table at 1.8 m may rise to about 1.2 m for a period of 30 days. No general rise.

Classification	**Range in Water-Table Depths, continued** Range in Water-Table Depths
Poor	Some alkali on surface; water table 1.2 to 1.8 m, may rise to 0.9 m for a period of 30 days.
Bad	Water table less than 1.2 m and rising. Natural and artificial drains too far away to drain land.

Generous long-time loans are made by land banks and other lending agencies when good lands are mortgaged as security; restricted or short-time loans are made on fair lands, and no loans usually are made on poor and bad lands. Positive assurance of immediate and adequate lowering of the water table by thorough drainage may induce financial agencies to make limited loans on bad lands.

The above standards are applicable to large segments of irrigated agriculture and are accepted by agricultural authorities. However, proper management can produce good and even excellent results in farming land with a water-table depth between 1 to 2 meters. Special care is required, and careful application of irrigation water is essential. Sprinkler systems are well adapted to controlled applications. At least one adequate leaching irrigation must be applied each year, preferably during the dormant season when roots of perennial plants will not be seriously affected. When no crop is growing on the land, a heavier leaching irrigation should be applied at the beginning of the dormant season so that the water table will have the maximum time to return to its normal position before crops are planted in the spring.

In irrigated regions where irrigation water and application costs are high, another benefit will accrue from a water table higher than would normally be thought advisable. Crops may obtain a significant portion of their needed water from the saturated soils. As much as 50 percent can come from ground water without seriously reducing the rooting depth. Saving the cost of applying 20 to 50 percent of the normally needed irrigation water may more than offset the reduction in yield resulting from a somewhat higher water table.

Also by exercising care it may be possible to avoid the cost of drainage and still not seriously restrict the agricultural production. In general the water table, including the capillary fringe which is essentially saturated, should not be allowed to occupy more than the lower one-third of the normal root zone of the crop. Since rooting depths vary with crops and vary during the season, no definite depth can be recommended. Furthermore, hazards are present and careful management is essential. Nevertheless, shallower water-table depths are economically feasible under excellent management.

15.5 LOWERING THE WATER TABLE

In addition to eliminating or controlling sources of excess water, improving natural drainage facilities and providing man-made drainage systems will also be of substantial aid in lowering the water table.

Proper maintenance of natural drainage systems, usually feasible at low costs, protects irrigated lands from excess percolation of water from rains and melting snows and also from flood damages.

In many arid regions artificial drainage also is required to provide adequate lowering of the ground-water table and is accomplished by one of three methods: (a) open channel drains; (b) covered clay or concrete pipe; and (c) pumping ground water.

In open-channel and tile drains, gravity pulls the excess water from the wet soils into the drains, and gravity also causes the flow in the drains. When gravity provides the mechanical power, the drains are designated gravity drains. Pumping of ground water in some valleys in arid regions provides for both irrigation and drainage.

15.6 FIELD INVESTIGATIONS

The four elements of primary concern in making a field investigation for drainage are topography, soil, water table, and water source. Each must be studied carefully and interpreted wisely. Existing data should be assembled and reviewed before proceeding with detailed field studies. The complexity and thoroughness of field investigation will depend upon the complexity of the drainage problems and the extent and usefulness of existing data.

The topography should be studied first by a visual field inspection. Aerial maps viewed under a stereoscope will frequently yield valuable information. Unless the problem can be solved readily, a topographic survey should be made. The accuracy needed will depend upon the problem and should be adjusted to give greater detail in critical areas.

The principal objective of the soils investigation is to ascertain the character of the soil profile and how it varies from location to location. Field investigations of the soil begin after reviewing existing data. Permeability is the major soil property of concern in a drainage study. Soil samples can be obtained and observa-

SOIL PROFILE CHART
FOR DRAINAGE INVESTIGATIONS

1—Clay
2—Silty clay loam
3—Clay loam
4—Silt loam
5—Loam
6—Sandy loam
7—Loamy fine sand
8—Sand
9—Coarse sand and fine gravel
10—Coarse gravel
11—Cobble

Property _John Jones_ ——————— Boring No. _A-1_
Physiography _Bench_ ———————— Relief _0.003 Slope_
Location _15-21-31_ — Elevation _3300_ —— Soil Symbol _1H34T_
Land Use _Pasture_ ————————— Crop Condition _Poor_
Technician _Y.B.B._ ————————— Date _3/1/55_

DEPTH IN FEET	TEXTURE	STRUC-TURE	SPECIAL FEATURES	HYDRAULIC CONDUCTIVITY INDEX						
				SLOW 1 2	MOD. 3 4	RAPID 5 6 7	VERY RAPID 8 9 10 11			
0.0	Cl	gr	pH 9.6							
			wet							
			pH 8.6							
			pH 8.0							
1.0										
	1	m								
2.0			pH 8.0 Free water							
3.0	sil	bk								
4.0			pH 8.0							
			Old tulle roots							
5.0	Sl	pl								
			pH 8.2							
6.0	sicl	m	Dark strains							
	S	Sq	Few small rocks							
7.0			pH 8.0							
8.0										
9.0	C	m	Very imperm.							
10.0										

Scale markings: <0.008, 0.06, 0.1, 0.2, 0.4, 0.5, 1.0, 2.0, 3.0, 5.0, 7.0, 75, 125, >150 — INCHES PER HOUR

Table 15.1 Soil profile chart developed for drainage investigations by the Soil Conservation Service, U.S.D.A.

FIELD HYDRAULIC CONDUCTIVITY TEST
PIEZOMETER METHOD
FOR DRAINAGE INVESTIGATIONS

Piezometer Number _A-1_

Estimated "HC" _0.2 in. hr_ Calculated "HC" _0.35 in. hr_

Location _John Jones 15-21-31_ Date _3/1/55_ Technician _L.B.B._

Stratum Thickness _2.0 ft_ Texture _sil_ Structure _bk_

Depth Piez. _4.2 ft_ Auger Dia. _1 3/16 in._ Piez. Dia. _1/4 in._

Length Cavity _4 in._ Sloughing _very slight_ Times Cleaned _4_

pH (Soil _8.0_ Water _8.1_) Salinity (Soil _2 X 10⁻³_ Water _250 X 10⁻⁶_)

TIME	ELAPSED TIME	Δt	DISTANCE TO WATER SURFACE FROM REFERENCE POINT			Δh	RESIDUAL DRAWDOWN
			BEFORE PUMPING	AFTER PUMPING	DURING RECHARGING		
			B	A	R	A–R	R–B
	Minutes	Minutes	Feet	Feet	Feet	Feet	Feet
11:00			2.00				
:01	0			4.05			2.05
:06	5		•		3.70		1.70
:11	10				3.35		1.35
:16	15				3.00		1.00
:21	20	20			2.63	1.42	0.63
:26	25				2.35		0.35
:31	30				2.23		0.23
:36	35				2.15		0.15
:41	40				2.10		0.10

COMMENTS

The test was very good

Table 15.2 Procedure and form used to obtain hydraulic conductivity in the field, developed by the Soil Conservation Service, U.S.D.A.

FIELD HYDRAULIC CONDUCTIVITY TEST: PIEZOMETER METHOD

The piezometer method is used to obtain the hydraulic conductivity of a given stratum or area in a soil profile. (Hyd. cond. is a permeability figure dependent on properties of the ground water as well as the soil profile.) This is possible because the hole which is bored into the soil for the conductivity measurement is cased, except for a small cavity at its end. The rate of entry into this cavity is a measure of the hyd. cond. of the soil around the cavity.

EQUIPMENT: A 1½- to 2-inch worm auger with a square bit end is used for the test. Electrical conduit 1¼- to 2-inch inside diameter, sharpened on one end, is used as the piez. The auger is ground to about ¹⁄₁₆ inch smaller than the inside dia. of the piez. A driving head on the piez. top prevents damage during driving. An electrical device sounding bell or blow tube can be used to measure the water level. A soil tube jack or piez. removal equipment can be used to remove the piez.

METHOD: An auger hole is bored to a depth of 6 inches. The piez. is then driven into the hole about 5 inches with light blows from a maul. The hole is again augered to a depth of 6 inches below the piez. This procedure is continued until the piez. reaches the desired depth. A cavity 4 inches long is carefully augered below the end of the piez. A stop on the auger handle helps make this length precise. The auger should be removed very slowly to prevent sloughing of the cavity wall. A hollow auger or small tube to the auger bit may be required to permit air to break the suction and prevent sloughing of the cavity. The piez. is pumped or bailed out, with a pitcher pump or bail bucket, to permit the pores in the cavity wall to be flushed out. Flushing is repeated until the rate of rise in the piez. is the same as a previous pumping.

TEST: The water level is lowered in the piez. a distance dependent upon the sloughing tendency of the profile. The water levels and times of observations are recorded and used in the following Kirkham Piezometer formula to calculate the hyd. cond.

$$HC = 377 \times \frac{r^2}{c} \times \frac{\Delta h}{\Delta t} \times \frac{1}{A+R-2B}$$

Piezometer Profile

HC—Hydraulic conductivity in inches per hour
r—Inside radius of piezometer in inches
c—Function from figure on this page
Δh—Raise of water level in feet in t timed interval (A–R)
Δt—Time required to give h in minutes
A—Depth to water level at start of test
R—Depth to water level at end of test
B—Depth to static water level in feet

r = _0.63_

c = _12.5_

Δh = _1.42_ $HC = 377 \times \dfrac{0.63^2}{12.5} \times \dfrac{1.42}{20} \times \dfrac{1}{(4.05+2.63-4.0)}$

Δt = _20_

A = _4.05_

R = _2.63_   HC = _0.35 in. hr_

B = _2.00_

C Function for Cavity
Four Inches Long

C FUNCTION IN INCHES

DIAMETER OF CAVITY IN INCHES

RESIDUAL DRAWDOWN (R–B) IN FT.

TIME IN MINUTES

An estimate of hydraulic conductivity can be determined by plotting the residual drawdown at various recharge times. The shape of the curve can also be used in evaluating characteristics of the soil strata.

tions made with hand-powered augers, jetting rigs, and piezometers discussed in Chapters 4 and 5. Kind of drainage system and size, depth, and capacity of the system may be determined primarily from the soil survey. A soil profile chart developed for drainage investigations by George B. Bradshaw is shown in Table 15.1. A method and form for determining hydraulic conductivity in the field is shown in Table 15.2. These tables can be modified readily for tabulating data in metric units.

Observing the depth of the water table and its variations yields valuable data affecting the drainage design. Interpretations of differences of water-surface elevation in terms of direction and magnitude of the flow velocity of the ground water are discussed in Chapter 5. Frequently drainage can be made most effective at a minimum cost by intercepting flow before it enters the problem area. Also drainage lines should be constructed perpendicular, not parallel, to the flow to be most effective. Upward flowing water cannot generally be effectively drained by open or tiled drains unless the drains can be placed within a very permeable aquifer. Usually drainage wells relieving the pressure causing upward flow are the most effective methods of drainage, providing the well can penetrate a permeable formation. Hence, knowing the direction of ground-water movement can be very helpful and is generally essential.

Knowing the source of water often suggests ways of reducing the quantity at the source. Also when the source is known, it is often possible to estimate at the source the quantity of water to be drained.

The feasibility of intercepting the drainage water in a location where it can be readily used for irrigation is enhanced when the source and direction of flow are known.

15.7 DESIGN OF OPEN DRAINS

On some drainage projects open drains are used largely to convey water to distant outlets. Water may flow into open drains directly from the ground water and also from collecting tile lines. The designing engineer selects the drain outlet and determines appropriate elevation of bed of drain and water surface in the drain at times of maximum flow. Then he decides the bed slope, which ranges as a minimum from 0.05 to 0.15 percent. The slopes of open drains in the lower nearly level lands should be as large as the ground-surface slope provided this will not cause excessive water-flow velocities and channel erosion. Uninformity of drain-bed slope is usually advantageous even though this causes some differences in depth of drain. Design of side slopes of open drains depends largely on the soil formation, with a range from the steep slope of ½ horizontal to 1 vertical in very stiff, compact clays, to flat slopes of 3 to 1 in loose, open sandy formation. Depths of open drains range from 1.5 to 3.5 meters or more.

The trend had been toward deeper drainage systems. Drains of 3.5 to 4.5 meters in depth were not uncommon. The high costs of very deeply constructed gravity drains and the value of land removed from production justify special effort toward pumping ground water for drainage in areas where the power costs and soil formations make pumping feasible or of installing pipe drains made of concrete, plastic, or clay and covered so that the ground surface is not obstructed.

15.8 CONSTRUCTION METHODS AND COSTS

In modern construction of open drains, large power-driven drag-line excavators are used as illustrated in Fig. 15.1. Deep open drains require strips of land 22 to 30 meters or more in width, and this is one serious objection to the use of open drains in areas of high priced lands. A typical open drain of average depth is shown in Fig. 15.2.

Nitroglycerine dynamite is used in the construction of open drains in areas too wet for earth-moving machinery to work.

Figure 15.1.   Excavator in operation constructing an open drain.

Figure 15.2 Typical open drain in New Mexico, bed width approximately 6-m capacity, 3.5 m³/s.

When the soil is saturated and sticks of dynamite are uniformly spaced, the explosion will "propagate" from one charge to the next. This method requires only one blasting cap and is economical. Dynamiting affords an easy means of opening silted creeks and ditches as well as for making new open ditches. The soft bottom resulting from the blast is unsuitable for laying tile drains.

15.9 TILE DRAINAGE SYSTEMS

The two most common tile-drainage layouts are: (1) relief drains, nearly uniform in depth and spacing, for fairly level lands, and (2) intercepting drains on irregular slopes and near sidehill lands.

In the relief system, spacing of the drains is much influenced by the texture and permeability of the soil. In clay soils of low permeability and drain depths of 1.5 meters, close spacing of 60 meters may be essential to satisfactory drainage; in average loam soils, 120 to 180 meters is good spacing provided the pipe is placed to depths of 1.8 meters or more; in sandy and gravelly soils, spacing at 250 meters or more represents the more general practice. Long main drains with short collecting laterals are called the herringbone system. The gridiron system consists of long parallel laterals connected to a short main drain. Manholes, sand traps, and observation wells at convenient points 200 to 400 meters along the lines facilitate essential inspection, cleaning, and maintenance of the lines.

The depths and locations of irregular cutoff drains to intercept seepage water and prevent it from flowing from sidehill lands toward the lower flat lands depend largely on the surface topography and soil formations. Intercepting drains must cut off the water flow in the sandy and gravelly soil strata, because the rate of flow in these soils is very high compared with the flow in loams and clays.

Pipe-drain sizes should be designed using the formulas presented in Chapter 12 for pipe flow. Allowance will need to be made for somewhat increased resitance due to the frequent joints.

15.10 INSTALLATION OF PIPE DRAINS

Trenching machines, like the ones shown in Figs. 15.3, 15.4, and 15.5, have replaced hand labor on nearly all drainage projects that are large enough to warrant moving the heavy trenching machines to and from the fields that need drainage.

Concrete, plastic, or clay drain pipes are hauled, usually direct from the factory to the field, and placed along the proposed drain lines. As the trenching machine moves forward in the field, the drain pipe is laid with each new length of pipe being placed securely against the one just laid. Some machines are equipped with hydraulic pressure devices to press the pipe firmly together.

Figure 15.3.   A trenching machine with digging wheel and retainer elevated.

Figure 15.4. Laying shallow drainage pipe.

Figure 15.5. Drainage machine for laying perforated plastic drain pipe.

Caving of soil into the trench near the excavator is prevented by a large two-walled steel shield. Water flows from the saturated soils into concrete and clay pipes through the joints, not through the walls. Plastic pipe is perforated to facilitate water entry. To keep soil materials from entering the pipe it is good practice to place over the top of the joints a strip of tar paper and a screened-gravel envelope about 7.5 centimeters thick around the pipe. For soils of high stability, moderate moisture content, and no caving of banks at the time of placing the tile, the gravel may well be placed 3 to 5 meters behind the trencher. Immediately after the gravel is placed around the pipe, soil should be carefully placed over the gravel envelop to a depth of about 30 centimeters. This process protects the pipe from displacement, and from damage when large volumes of earth material are forced into the trench by heavy machinery for backfilling.

Pipe-laying machines can place about 300 meters of pipe a day when a tractor is used to load the gravel into the chute and backfill the trench.

15.11 MAINTENANCE OF DRAINS

The principal need in the maintenance of drainage systems is the removal of soil and vegetation from drains. Maintenance of open drains is the same as for other water conveyance channels and is discussed in Chapter 12.

Several types of trees and plants extend their root systems many feet to obtain water. Among these are greasewood, willows, and poplars. The roots enter the joints of closed drains and continue to grow inside the pipe, eventually obstructing the flow of drainage water. Partial obstruction by roots may retard the velocity of flow sufficiently to allow soil particles to settle in the pipe, thus gradually sealing the drains. Chemicals should be added to the drain water periodically when roots tend to penetrate the drains.

It is very difficult to construct drains in fine sandy soils with joints through which ground water may enter the pipe, and yet exclude all soil particles. The volume of soil entering the drain is small when a suitable gravel envelope is provided for the joints.

There are large variations in the quantity and velocity of water flow in closed drains. The largest flows with the greatest velocities occur during or after storms and irrigations. Soil particles entering the drains during the periods of high-velocity flow may largely be carried in suspension to the sand boxes or to the drain outlet, but as the quantity and velocity of flowing water decreases the soil particles settle to the bottom of the pipe. They may be rolled along with the low-velocity flow of water, but with a further reduction in the velocity the soils come to rest. The soil may become sufficiently stable to resist the scouring action of subsequent high-velocity flows, after which more soil accumulates, thus gradually filling the pipe unless the soil is removed.

The term "wash-ins" refers to holes formed in the land where irrigation water flows downward through the backfilled trench, washing large volumes of soil into the closed drain through the joints. Wash-ins have occurred most frequently soon after construction, and less frequently as the backfill material settled in the trench, thus becoming more compact.

Wash-ins have been attributed to careless irrigation methods. Irrigation water applied excessively and allowed to pond over the drains is especially conducive to wash-ins. The damage caused by wash-ins is twofold: the soil washed in obstructs the drains, and the holes formed in the land surface render that part of the land unproductive until repaired.

In sandy soils drain pipes have been found shifted as much as 90° from alignment. Loosely placed or unmatched joints cause this condition. Drainage water flowing over a wide or unmatched joint erodes the soil from under the joint, thus allowing it to settle. As the joint settles and the pipe ends separate, the eddies in the hole forming under the joint become greater, thus increasing the eroding effect of the water. This continues until the flow of water in the drain is completely blocked by soil entering the joint or by silt collecting in the lower reaches of the drain.

15.12 NEED FOR A GOOD QUALITY PIPE

Good quality pipe is essential if drainage systems are to operate over extended periods of time. Pipe must be highly resistant to deterioration from freezing and thawing, resistant to attack by acids and sulfates, and dense enough to be relatively impermeable. Modern processes of manufacture and availability of new high quality materials remove any excuse for using poor quality pipe.

15.13 GRAVEL ENVELOPES FOR PIPE DRAINS

Gravel envelopes are placed around pipe drains for two reasons: (1) to prevent soil particles from moving into the drain, which may clog the line and also cause "wash-ins" as discussed above and (2) to provide a more permeable material adjacent to the tile, thus increasing the effective diameter of the tile.

Design of gravel envelopes for drains follows the same procedure outlined in Chapter 13 for wells. Because of the rather large
quantities of gravel frequently used on a given drainage project,
it becomes rather costly to adjust the gravel mix to fit the theoretical considerations. In practice, naturally occurring gravels
which are properly graded have been used successfully. Gravels
used for filters should contain a predominance of coarse- and
medium-sized sand.

15.14 DRAIN DEPTHS, SPACING, AND GROUND-WATER FLOW

Two types of soil profile are considered to illustrate the
influence of drain depth, spacing, and other factors on the quantity of flow of ground water toward and into drains.

In highly permeable sandy soils, underlain by compact clay of
low permeability 1.7 to 3.0 meters below the land surface, as illustrated in Fig. 15.6, the ground-water flow is essentially horizontal toward the drains.

To simplify the illustration, the source of water flowing toward
the drain is considered a reservoir as illustrated on the left of
Fig. 15.6. The water surface is maintained in the reservoir and
adjoining soil a distance of H meters above the clay. Flow from
the reservoir to the drain is steady, it being assumed for simplicity that the reservoir is the only source of water.

Ground water actually flows to the drain from both sides with
similiar conditions existing on both sides of the drain. Let $2q$ rep-

Figure 15.6.   Illustrating for sandy soils over clay the linear flow of
ground water toward drains spaced $2R$ ft in which the water table
midway between drains is $H - h$ ft above the water surface in the
drain.

resent the flow into a drain in length L. Then the ground-water flow from one side to the drain is

$$q = av$$

Where a is cross-sectional area and v is velocity.

From Darcy's law,

$$v = k \, \frac{h_f}{R} = k \, \left(\frac{H - h}{R} \right)$$

Consider the depth of saturated sand about midway between the reservoir and the drain as average; then the average area of saturated soil, in drain length L, through which the ground water flows is:

$$a = \left(\frac{H + h}{2} \right) L$$

and the quantity of flow from the reservoir to the drain

$$q = \left(\frac{H + h}{2} \right) L \times k \left(\frac{H - h}{R} \right) = \frac{kL(H^2 - h^2)}{2R} \qquad (15.1)$$

The quantity of flow to the drain from reservoirs on both sides would be:

$$Q = 2q = \frac{kL(H^2 - h^2)}{R} \qquad (15.2)$$

from which

$$R = \frac{kL(H^2 - h^2)}{Q} \qquad (15.3)$$

The lengths H, h, and L can be accurately measured at any time, and q can be measured within 5 percent accuracy. However, computations of drain spacing using Equation 15.3 must be regarded as only approximations because of the fact that permeability k varies greatly in field soils, and assumed boundary conditions are approximations.

Rainfall and irrigation water which percolate below the water table may be the source of drainage water rather than essentially horizontal flow from a distant source such as a reservoir. Since the flow entering the ground water is essentially vertical, a modification must be made in the above analysis.

In the previous development, flow midway between the reservoir and the drain was assumed to be q. However, with uniform vertical flow the discharge past this midsection will be only $q/2$. Hence Equation 15.1 becomes

$$q = \frac{kL(H^2 - h^2)}{R} \qquad (15.4)$$

and the quantity of flow into the drain receiving percolating irrigation or rain water will be

$$Q = \frac{2kL(H^2 - h^2)}{R} \qquad (15.5)$$

from which

$$R = \frac{2kL(H^2 - h^2)}{Q} \qquad (15.6)$$

For soils of great depth and approximately uniform permeability, as illustrated in Fig. 15.7, ground water flows radially toward the drain. With these soil conditions and designating radial flow through a semicircular area (a little less than the actual area) it is essential to use the calculus to derive the rational equation for radial flow

$$Q = \frac{\pi kL(H - h)}{2.3 \log_{10} R/r} = \frac{\pi kL(H - h)}{2.3 \log_{10} S/d} \qquad (15.7)$$

in which all the symbols but S and d have the same meaning as in Equation 15.2. S is the spacing between drains, and d is the diameter of the drain.

The basic differences of the two flow conditions of special importance are: for the first condition for essentially horizontal flow,

Figure 15.7.   Illustrating for deep, uniform soils the radial flow of ground water toward drains spaced $2R$ ft, in which the water table midway between drains is $H - h$ ft above the water surface in the drain.

if h is small as compared to H, the flow to the drain is proportional approximately to the square of the effective depth $(H - h)$, whereas in the second condition for essentially radial movement the flow is proportional to the first power of the effective depth.

15.15 LIMITATIONS OF EQUATIONS

Of necessity Equations 15.2 to 15.7 are based on uniform soil conditions and a uniform source of water. Also, the capillary range above the water table and the seepage surface at the drain have been neglected. Simplified boundary conditions have been assumed. Hence, the formulas shown are useful as approximations, to be modified by experience and applied with wisdom. The method of developing the equations and the assumptions made will be helpful in visualizing the physical processes involved in drainage and the degree to which variation from assumed conditions will influence results. Further discussions of these principles are contained in Chapters 5 and 13.

15.16 PUMPING FOR DRAINAGE

The main physical defect of gravity drainage systems is failure to lower the water table to an adequate depth. Many gravity drains are either too shallow, spaced too far apart, or both. Pumping ground water in some areas where a suitable permeable aquifer exists is a more effective means of lowering the water table.

The influences on well discharge of the soil permeability, effective well depth, and diameter of well when pumping from ground water are considered in Chapter 13. In pumping from confined ground water in an artesian aquifer of depth D the water flows radially to the well through cylindrical surfaces having a vertical axis and the following well-discharged formula applies:

$$Q = \frac{2\pi k D(H - h)}{2.3 \log_{10} R/r}$$

(15.8)

Comparisons of Equations 15.7 and 15.8 show that if the length of drain L in Equation 15.7 is equal to the depth of aquifer D in Equation 15.8, and all other items are the same, then the flow to the well through "vertical cylindrical" surfaces would be twice that to the drain through "horizontal semicylindrical" surfaces.

Pumping ground water for drainage is influenced favorably by adequate depths and permeabilities of the water-bearing formations, by high values of pumped water for irrigation, and by low power costs. The experiences of one California irrigation district and of the Salt River Valley Water Users Association in Arizona indicate the feasibility of ground-water pumping where subsoils are favorable and the pumped water can be used for irrigation.

From 1907 to 1922, the Modesto Irrigation District in California spent $356,000 for construction and maintenance of gravity drains for 18,000 hectares. Subirrigation had prevailed for several years. In many locations where the rich soil had previously produced abundant crops, the yields decreased, orchard trees died, and vines withered, for the alkali salts had become sufficiently concentrated to render the soil unfit for plant growth. In 1922, the Modesto group drilled the first drainage well, and by 1939 had put into operation 77 pump wells, reaching a combined capacity of 6 cubic meters per second. The pumping costs on 20,000 hectares subject to Modesto's high water table with 36 percent for construction, maintenance, and operation, and 64 percent for power cost, was a third more than the cost for gravity drains. However, during the period in which the district operated the pumps, a total of 74,000 hectare-meters of water was pumped and about 75 percent of that water was utilized for irrigation. The pumped irrigation water had a value considerably in excess of all drainage-pumping costs. The operation of deep-well pumps was not only a satisfactory method of subsurface drainage but also a profitable method.

In the Salt River Valley, Arizona, irrigation was greatly advanced in 1911 by completion of the Bureau of Reclamation Roosevelt dam and reservoir. Drainage did not become a problem there until about 1918; then the Water Users Association decided to pump ground water. The association pumped 6000 hectare-meters in 1920 and the same amount in 1921. In 1922, it increased the drainage to 12,000 hectare-meters, and the menacing water table began to go down. Since then, the volumes of water pumped annually have increased, and the water-table depth has also greatly increased, thus solving the drainage problem. The association operates about 250 pumped wells, drawing nearly one-half of its irrigation water supply from the ground water.

15.17 REUSE OF DRAINAGE WATER

With the increasing shortage of water, drainage designs should include provision for maximum reuse of drainage water for irrigation and other beneficial purposes. The practice by Modesto Irrigation District and Salt River Valley Water Users, discussed in the preceding paragraphs, shows what can be done. Many other irrigated areas having drainage problems could do likewise. Others may not be able to drain by pumping, but drainage water could be combined with surface water to advantage. Public and company officials have an obligation to see that drainage water is reused to its fullest potential.

CHAPTER
16
WATER MEASUREMENT

Efficient use of water for irrigation depends largely on measurement of water. Increasing utilization and value of available water and the growing tendency among irrigation companies to base annual water charges on the volume of water used make an understanding of the principles and methods of water measurement necessary. Information concerning the relationships between water, soils, and plants cannot be utilized in irrigation practice without the measurement of water.

16.1 UNITS OF WATER MEASUREMENT

Units of water measurement are considered in two classes: first, those expressing a specific volume of water at rest; and second, those expressing a time rate of flow. The commonly used units of volume of water at rest are the liter, cubic meter, hectare-centimeter, and hectare-meter. A hectare-centimeter is a volume of water sufficient to cover 1 hectare 1 centimeter deep, which is 100 cubic meters. A hectare-meter of water will cover 1 hectare 1 meter deep, and is equal to 10,000 cubic meters.

The commonly used units of flow are liters per second, cubic meters per second, hectare-centimeters per hour, and hectare-meters per day.

16.2 FLOW THROUGH AN ORIFICE

When the water-pressure intensity inside the pipes of a domestic water system is high, water flows out of an open tap at a high velocity, and when the pressure is low, it flows out slowly. If water pressure within the pipe were exactly equal to the pressure of the air outside the pipe, there would be no flow. Pressure intensity at any point within a static body of water with a free surface is proportional to the depth of the point below the water surface. The velocity of water flow through an opening in a vessel (or in a wall built across a stream) which is far below the water surface is greater than the velocity through an opening near the water surface. In irrigation practice it is important to know just what the velocity will be through an orifice at any vertical distance below the water surface. The basic physical law which determines the velocity of water through an orifice is the same as the law which determines the velocity of a falling body at any vertical distance below the point from which it begins to fall.

The velocity of a falling body, ignoring atmospheric friction, may be determined from the vertical distance through which the body falls from rest. Likewise, the velocity of water escaping from an opening (orifice) in a vessel, ignoring friction, may be determined from the height of water in the vessel above the opening. Stated as an equation, this very important law of falling bodies, as applied to the flow of water, is:

$$V = \sqrt{2gh} \qquad (16.1)$$

where V = velocity, in cm/s
   g = the acceleration due to gravity (or the force of gravity per unit mass of water), which is 981 cm/s
   h = the depth of water, in cm, or pressure head, causing discharge through an orifice

If the vertical dimension of the orifice opening is very long, the velocity of flow through the orifice will be appreciably greater near the bottom of the orifice than near the top. For the purpose of this discussion it is assumed that the orifice height is so small as compared to the pressure head causing the discharge that the difference between velocity near the top and near the bottom of the orifice is negligible. To illustrate the use of Equation 16.1 assume that h in Fig. 16.1 = 120 cm. Then:

$$V = \sqrt{2 \times 981 \times 120} = 485 \text{ cm/s}$$

Therefore, theoretically, water should flow through an orifice which is 120 centimeters below the surface at a velocity of approximately 485 centimeters per second. Owing to frictional resistance, the actual velocity is somewhat less than the theoretical velocity.

The fact that the quantity of water that flows through an opening is directly proportional to the cross-sectional area of the opening and to the velocity of flow was given as Equation 12.1, identified as the continuity equation. This basic rational equation is repeated here for convenience as:

$$Q = AV \qquad (16.2)$$

where Q = quantity of flow
   A = cross-sectional area of water
   V = the mean velocity of perpendicular to the cross-sectional area

The theoretical discharge through an orifice may be determined by substituting the value of V from Equation 16.1 in Equation 16.2, i.e.,

$$Q = A \sqrt{2gh} \tag{16.3}$$

If the orice opening in Fig. 16.1 is 10 centimeters high by 45 centimeters long (perpendicular to the plane of the paper), the area would be:

$$A = 10 \times 45 = 450 \text{ cm}^2$$

Experiment has shown that the actual discharge for a standard orifice is approximately six-tenths the theoretical discharge, so that the actual Q would be computed thus:

$$Q = 0.6 \times 450 \times 485 = 130{,}950 \text{ cm}^3/\text{s}$$
$$= 0.13 \text{ m}^3/\text{s}$$

Finally, the equation for actual discharge through an orifice is

$$Q = CA \sqrt{2gh} \tag{16.4}$$

in which C is a coefficient of discharge determined by experiment. The coefficient C ranges from 0.027 up to 0.035 or more, depending on the position of the orifice relative to the sides and bottom of the vessels or of the water channel and also on the degree of roundness of the edges of the orifice.

Submerged Orifices

The water cross-sectional area of a submerged orifice, Fig. 16.2, is the length of the opening times the height, or $A = L \times H$. The loss of head as water flows through a submerged orifice is a

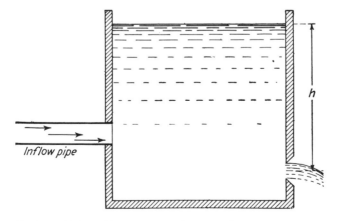

Figure 16.1. Illustrating the discharge of water through an orifice under head, h.

difference in elevation of the water surface upstream and downstream, as shown by h in Fig. 16.2. Hence, from Equation 16.4 and from experiments, the discharge of a standard submerged orifice is:

$$Q = 0.027A\sqrt{h} \qquad (16.5)$$

where Q = discharge, in l/s
     A = area of orifice, in cm^2
     h = head, in cm

The submerged orifice is used both in its standard form and as a combination head gate measuring device.

Types of Orifices Used

Types of orifices used to measure irrigation water are: (1) submerged orifices with fixed dimensions, (2) adjustable submerged orifices, and (3) calibrated gates.

Submerged orifices with fixed dimensions are used where the available head is insufficient for weirs. For best results they are usually rectangular with the horizontal dimension from two to six times the height. They are generally installed in sufficiently large channels so that the contractions[1] are complete or very nearly so. The coefficient of discharge will then be approximately 0.027.

[1]Full contraction is secured on an orifice or weir whenever the sides, bottom, and water surface are far enough from the opening that no change in discharge would occur if they were further removed, other conditions remaining constant.

Figure 16.2. Submerged orifice.

There are several kinds of adjustable submerged orifices. Some resemble submerged orifices with fixed dimensions except that their height is adjustable to accommodate a wide range of flow without excessive head loss. The more usual type of adjustable submerged orifice is a combination head gate, or turnout, and measuring device. Such structures generally have one or two slide gates that may be held open in any desired position. Two gates can be used together to provide a constant head orifice with greater accuracy. One gate is used to control the water level. Single submerged orifice head gates are generally not accurate measuring devices and are gradually being replaced by better types of structures.

A development of recent years has been the calibrated commercial gates for water measurement. Tests have been conducted on gates of various size, and curves and tables prepared giving the flow in cubic meters per second for different gate openings measured on the rising stem. The head is the difference in elevation of the water surface in the supply canal and the outlet ditch. A discharge chart and table are furnished with commercial gates at the time of purchase.

For accurate measurements with free-flowing orifices, the height of the orifice should not be greater than the head. The pressure head is the most difficult to regulate. The means generally employed is either an adjustable regulating gate or a spill crest at the desired level allowing excess water to overflow. Some orifice installations employ both principles to ensure accurate measurement. With complete contraction, velocity of approach is usually negligible.

16.3 FLOW THROUGH A WEIR

Suppose that the water surface is lowered until it drops below the upper edge of the orifice, Fig. 16.3. Then the pressure head which causes the average velocity, as represented by h of Fig. 16.3, is one-half the total depth of water over the bottom edge of the orifice. Representing the length of orifice by the symbol L, the area becomes:

$$A = 2hL$$

Substituting this value of A in Equation 16.4

$$Q = 2CLh \sqrt{2gh} = 2CLh^{3/2} \sqrt{2g} \tag{16.6}$$

Since the acceleration due to gravity g is nearly constant, it is convenient to represent the product $2C\sqrt{2g}$ by a single symbol, say C'. Then it follows that

$$Q = C'Lh^{3/2} \tag{16.7}$$

Figure 16.3. Showing that discharge through a partly filled orifice is similar to the discharge over a weir.

Equation 16.7 gives the theoretical discharge of an orifice when the top edge of the orifice is above the water surface.

The term weir, as used in measurement of water, is defined as a notch in a wall built across a stream. The notch may be rectangular, trapezoidal, or triangular in shape. The orifice in Fig. 16.3, when flowing partly full, is a weir shown in Fig. 16.4 according to the above definition. In measuring the flow of water over weirs, it is convenient and customary to measure the total depth; i.e., $2h$ of Fig. 16.3. Although the total depth, as represented in Fig. 16.4 by the symbol H, does not represent the point in the stream of average velocity, it can be used in Equation 16.7 by changing only the coefficient C'. Substituting for h its equivalent $H/2$ in Equation 16.7 there results:

$$Q = C'L \left(\frac{H}{2} \right)^{3/2} = \frac{C'}{2^{3/2}} LH^{3/2}$$

or

$$Q = C''LH^{3/2} \tag{16.8}$$

in which

$$C'' = \frac{C'}{2^{3/2}}$$

Equation 16.8 is the general form for discharge of rectangular and trapezoidal weirs. Although, as shown above, it is based on the fundamental equation $Q = AV$, the only measurements es-

Figure 16.4. A rectangular weir with complete end contractions.

sential in using Equation 16.8 are those of length of weir crest, L, and depth of water flowing over it, H. The velocity need not be measured directly. The coefficient C'', ordinarily represented by C, has been determined by experiment by many workers. For rectangular weirs it was early found by Francis to be 0.0184 and hence, the widely used sharp-crested weir discharge equation

$$Q = 0.0184 \, LH^{3/2} \tag{16.9}$$

where Q = discharge, in l/s
$\quad\quad L$ = length of weir crest, in cm
$\quad\quad H$ = total head, in cm

Equation 16.9 without modification applies accurately only to rectangular weirs in which the length of weir is the same as the width of the rectangular channel immediately above the weir; i.e., weirs having suppressed end contractions. For weirs having complete end contractions, such as represented in Fig. 16.4, the effective length of weir crest, L, is found from the relation

$$L = L' - 0.2H \tag{16.10}$$

in which L' = the measured length of weir crest.

The Italian engineer, Cipolletti, long ago designed a trapezoidal weir with compete contractions in which the discharge is believed to be directly proportional to the length of weir crest, making it unnecessary to correct for end contraction. It has been widely used because of its advantages. The equation giving the discharge is:

$$Q = 0.0186 \, LH^{3/2} \tag{16.11}$$

where Q is in liters per second and L and H are in centimeters. In this trapezoidal weir the sides have a slope of 1 horizontal to 4 vertical, Fig. 16.5.

Figure 16.5.   A trapexoidal or Cipolletti weir.

For the 90° triangular weir shown in Fig. 16.6, the water cross-sectional area is $H \times H$, or H^2, and therefore, from Equation 16.4

$$Q = CH^2\sqrt{2gh} = C'H^{5/2}$$

The actual discharge has been found by experiment to be approximately

$$Q = 0.0138H^{5/2} \qquad (16.12)$$

where Q is in liters per second and H is in centimeters.

Summary of Weir and Orifice Formulas

To show the similarity between the discharge equations and to summarize the material of the preceding section, Table 16.1 has been prepared. When the orifice is no longer flowing full with the result that h becomes equivalent to H then $H \times H^{1/2} = H^{3/2}$, or the form of Equation 16.9 is obtained to describe the rectangular weir without side contraction. However, when the sides are contracted, the effective width of the crest L is reduced on each side by $0.1H$. The 1 to 4 side slopes of a trapezoidal weir nearly produce the same flow as a rectangular weir, the only difference

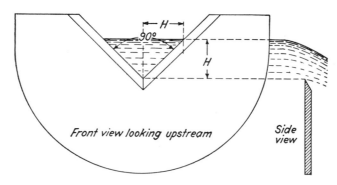

Figure 16.6.   A 90° triangular notch weir.

Measuring Device (all sharp crested)	Views	Formula	Equation Number
Orifice	Front view; Side view	$Q = 0.61 A \sqrt{2gh}$	(6.5)
Rectangular Weir (without contraction)	Top view	$Q = 3.33 L H^{3/2}$	(6.9)
Rectangular Weir (with contraction)	Top view $(L - 0.2H)$; Side view	$Q = 3.33 (L - 0.2H) H^{3/2}$	(6.9) and (6.10)
Trapezoidal Weir	End view	$Q = 3.37 L H^{3/2}$	(6.11)
90° Triangular Weir	End view	$Q = 2.49 H^{5/2}$	(6.12)

Table 16.1   Summary of Orifice and Weir Formulas.

being a slight change in coefficient of discharge from 0.0184 to 0.0186. A 90° triangular weir has a flow proportional to $H^{5/2}$ because the length also increases directly as the head increases; and L becomes equal to H, and $H^{3/2} \times H^{3/2} = H^{5/2}$. The coefficient of discharge must be varied slightly to take care of changing contractions and losses when boundaries are changed.

Effect of Boundary Form on Coefficient of Discharge

When silt deposits in front of an orifice or weir, the flow is changed. This causes one of the major sources of error in these measuring devices. Orifice and weir formulas presented in the preceding sections are for average conditions and do not reflect changes in boundary form. Figure 16.7 can be used to determine a more accurate discharge coefficient for orifice and weir formulas when the installations do not conform to standard conditions.

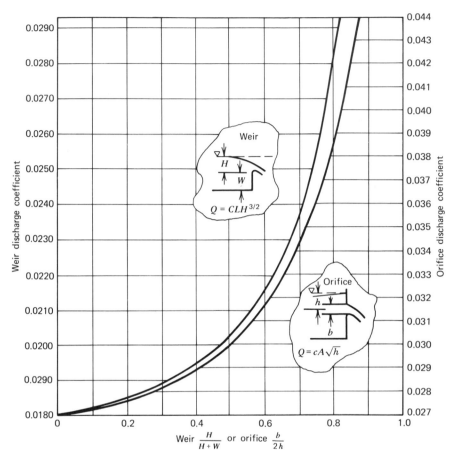

Figure 16.7.   Variation of discharge coefficients with boundary form for a rectangular weir and for an orifice.

For orifices the coefficient of discharge C_d in the following formula is a function of the height of opening b and the depth of water h over the center of the opening:

$$Q = C_d A \sqrt{2gh} \tag{16.13}$$

From Fig. 16.7, a coefficient of 0.027 corresponds to conditions where $b/h = 0$, which is not always the case.

For weirs, the coefficient 0.0184 in Equation 16.9 corresponds to a value of $H/(H = W)$ of 0.22—essentially the conditions shown in Fig. 16.4, where the depth over the crest is one-third the distance from the bottom to the crest. Hence, when the bottom is other than $3H$ below the crest, the coefficient of discharge given in Fig.

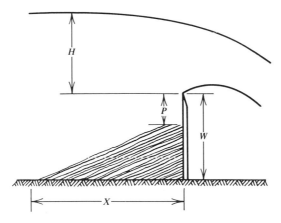

Figure 16.8. Definition sketch for a change
in geometry or silting behind a weir to be
used with Table 16.1.

16.7 should be used in the following formula for more accurate
results

$$Q = C_d L H^{3/2} \tag{16.14}$$

The same coefficient can be used for a contracted weir or a
trapezoidal weir with very little error.

Often the bottom of the channel in front of the weir is sloped
and consequently effects the value of the weir coefficient. Also
silt settles behind a weir changing the geometry of approach.
The discharge for a silted weir is always more than the flow calculated from the measured head. Figure 16.8 shows the necessary dimensions for determining from Table 16.2 the increase in

TABLE 16.2   Percent Increase in Discharge Caused by a Change
in Geometry or Silting Behind a Weir

$\dfrac{P}{W}$	$\dfrac{X}{W}$					
	0	0.5	1.0	1.5	2.0	2.5[a]
0	0	10	13	15	16	16
0.25	0	5	8	10	10	10
0.50	0	3	4	5	6	6
0.75	0	1	2	2	3	3
1.00	0	0	0	0	0	0

[a]For values of X/W greater than about 3.0 the increase in discharge
can be ignored.

discharge occurring as a result of the upstream geometry. For example, note that if $P/W = 0.25$ and $X/W = 2.0$, an increase of 10 percent in the flow will result. Note that this is the increase over the discharge given by use of the revised coefficient obtained from Fig. 16.7, not the discharge calculated from Equation 16.9 using a coefficient of 0.0184

The error in orifice and weir measurement is generally not greater than 5 percent unless approach conditions are not correct. However, it should be clearly understood that boundary form does have an influence on discharge. Therefore, when accuracy greater than 5 percent is necessary, measuring devices should be carefully constructed to the specifications of the researchers who have developed the formula to be used. Many formulas are available, and each one applies to a rather narrow range of boundary conditions. Only the simpler formulas which are sufficiently accurate for most irrigation work have been presented herein.

Properties of Weirs and Orifices

It is important to note the influence that change of depth of water H has on the discharge Q for the various weirs and orifices. For example, when the stream over a rectangular weir increases until the depth H is doubled, the cross-sectional area A is doubled, and the discharge Q increased by 2.8 times; whereas doubling H over a trapezoidal weir slightly more than doubles the area, and increases the discharge 2.88 times. When the H on a triangular weir is doubled, area is increased 4 times and discharge is increased nearly 5.66 times. When the depth of water H causing the discharge through a submerged orifice is doubled, the area remains unchanged, and the discharge Q is made only 1.4 times the original discharge. Comparisons similar to the above may readily be made for any change in depth by remembering that discharge Q varies with the three-halves power of depth H for the rectangular and trapezoidal weirs, with the five-halves power for the triangular weir, and with the one-half power of the head h for the submerged orifice. The variations of discharge with depth of water have a very important bearing on irrigation practice and should be clearly understood. In order to illustrate further the influence of change of depth on the discharge, the curves of Fig. 16.9 are presented.

Advantages and Disadvantages of Weirs

The advantages of weirs for water measurement are: (1) accuracy, (2) simplicity and ease of construction, (3) nonobstruction by moss or floating materials, and (4) durability.

Figure 16.9. Curves showing the relation between the discharge of water and the depth of water on the weir crest for a 30-cm rectangular weir with suppressed end contractions, a 30-cm trapezoidal weir, a 90° triangular notch weir, and a submerged orifice of 30 × 30 cm cross-sectional area.

The disadvantages of weirs are: (1) requirement of considerable fall of water surface, which makes their use in areas having level land impracticable, and (2) deposition of gravel, sand, and silt above the weir prevents accurate measurements.

The principal advantage for the submerged orifice is the small loss of head, or difference in elevation of the water surfaces on the upstream and the downstream sides of the orifice, making it suitable for use in canals and ditches having very small slopes, where it is difficult to obtain fall enough to use weirs. Orifices have in addition most of the advantages enumerated for weirs.

Most serious disadvantages of orifices are: (1) collecting of floating debris, and (2) collecting of sand and silt above the orifice, thus preventing accurate measurement.

Weir Box and Pond

When using weirs, the ditch or canal must be made wider and deeper than average for some distance upstream from the weir. This is to make the water approach the weir at a low velocity (usually less than 15 centimeters per second) by flowing through a relatively large channel. The enlarged section of the ditch should be gradually tapered to natural size. Cross currents upstream from and near the weir must be prevented. This may be done by placing baffle boards across the weir channel. The weir may be placed in a weir box built of concrete or lumber, Fig. 16.10, or it may simply be placed in an enlargement of the ditch, Fig. 16.11.

Less room is required when a box is used, but cleaning is made more difficult. For temporary use, the placing of a weir in an open ditch as in Fig. 16.11 is the more economical method.

Cleaning is also less expensive in an open ditch, since a scraper may be used. The ditch downstream must be protected with loose rock or other material to prevent erosion by the falling water.

Table 16.3 gives the sizes of weirs best adapted to measuring streams of water varying from 15 to 625 liters per second, and also the proper dimensions for each size of rectangular, trapezoidal, and 90° triangular notch weirs.

Figure 16.10.   Plan of weir box. (Courtesy U.S.D.A.)

Figure 16.11. Weir notch and bulkhead in weir pond.

Weir dimensions, listed in Table 16.3 and illustrated in Fig. 16.10 are a little smaller than would be necessary to obtain rigid accuracy, but boxes of these sizes generally will give results within 5 percent of the correct values if care is taken to maintain proper flow conditions.

For temporary wooden weirs, the wood of which the weir is constructed may well be used also to form the weir crest and sides. However, since wood warps easily and its sharp edges become worn and splintered, its use on permanent weirs for crests and sides is seldom desirable.

Measurement of Head or Depth on Weir Crest

Measurement of head or depth of water on a weir crest is obtained with a specially constructed scale or a carpenter's rule. A scale, called a weir gauge, must be set upstream from the weir a distance no less than four times the depth of water H flowing over the crest, or upstream in the corner near the bank where the velocity is essentially zero. This is made necessary by the downward curvature of the water surface near the crest. The zero point on the scale must be set level with the crest of the rectangular or trapezoidal weir, or with the vertex of the triangular weir. The scale, or a lug upon which to place a rule, may be fastened to the bulkhead at a lateral distance from the end of the notch of not less than twice the greatest depth of water H over the crest. To get the zero point of the scale or the lug level with the crest, a carpenter's level may be used. Allowing the water to flow into the pond and slowly rise until it flows over the weir crest is inaccurate, since the water surface will rise appreciably above the crest before flow over the crest begins. Small errors in reading H cause relatively large errors in discharge determination. The scale or gauge may be marked to read discharge directly.

TABLE 16.3 Weir-Box Dimensions for Rectangular, Cipolletti, and 90° Triangular Notch Weirs[a]

Rectangular and Trapezoidal Weirs with End Contractions

Flow (s-cm)	H Maximum Head	L Length of Weir Crest	A Length of Box Above Weir Notch	K Length of Box Below Weir Notch	B Total Width of Box	E Total Depth of Box	C End of Crest to Side	D Crest to Bottom	F Hook Gage Distance Upstream	G Hook Gage Distance Across Stream
15 to 35	30	30	180	60	165	105	68	60	120	60
60 to 150	33	46	210	90	210	120	82	75	135	60
115 to 225	36	60	240	120	256	135	98	82	150	75
170 to 400	39	90	270	150	360	150	135	98	165	90
280 to 625	45	120	300	180	420	165	150	105	180	90
90° Triangular Notch Weir										
15 to 70	30	—	180	60	150	90	75	45	120	60
55 to 120	38	—	195	255	195	90	98	45	150	75

[a] All dimensions are in cm. The letters at the heads of the columns in this table refer to Fig. 16.10.

Hook gauges are widely used and considered the most accurate for determining water depths or stages. They consist of two essential parts; a moveable scale on which is fastened a hook; a fixed part containing an index mark and usually a vernier scale. The moveable part is raised until the point produces a slight pimple on the water surface, and the gauge height is read opposite the index. A blunt point is preferable to a sharp point.

Recording gauges, called water-level recorders, are used to obtain a continuous graph of gauge height. Essential parts of a recording gauge are: (1) float or pressure-indicating device, (2) recording mechanism, and (3) clock. Several different kinds of recording gauges are available.

Stilling wells for measuring water elevation are essential if accuracy is desired. A box or large-diameter pipe set vertically at one side of a connected stream or channel can be used as a stilling well. Stilling wells are used to eliminate wave action and provide a still water surface. To function properly, the cross-sectional area of a stilling well should be about 100 times the area of the inlet pipe or opening. Care should be taken to prevent the inlet pipe from clogging, and a convenient means of cleaning both the inlet pipe and stilling well should be provided. Flow of high velocity past a protruding stilling-well pipe should be avoided.

Portable Weirs

It is sometimes desirable to measure small streams at points where the cost of installing permanent weirs would not be warranted. For example, the occasional measurement of surface runoff from various fields, though desirable, would hardly warrant the installation of a permanent weir. In situations like this, a small steel plate cut like a half-circle and having a weir notch serves well, such as shown in Fig. 16.6. The notch may be cut as a rectangle, trapezoid, or triangle, depending on the type of weir desired. Bulkheads of plastic or canvas attached to weirs or orifices make excellent portable measuring devices.

Weirs Without End Contractions

A standard rectangular weir without end contractions consists of a bulkhead having a sharp crest, built across a rectangular channel, high enough to cause a complete deflection of water filaments as the stream passes over the weir. The conditions for accuracy are the same as for the standard rectangular weir with contractions, except for those relating to side contractions. This type of weir can be used only in channels having a uniform rectangular cross section. Air holes must be made through the weir box just below the weir crest to fully admit air under the sheet of over-falling water. Lack of an adequate air supply at atmospheric pressure under the sheet of falling water will cause a suction, drawing down the water surface and increasing the discharge. Equation 16.9 or 16.14 can be used to determine the discharge, providing atmospheric pressure exists under the sheet of falling water.

Rules for Setting and Operating Weirs

1. The weir should be set at the lower end of a long pool sufficiently wide and deep to give an even, smooth current with a velocity of approach of not over 15 centimeters per second, which means practically still water.

2. The center line of the weir box should be parallel to the direction of the flow.

3. The face of the weir should be vertical, i.e., leaning neither upstream nor downstream.

4. The crest of the weir should be level, so the water passing over it will be of the same depth at all points along the crest

5. The upstream edge should be sharp so that the over-falling water touches the crest at only one point.

6. The distance of the crest above the bottom of the pool should be about three times the depth of water flowing over the weir crest.

7. The sides of the pool should be at a distance from the sides of the crest not less than twice the depth of the water passing over the crest.

8. For accurate measurements the depth over the crest should be no more than one-third the length of the crest.

9. The depth of water over the crest should be no less than 15 centimeters, as it is difficult with smaller depths to get sufficiently accurate gauge readings to give close results.

10. The crest should be placed high enough so water will fall freely below the weir, leaving an air space under the over-falling sheet of water. If the water below the weir rises above the crest, this free fall is not possible, and the weir is then said to be submerged. Unless complicated

corrections are made, measurements on submerged weirs are unreliable.

11. The gage or weir scale may be placed on the upstream face of the weir structure and far enough to one side so that it will be in comparatively still water, Fig. 16.10, or it may be placed at any point in the weir pond or box, so long as it is a sufficient distance from the weir notch as to be essentially beyond the downward curve of the water as it flows over the weir crest. The zero of the weir scale or gauge should be placed level with the weir crest. This may be done with a carpenter's level, or where greater refinement is desired, with an engineer's level.

12. To prevent erosion by the falling water, the ditch downstream from the weir should be protected by loose rock or by other material.

16.4 PARSHALL FLUMES

Parshall has designed a measuring device with which the discharge is obtained by measuring the loss in head caused by forcing a stream of water through a throat or converged section of a flume with a depressed bottom. The disadvantages of weirs and submerged orifices are largely overcome by the Parshall flume. Since the head H is small on which measurement is based, care must be exercised in determining the differences in water level to get accurate measurements. The Parshall flume is illustrated in Figs. 16.12 and 16.13.

The Parshall measuring flume is a product of many years of painstaking research. It was first known as the venturi flume, being similar to the venturi tube or meter early designed to measure the flow of water in pipes.

The accuracy of the Parshall measuring flume is within limits that are allowable in irrigation practice, ordinarily within 5 percent. Flumes ranging from 2.5 centimeters to 3 meters in throat width are used to measure flows from 0.3 liter per second to 5.5 cubic meters per second or larger. Small flumes are well suited to the requirement of measuring farm water deliveries and flows in furrows. The Parshall flume operates successfully with less loss of head than required for weirs.

Silt will not deposit in the structure where it will affect the accuracy, because velocity is higher in the flume than in the channel. Ordinary velocities of approach have little or no effect on the measurement. The flume may be used with recording or register-

Figure 16.12.   Finished 9-in. Parshall flume with staff gauge. One-in. angle irons at the upstream end, crest, and downstream end of the structure serve as guides for striking off the floors at exact elevations. (Courtesy Soil Conservation Service, U.S.D.A.)

Figure 16.13.   Parshall measuring flume constructed with reinforced concrete. (Courtesy Soil Conservation Service, U.S.D.A.)

ing instruments when continuous records of flow are desired, or with an indicating gage graduated to give the flow in any unit desired.

The Parshall flume cannot readily be combined with a turnout. For free-flow conditions the exit velocity is relatively high, and channel protection is generally necessary downstream from the flume to prevent erosion.

Only a single head need be measured for free-flow conditions which exist when the head at the lower gauge is less than about 60 percent of that at the upper gauge. Free-flow is determined from a measurement of the head at the upper gauge by use of Table 16.4. When the head at the lower gauge is greater than 70 percent of that of the upper gauge, the upper gauge reading is affected and submerged flow results. Fairly accurate measurements can be made with a submergence of 90 percent, provided that the heads at both places are measured and corrections made for submergence.

The successful operation of the Parshall flume depends largely upon the correct selection of size and proper setting of the flume. The probable maximum and minimum flow to be measured is estimated, and maximum allowable head is determined. The maximum allowable head will depend on the grade of the channel and the freeboard (distance from normal water surface to top of banks) at the place where the flume is to be installed. Whenever possible, the selection should be made so that free-flow will always result. For economy the smallest flume that will satisfy the conditions should be selected.

Construction of Parshall Flumes

Parshall flumes may be built of wood, concrete, concrete blocks, concrete and brick, or, in the smaller sizes, of heavy sheet metal. The dimensions of flumes ranging from 2.5 centimeters to 3.05 meters are shown in Fig. 16.14 and tabulated in Table 16.5 and 16.6.

To secure accuracy in measurement, these flumes must be built to exact dimensions, especially the converging and throat sections. The flow of the upstream converging section, especially the crest, must be level. Where the flume is more than 15 centimeters above the channel bottom, a short inclined floor should be provided as a reasonably uniform approach.

The diverging section downstream from the throat can be omitted without affecting the discharge. The function of the diverging section is to reduce the head loss through the flume and to minimize erosion downstream.

TABLE 16.4   Free Flow Through Parshall Flume

Upper Head, H_a (cm)	Throat Widths (cm)									
	7.5	15	23	30	61	91	122	152	183	244
	Flow (l/s)									
1										
2										
3	.78	1.4	2.5							
4	1.2	2.3	4.0							
5	1.6	3.1	5.4							
6	2.3	4.5	7.3	9.8	18	27	35			
7	2.9	5.7	9.0	12	23	34	45			
8	3.5	7.1	11	15	28	41	54	67	79	
9	4.3	8.6	13	18	35	51	67	83	99	130
10	5.0	10	15	21	41	60	79	97	116	152
11	5.8	12	18	24	46	69	91	112	133	176
12	6.5	13	21	27	52	78	103	127	152	200
13	7.5	15	24	31	61	90	119	148	176	233
14	8.5	17	27	35	68	101	133	165	190	260
15	9.6	19	29	38	75	111	147	183	218	288
16	10.3	21	32	42	82	122	162	201	240	316
17	11	23	35	47	92	137	181	225	270	356
18	12	25	38	51	100	149	197	246	295	388
19	13	27	42	55	108	161	213	266	317	421
20	14	29	45	59	117	173	230	286	342	454
21	16	32	49	64	127	190	252	312	374	495
22	17	35	52	69	136	204	270	336	404	531
23	18	37	56	73	145	216	288	358	430	570
24	19	40	60	78	155	234	310	383	464	615
25	21	43	64	84	166	248	332	414	495	660
26	22	45	68	89	176	264	350	440	525	698
27	23	48	72	94	186	278	370	463	555	730
28	25	51	76	100	199	298	398	496	595	780
29	26	54	80	105	209	313	418	522	625	835
30	27	57	84	110	220	330	440	550	660	880
32	30	63	93	122	244	368	488	612	734	980
34		70	103	134	270	400	540	680	810	1060
36		76	110	146	290	440	590	740	880	1180
38		83	121	157	320	480	640	810	970	1300
40			131	170	350	520	690	880	1050	1400
42			142	184	380	560	750	940	1140	1520
44			152	198	400	600	810	1010	1210	1630
46			163	210	430	650	870	1090	1310	1750
48			174	230	460	690	920	1160	1400	1870
50				240	490	740	990	1240	1490	2000
52				260	520	790	1060	1320	1590	2130

Table continues on next page

TABLE 16.4   Free Flow Through Parshall Flume (continued)

Upper Head, H_a (cm)	7.5	15	23	30	61	91	122	152	183	244
			Throat Widths (cm)							
				Flow (l/s)						
56				290	580	880	1180	1490	1790	2410
58				300	610	930	1260	1580	1890	2540
60				320	640	980	1320	1660	2000	2690
62				340	680	1030	1390	1750	2120	2840
64				350	710	1080	1460	1840	2220	2980
66				370	740	1130	1510	1920	2320	3120
68				390	780	1190	1580	2020	2440	3280
70				400	820	1250	1670	2100	2560	3440
72				420	850	1290	1740	2180	2680	3580
74				440	890	1350	1820	2300	2780	3760
76				460	940	1420	1920	2420	2920	3940

Source: Water Measurement Manual, U.S. Dept. of the Interior, Bureau of Reclamation, 1953.

TABLE 16.5   Standard Dimensions of Parshall Measuring Flumes from 7.5- to 23-cm Throat Width

Dimension Letter (Fig. 16.4)	2.5 cm	5 cm	7.5 cm	15 cm	23 cm
	Dimensions (m) for Throat Widths (W) of				
A	0.36	0.42	0.47	0.62	0.87
2/3A	0.24	0.28	0.31	0.41	0.60
B	0.36	0.41	0.46	0.61	0.86
2/3B	0.24	0.27	0.30	0.40	0.57
C	0.09	0.14	0.18	0.39	0.38
D	0.17	0.21	0.26	0.39	0.57
E	0.23	0.25	0.38	0.45	0.61
F	0.08	0.11	0.15	0.30	0.30
G	0.20	0.25	0.03	0.61	0.45
K	0.02	0.02	0.03	0.08	0.08
N	0.003	0.017	0.06	0.12	0.12
X	0.008	0.016	0.03	0.05	0.05
Y	0.01	0.025	0.04	0.08	0.08

PLAN

SECTION A-A

Figure 16.14. Plan and longitudinal section of Parshall measuring flume. Dimensions are given in Table 16.5 for flumes having a throat width from 2.5 to 23 cm and in Table 16.6 for throat widths from 0.30 to 3.05 m.

TABLE 16.6 Standard Dimensions of Parshall Measuring Flumes from 0.30 m to 3.0 m

Throat Width, W (m)	Dimensions (m)[a]					
	A	$2/3A$	B	$2/3B$	C	D
0.30	1.37	0.9	1.32	0.9	0.61	0.84
0.61	1.52	1.0	1.47	1.0	0.91	1.21
0.91	1.65	1.1	1.62	1.12	1.22	1.57
1.22	1.83	1.22	1.77	1.22	1.52	1.94
1.52	1.98	1.32	1.93	1.30	1.83	2.29
1.83	2.13	1.42	2.08	1.40	2.13	2.67
2.13	2.28	1.53	2.25	1.50	2.44	3.04
2.44	2.44	1.63	2.39	1.60	2.75	3.40
3.05	2.75	1.83	2.65	1.79	3.35	4.12

[a]Letters refer to Fig. 16.14, in which other dimensions for these flumes are shown.

338

16.5 TRAPEZOIDAL FLUMES

Trapezoidal flumes somewhat similar to Parshall's rectangular flumes have been developed, Fig. 16.15. The head-discharge relationships are shown in Fig. 16.16. Based upon the study of Robinson and Chamberlain[2], the characteristics are listed as follows: (1) extreme approach conditions seem to have a minor ef-

[2]A. R. Robinson and A. R. Chamberlain, "Trapezoidal Flumes for Open Channel Flow measurement," *Trans. Am. Soc. Agr. Eng.*, Gen. Ed., Vol. 3, No. 2, pp. 120–124, 1960.

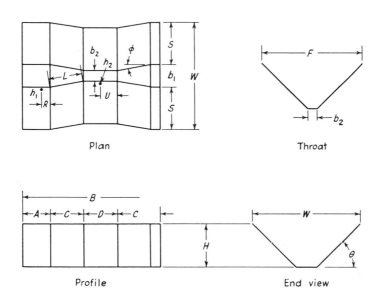

Flume no.	Description	b_1	b_2	A	L	C	D	E	F	H	B	R	S	U	W	θ	ϕ
1	Large 60°-v	5.1	0	17.8	17.8	17.6	17.8	7.6	20.0	17.1	78.4	3.8	10.2	8.9	25.4	60°	8.25°
2	Small 60°-v	5.1	0	12.7	10.5	10.3	12.7	5.1	12.1	10.2	51.4	2.5	6.0	6.4	17.2	60°	11.20°
3	5.1 cm-60° wsc	12.4	5.1	20.3	21.6	21.4	21.6	7.6	35.6	34.3	92.2	3.8	15.2	10.8	42.9	60°	9.22°
4	5.1 cm-45° wsc	12.4	5.1	20.3	21.6	21.3	21.6	7.6	56.4	26.9	92.1	3.8	26.9	10.8	66.2	45°	9.91°
5	5.1 cm-30° wsc	12.4	5.1	20.3	21.6	21.3	21.6	7.6	93.1	25.4	92.1	3.8	44.0	10.8	100.4	30°	9.80°
6	10.2 cm-60° wsc	20.3	10.2	22.9	25.4	23.9	25.4	7.6	50.8	35.2	105.7	3.8	20.3	12.7	61.0	60°	11.20°
7	5.1 cm-30° csu	25.4	5.1	25.4	27.3	25.4	25.4	7.6	90.1	24.5	109.2	3.8	42.5	12.7	110.4	30°	21.80°

Note: Dimensions in centimeters except as shown. See Fig.16.16 for the relationship of Q to depth in each of the 7 flumes.

Figure 16.15.   Details of designs of trapezoidal flumes. (After Robison and Chamberlain.)

Figure 16.16.   Calibration curves for trapezoidal flumes under free-flow conditions.

fect upon head-discharge relationships; (2) material deposited in the approach did not change the head-discharge relationships noticeably; (3) a large range of flows can be measured through the structure with a comparatively small change in head; (4) the flumes will operate under greater submergence than rectangular-shaped ones without corrections being necessary to determine the exact discharge; (5) the trapezoidal shape fits the common canal section more closely than a rectangular one; (6) construction details such as transitions and form work are simplified; (7) the relationship between head and discharge is not as easily expressed in the form of an equation as is the rectangular-

Figure 16.17.
Current meter showing rod suspension with double-end hanger and round wading base. (Courtesy The A. Leitz Co.)

shaped flume; (8) since a small change in head results in a comparatively large change in discharge, the sensitivity of the flume to changes in discharge is less than that for the rectangular shaped ones.

16.6 THE CURRENT METER

A device widely used by engineers for measuring flowing water is the current meter. Meters of three general types are used: the cup type with vertical axis shown in Fig. 16.17, the propeller type with horizontal axis shown in Fig. 16.18, and the electromagnetic probe shown in Fig. 16.19.

Propeller and Cup Type Meters

Cup meters are generally used by the U. S. Bureau of Reclamation and the U. S. Geological Survey. Propeller meters are favored in European countries. Both have their advantages. The cup meter is more sensitive to disturbances such as being

Figure 16.18.   Propeller type current meter. (Courtesy A. Ott.)

Figure 16.19.
Electromagnetic probe showing the tear-drop shaped sensing unit, meter, and rod used in measuring the velocity of flowing water. (Courtesy Marsh-McBirney.)

brought close to the bank and not being held stationary. The cup meter also offers greater resistance to the flow of water and is thus swept farther downstream when suspended from a cable than is the propeller meter. One distinct advantage of the cup meter, and one of the principal reasons why it is used extensively in the United States, is that it is a more rugged instrument than the propeller meter and thus can be used by relatively unskilled technicians. The propeller meter has been used for higher ranges of velocities than the cup meter (up to 6 to 9 meters per second as against 3 to 5 meters per second for the cup meter). The propeller meter can also be made much smaller than the cup meter. The small size is sometimes an advantage when measurements must be taken near the walls of pipes and the edges of channels where there is rapid change of velocity. Another major advantage of the screw-type propeller meter is that it is less likely to be fouled by floating weeds and debris. Hence, the choice of meter depends upon conditions under which the meter will be used. However, for field observations the cup-type meter has generally been found superior; whereas for pipe measurements and laboratory observations the propeller meter has been found to be more useful.

Electromagnetic Probe

Figure 16.19 shows the component parts of an electromagnetic probe. When the electromagnetic probe is immersed in flowing water, a voltage is created around the probe. This voltage, sensed by electrodes imbedded in the probe, is transmitted through the cable to the meter box. The voltage created by water flowing through the magnetic field is proportional to the velocity of the water. These small voltages are electronically processed and displayed on the panel meter. Accuracies of 2 percent are attainable over a range from 1.5 centimeters per second to 3 meters per second.

Measuring Technique

The accuracy of current meter measurements will depend largely upon the skill of the operator and conditions at the location of the measurement. Under ideal conditions the error may be less than 2 percent and is usually not more than 5 percent, providing the flow is steady.

The meter is calibrated by passing it through still water at a known speed and noting the number of revolutions per second. When the calibrated meter is held still in running water at the proper depth, it is thus possible to determine the average velocity of the water by observing the number of revolutions per second. The average velocity in streams not over 45 centimeters in depth is at about 0.6 of the depth from the surface[3]; in streams over 45 centimeters in depth the average velocity is better represented by the average of the velocities at 0.2 and 0.8 of the depth. In the measurement of flowing water it is essential that the current meter be placed at the point or points of average velocity.

Another method of determining the average velocity in a stream is the integration method, in which the current meter is raised and lowered slowly and at a constant rate from the bottom to the top of the stream. On practically all the larger canals and

[3]Some authorities have found that velocities measured at 0.6 of the depth from the surface in shallow streams usually range from 4 to 6 percent higher than the true average velocities.

on rivers discharge measurements are computed from current-meter readings of velocity and measured cross-sectional areas. The cross section of a stream should be divided into twenty or more parts, except for a very small stream. Not more than 10 percent, and preferably not more than 5 percent, of the discharge should occur in any one section. The velocity is then measured in each section and multiplied by the area to obtain the discharge of the section. The flow in the channel is obtained by adding the flow in each section.

By measuring the discharge of a canal or river at several different stages (or depths), data was obtained from which a relationship can be derived between the depth of the water and the discharge of the stream. The changes in depth are usually referred to a permanent bench mark, or elevation datum, and distances vertically above the datum are designated "gauge heights." After measuring the discharges at various gauge heights, a rating curve can be plotted, Fig. 16.20. This figure shows discharges ranging from zero discharge at zero gauge height to 2.5 cubic meters per second at a gauge height of 104 centimeters.

The major advantages of current meters are that they require no obstruction of stream flow and are suited to large streams. Water commissioners whose responsibility it is to distribute the public waters to those entitled to their use depend very largely on rating curves for their measurements. The gauge height may be read by nontechnical men, but the actual use of the meter and

Figure 16.20.   Typical rating curve.

the making of rating tables and rating curves are tasks for a trained and experienced hydrographer or for an engineer.

16.7 MECHANICAL MEASURING AND RECORDING DEVICES

A number of mechanical devices for water measurement have been designed, most of which measure the rate of flow and also automatically register the total amount of water passing in any given period of time. Canal companies that base water charges on the amount of water delivered to individual irrigators find self-recording devices serviceable and convenient. More extended use of mechanical-automatic registering devices will be made as water increases in value and more irrigation companies in order to stimulate economy among their irrigators find it necessary to base water charges on actual amounts of water delivered to encourage more efficient use of water.

Meters, although desirable under certain conditions, involve numerous practical difficulties. Therefore, one should understand the limitations of any meter before purchasing it for a particular purpose. The most common mechanical meters operate by the flow of water and may be classified as displacement, velocity, or by-pass type.

The displacement-type meter measures volumetrically. Water passing through the meter displaces a vane or disk, which in turn operates the recording mechanism. The operation is positive, and the measurement is fairly accurate. Such meters generally require a greater loss of head than velocity meters and are more expensive. Their use for measuring irrigation water is limited to pressure pipe lines in localities where water is rather expensive and stream flows less than 6 liters per second.

Velocity meters are operated by the kinetic energy of moving water. They usually contain an impeller vane turned by the water; and their operation is similar to that of a current meter. For high velocities, the impeller vane rotates at a rate almost directly proportional to the velocity of water; for low velocity it may turn more slowly, and below certain velocities it may not move at all. Velocity meters are less expensive and can accommodate larger flows than displacement meters.

The by-pass meter operates differently from either the velocity or displacement meters. Only a small part of the flow passes through the registering element, but the meter is calibrated to register the volume that passes. The percentage of water passing through it is not constant but usually varies slightly with the flow.

Mechanical meters have the advantage of eliminating computations in volume determination. This convenience often justifies the expense involved. Meters which are subject to clogging should never be used on pipe lines receiving water from open ditches unless the entrances are adequately screened to keep out debris.

16.8 VENTURI TUBES AND SIMILAR DEVICES

The venturi tube is a convergent-divergent tube which has been used extensively for measuring flow in pipe lines of diameters ranging from a few centimeters to several meters, but it has been used only to a limited extent for irrigation water. The standard venturi tube has been considered too expensive for irrigation purposes.

Modified venturi sections have been developed and used to measure irrigation water in open channels. They are used primarily where insufficient slope exists to permit accurate measurement by Parshall flume or weir. Losses of energy through the measuring section are minimized by using a movable diverging section over the top of the channel to recover most of the kinetic energy of the flow after it passes through the restricted section. The recovery of the kinetic engery is accomplished by increasing the cross-sectional area and thereby reducing the velocity. When this can be accomplished without excessive turbulence, most of the kinetic energy is transformed into potential energy.

The sides and bottom of the channel are straight while the top is a plate with a typical venturi shape. This top confining plate is moved vertically to accommodate the flow and thereby give a variable head loss and a variable accuracy of flow, greater head loss ensuring greater accuracy. When greater accuracy is desired, the throat section is made smaller with a resulting increase in velocity and head loss.

These modified venturi tubes are calibrated individually, and discharge charts and tables are provided for each tube. They have been in use for several years and have proved satisfactory. With them less energy is lost than with most other practical devices. Their principal disadvantages are: the lack of standardized sizes and shapes, the lack of information regarding the coefficients of discharge, and they are relatively expensive except when made in large quantities.

Flow nozzles, the thin plate orifice in pipes and the thin plate orifice on the end of a pipe, are being used for measuring irrigation water. These methods are particularly useful for testing pumps and measuring flow from wells.

16.9 THE DIVISION OF IRRIGATION WATER

Irrigation companies divide their streams according to the number of shares of stock owned by individuals or groups of individuals. Some users are less interested in the measurement of water than in the division of the stream. For example, one company may be entitled to 40 percent of the stream and the other company to 60 percent. Many times a division must be made where it is impracticable to make a measurement.

For satisfactory division, a few principles must be observed. The water must approach the divider in parallel paths; i.e., there must be no cross currents. To secure this condition the divider box must be placed at the lower end of a long flume or straight open channel. The floor of the channel immediately above the divider should be level transversely. If the water is reasonably free from silt, it is desirable to have it approach the divider at a low velocity. For streams carrying considerable silt and gravel there should be no obstruction in the channel in the form of a bulkhead, and the velocity with which the water enters should be maintained through the structure. Any gravel or debris allowed to collect in the channel of approach will cause cross currents and interfere with proper division.

The flow over a weir can be easily divided by placing a sharp-edged partition below the weir to divide the stream as it falls over the crest. The crest of this partition should be placed a sufficient distance below the weir crest to permit a free circulation of air between the divider and the sheet of water falling over the weir.

The discharge over a weir is not exactly proportional to the length of the crest; the error in considering it so is slight for low heads. Trapezoidal weirs are the most desirable type for dividers. The flow over this weir is very nearly proportional to length of crest. If it is desired to divide the stream into two parts, one taking 30 percent and the other 60 percent of the flow, the divider should be placed 30 percent of the distance from the end of the weir. Figure 16.21 shows a trapezoidal weir divider fixed to divide a stream into three parts.

If it is desired to divide a stream into two equal parts, the rectangular weir, either with or without end contractions, is entirely satisfactory. In localities where water is distributed to several

Figure 16.21.   Adjustable divider. (Courtesy Utah Agr. Exp. Sta.)

parts of a farm in underground pipes, it is essential to provide special boxes for making an appropriate division of the water. A typical concrete proportional division box connected to a pipe line is illustrated in Fig. 16.22. Any rotational motion of the water as it emerges from the vertical pipe and approaches the outlet will cause uneven flow.

16.10 FLOAT MEASUREMENTS

Quick, inexpensive measurements of discharge are frequently made by timing the velocity of an object on the surface of the water. The observed surface velocity must be reduced, since it is greater than the average velocity. The correction factor is influenced by roughness, shape of channel, and depth of flow. The observed surface velocity should be multiplied by about 0.85 to obtain the mean velocity. The correction factor may be as low

Figure 16.22. Proportional division box. (Courtesy U.S.D.A.)

as 0.8 and as high as 0.95. It increases as depth increases and diminishes as roughness of channel bed increases. Float measurements can be expected to involve errors of at least 10 percent.

16.11 FLOW FROM A VERTICAL PIPE

When water flows vertically out of an open pipe, the height to which it will rise above the pipe is proportional to the flow. Lawrence and Braunworth[4] made careful measurements of the flow and found that when the height of the jet was less than 0.37D, where D is the inside diameter of the pipe, the flow is similar to the flow over a weir and can be expressed by the following empirical equation, when the lower nappe springs free from the pipe:

$$Q = 3.1 \times 10^{-7}D^{2.5}H^{3.5} \qquad (16.15)$$

When the height of the jet exceeds 1.4D the flow is comparable to jet flow and can be expressed by the empirical equation:

$$Q = 0.029D^{1.99}H^{0.53} \qquad (16.16)$$

In both equations Q is the flow in liters per second, D is the inside diameter, and H is the average height of rise of the center of the jet, with both D and H measured in centimeters. For heights between 0.37D and 1.4D, the flow is somewhat less than the flow given by either equation. Using these equations, accuracies of 10 percent can be expected; however, the accuracy increases as H increases.

[4]F. E. Lawrence and P. L. Braunworth, "Fountain Flow of Water in Vertical Pipes," *Trans. Am. Soc. Civil Eng.*, 1906.

16.12 FLOW FROM A HORIZONTAL PIPE

The trajectory of a stream of water from a horizontal pipe can be used to estimate the discharge. Such a procedure is rapid, inexpensive, and convenient. Measurements are made of the X and Y coordinates, where X is measured parallel to the pipe and Y is measured vertically. The formula is obtained by combining the following three equations:

$$Y = \tfrac{1}{2}gt^2 \qquad X = Vt \qquad Q = AV$$

$$Q = C' \sqrt{\frac{g}{2}} \frac{AX}{\sqrt{Y}} = 0.022CA \frac{X}{\sqrt{Y}} \qquad (16.17)$$

where Q = discharge, in l/s
$\quad\ C$ = Coefficient of discharge (see Table 16.7)
$\quad\ g$ = acceleration of gravity
$\quad\ A$ = cross-sectional area of water at end of pipe, in cm^2
$\quad\ X$ = coordinate of the point on the surface measured, in cm to the pipe.
$\quad\ Y$ = vertical coordinate measured in cm

TABLE 16.7   Approximate Coefficients of Discharge for Use in the Coordinate Method for Measuring the Rate of Water Flow from Horizontal Pipes

d/D	A. Flowing Partially Full at the End X/D							
	1.00	1.50	2.00	2.50	3.00	4.00	5.00	8.00
0.2	1.02	1.01	1.00					
0.3	1.11	1.06	1.03	1.02	1.01	1.00		
0.4	1.17	1.10	1.06	1.03	1.02	1.01	1.00	
0.5	1.22	1.13	1.07	1.04	1.03	1.01	1.01	1.00
0.6	1.26	1.15	1.09	1.05	1.03	1.02	1.01	1.00
0.7	1.30	1.17	1.10	1.06	1.04	1.02	1.01	1.00
0.8	1.33	1.18	1.11	1.07	1.04	1.02	1.01	1.00

Y/D	B. Flowing Completely Full at the End X/D							
	1.00	1.50	2.00	2.50	3.00	4.00	5.00	8.00
0.5	1.44	1.28	1.18	1.13	1.10	1.06	1.03	1.00
1.0	1.37	1.24	1.17	1.12	1.09	1.06	1.03	1.00
2.0		1.11	1.09	1.08	1.07	1.05	1.03	1.00
3.0			1.04	1.04	1.04	1.04	1.03	1.00
4.0			1.01	1.01	1.02	1.03	1.02	1.00
5.0			0.97	0.99	1.00	1.01	1.01	1.00

Note that the coordinates are measured from the surface of the water as it emerges from the pipe. The coefficient of discharge C varies greatly depending primarily upon the depth of flow in the pipe and the coordinates X and Y. Variation for a partially full pipe is shown in Table 16.7, where the depth of flow at the end of the pipe is designated by d and the inside diameter of pipe by D. The table indicates that the coefficient of discharge has the value of 1.33 for $d/D = 0.8$ and $X/D = 1.00$. Hence this coefficient is extremely important, if accuracy is to be obtained.

The error involved with the coordinate method will seldom be less than 5 percent. Errors of 10 percent can be expected, even when the coefficient is used, if the jet is turbulent and the free surface poorly defined. Wherever practical, X/D should not be less than 3.

16.13 DISCHARGE FROM SIPHON TUBES

When water is diverted from a ditch using a siphon tube, the head and diameter can be used to estimate the discharge, Fig. 16.23, with an accuracy of 5 to 10 percent. In essence, the orifice-discharge equation applies to this type of flow with the discharge coefficient depending upon the length of siphon and entrance and exit conditions of flow. The effective head causing flow is the difference between the water level in the ditch and the level of water in the furrow for a submerged outlet. When the outlet is not submerged, the effective head is the height of the water in the ditch above the center of the outlet end.

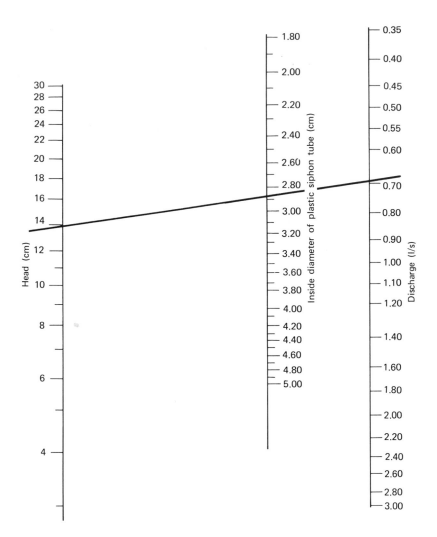

Figure 16.23.   Discharge versus head for plastic siphon tubes.

CHAPTER
17
LEGAL AND ADMINISTRATIVE ASPECTS OF IRRIGATION AND DRAINAGE

Since legal and administrative aspects are so diverse, it is impractical to present more than a brief summary of the important principles. To illustrate these principles, legal and administrative patterns in the United States and irrigation practices in the arid West will be used. The individual interested in irrigation should become intimately acquainted with legal and administrative structures used in his own area and then do his utmost to improve the existing patterns.

17.1 IRRIGATION AND DRAINAGE ENTERPRISES

The type of organization formed to deliver irrigation water and to drain land is vitally important. The legal powers and the relation between the organization and the water users are governed by the kind of organization. Good irrigation projects have failed because the wrong organization was adopted. The most successful enterprises have been those owned and operated by the water users. When the users are involved directly, their interest is keen and their services are rendered at a nominal cost. Most private associations formed to make a profit through the conveyance and delivery of irrigation water have become insolvent, or have been replaced by cooperative organizations formed by the water users. Most successful irrigation and drainage enterprises are operating on a nonprofit basis. Three types of organizations are used: private, quasipublic, and public.

Private enterprises include individual projects, and mutual and commercial company organizations.

Quasipublic enterprises operate under public laws which prescribe procedure and public agencies participate in the organization and management without assuming direct financial responsibility.

Public irrigation enterprises are organized under public laws, administered by public agencies, and financed with public funds.

17.2 PRIVATE ENTERPRISES

Commercial Companies

The irrigation enterprise which supplies water for compensation to irrigators who have no direct financial interest in the irrigation works, or who hold an equity which has not yet ripened into ownership and control, is designated a commercial irrigation company. Some commercial companies furnish water on an annual rental basis; others sell the prospective irrigator a water right and in addition charge an annual rental; and some sell a water right which carries with it a perpetual interest in the irrigation system. Commercial companies of the last type ultimately become mutual companies in which the irrigation works are owned and operated by the irrigators. Service rates of companies furnishing water on an annual rental basis are generally subject to public regulation as a result of dedication of the water to public use. The annual rentals charged by companies which sell water rights are not subject to public regulation if the contracts for sale of rights and charging of rentals are held to be private contracts.

Individual and Partnership Enterprises

Comparatively few large, but many small, irrigation projects are built by individuals working alone or in partnerships. Small streams adjacent to arable land in isolated sections favor individual effort in irrigation. Also, where ground water is available for pumping, or where other water sources may be best developed by small pumping plants, individuals build and operate their own irrigation projects. The advantages claimed for individual enterprises are that they permit the farmer to irrigate at any time he desires, so that he can regulate his own practices, and he is independent of the assessments, rules, regulations, and irrigation practices of his neighbors.

Individual irrigation activity is usually more expensive than the combined activity of groups who need irrigation water, and, moreover, it is rigorously restricted by nature since it is quite impossible, as a rule, for the farmer to build the storage and diversion works, and conveyance canals necessary to provide water for lands which are not near the sources of water supply.

Cooperative Enterprises

There are water resources and arable lands which can be brought together by small groups of individuals voluntarily forming an association for the purpose of constructing and operating irrigation systems. Cooperative irrigation enterprise is locally designated by a variety of names. All such enterprises are included in two main types of organization, incorporated, referred to herein as mutual irrigation companies, and unincorporated, referred to as mutual associations. The incorporated group units are far more numerous and extensive.

The success of an unincorporated mutual association, sometimes designated as a mutual company, rests largely on the fairness and congeniality of each member, because the association provides no means of legally enforcing the payments of dues or of enforcing contributions to the maintenance, betterment, renewal, and operation or expansion of the project. The major asset of the association is the labor of its members. Its activities are limited to small projects which require no difficult construction and but little capital.

Mutual Irrigation Company

A corporate body of irrigators, voluntarily organized for the purpose of supplying water to its stockholders, is known as a mutual company. It is a nonprofit organization for the delivery of water to its members only. It obtains its revenues by stock assessments, and its dividends consist of water delivered in proportion to the stock owned by each irrigator. It enforces payment of assessments by the sale of stock if necessary. The stockholders delegate the responsibility of management to a board of directors, from three to seven or more, elected by ballot. Each stockholder has as many votes as he owns shares of stock. The tenure of office of directors, fixed by the articles of incorporation, ranges from 1 to 3 or more years. The directors elect one of their members as president and appoint a secretary, treasurer, and watermaster, any of whom may or may not be directors. In the small irrigation companies the watermaster has charge of the project operation and maintenance, including the distribution of water to stockholders. The watermaster, with the aid of check gates, takeout gates, and diversion structures, is expected to distribute equitably a valuable commodity to numerous claimants.

The larger mutual companies sometimes employ an engineer-manager who is assigned the responsibility of water distribution and to whom the watermasters, one for each of several districts, are instructed to report. The mutual irrigation company has wide flexibility. It is especially suited to maintenance and operation of irrigation projects and is the dominant type of operating organization in many irrigated areas.

Mutual companies are exempt from general taxation as long as they are used for the service of their own stockholders only. Some local governmental units relieve mutual irrigation companies from license taxes assessed against corporations.

17.3 QUASIPUBLIC ENTERPRISES

Irrigation Districts

An irrigation district is a quasipublic corporation for providing water for lands within its boundaries. An irrigation district is a public corporation, a political subdivision of a State with defined geographical boundaries. It is created under authority of the State legislature through designated public officials or courts and with the consent of a designated fraction of the landowners or of the citizens, as the case may be, of the particular territory involved. Being public and political, the formation of a district is not dependent upon the consent of all persons concerned, but may be brought against the wishes of the minority. In this respect the district differs fundamentally from the voluntary mutual company and the commercial irrigation company.

It is a cooperative undertaking, a self-governing institution, managed and operated by the landowners or citizens within the district. Supervision by State officials is provided for to the extent of seeing that the laws are enforced, and in most States is extended in greater or less degree over organization, plans, and estimates prior to bond issues, and construction of works.

It may issue bonds for the construction or acquisition of irrigation works, which bonds are payable from the proceeds of assessments levied upon the land. Hence, it has the taxing power. Each assessment becomes a lieu upon the land. While the ultimate source of revenue is the assessment, an additional source frequently provided for is the toll charged for water. Other revenue may in some cases be obtained from the sale or rental of water or power to lands or persons outside the district.

Finally, the purpose of the irrigation district is to obtain a water supply and to distribute the water for the irrigation of lands within the district. Additional authority is granted irrigation districts, almost without exception, to provide for drainage. In some States districts may also develop electric power. These additional powers, however, are subsidiary and are intended to make more effective the principal function of the organization, which is to provide irrigation water.

The irrigation district has become an increasingly important irrigation agency. It is adapted to large-scale irrigation enterprise. Many of the large irrigation systems are managed by this type of organization.

17.4 PUBLIC ENTERPRISES

Many federal and state governments have found it advantagous and necessary to enact reclamation laws to provide direct use of federal or state funds without interest or at low interest for construction of large irrigation projects. These efforts have made outstanding contributions in the design and building of engineering structures of large magnitude.

The usual features of projects constructed by federal and state funding are:

(a) The settler has the use of public money for a period of sufficient length to permit repayment of construction costs.

(b) Public lands and water resources too costly to be developed by private capital are included in many projects.

(c) Until a substantial part of the construction charges are paid the project is under control of the federal or state government.

(d) Annual payments of construction charges and operation and maintenance costs are fixed by the government and paid by the settler to its representatives.

(e) Construction costs of multiple-purpose projects are prorated according to the benefits to irrigation, municipal water development, power development, flood control, navigation, recreation, and other purposes.

The essentials of success, as on irrigation districts and other irrigation enterprises, have proved to be productive land, sufficient water, reasonable construction costs, and adequate land settlement.

17.5 LEGAL ASPECTS OF IRRIGATION AND DRAINAGE

The legal as well as the administrative structure under which irrigation and drainage are practiced will often ensure the success or seriously retard, if not destroy, the undertaking. Therefore the discussion of technical aspects alone is not sufficient.

The student of irrigation and drainage must understand and appreciate the importance of legal and administrative aspects. The legal aspects focus upon two conflicting doctrines, riparian rights and appropriation.

Doctrine of Riparian Rights

Settlers along streams in humid regions used the water as needed under the common-law doctrine of riparian rights, each owner along the stream being entitled to have the water flow in its natural channel "undiminished in quantity and unpolluted in quality." Use does not create nor does disuse destroy the riparian right.

Since the doctrine of riparian rights does not contemplate or allow for consumptive use of water or pollution, changes in concept in arid regions were inevitable. Thus in some areas modification of doctrine has been made to allow each proprietor to divert water for use on his adjoining lands for "domestic" and "natural purposes." In some areas diversion for these purposes is permitted even though it may consume the water previously used by a lower riparian owner. Thus no right to the water is secured by use.

Use of water for irrigation is considered as "artificial" and is not normally allowed. Modifications have been made which result in the "natural flow" doctrine which permits a riparian owner to divert water for irrigation only if there is a surplus to the needs of lower riparian owners, and if such diversion does not appreciably lower the water level or quality. A "reasonable use" doctrine permits a more liberal removal of water for irrigation.

Under any of the riparian doctrines, beneficial use does not create priority. Usually land must be adjacent to the stream, and diversion as well as return flow must be made upon the same land to be irrigated.

Riparian and modifications of the riparian doctrine are used in the more humid regions. However, the increasing use of water for irrigation in humid areas is creating serious water-right problems which are being solved by adoption of a modified doctrine approaching the doctrine of appropriation.

Doctrine of Appropriation

When irrigation was begun in the arid West of the United States, the doctrine of riparian rights was readily seen to be inadequate. Agricultural and community developments depended upon a reliable supply of water. Frequently the existing waters were used fully. Development, therefore, depended upon establishing a right to the water by prior and by beneficial use. Out of this need came the doctrine of appropriation which asserts that all water rights are founded upon priority of use: that use creates the right, and that disuse destroys or forfeits the right. Beneficial use is declared "the basis, the measure, and the limit of the right."

17.6 GROUND-WATER LAWS

Laws governing surface water are generally applied in their essence to ground water. However, because of the difficulty of defining ground water, considerable confusion exists. The courts have attempted to distinguish between different underground streams; "underflow of surface streams," "percolating waters," "artesian water," etc. Since these distinctions cannot be supported by sound physical facts, much confusion has resulted.

Vigorous attempts are being made to establish good, sound laws which will permit maximum utilization of underground water reservoirs and resources.

The common law doctrine of riparian rights is still used in many states. However, an increasing number of states have some form of ground-water laws based on the doctrine of appropriation, and the reasonable-use concept is being applied increasingly to the riparian doctrine.

17.7 DRAINAGE ENTERPRISES

Group action is essential in the drainage of irrigated land. For one landowner to be able to provide adequate drainage for his land without cooperating with his neighbors is the exception rather than the rule. Group action in some areas is obtained by organizing drainage districts which are quasipublic corporations provided for by state laws and which have authority to tax irrigated land for drainage purposes.

Two principal purposes influence the organization of drainage districts, namely: (1) to consolidate into one drainage agency the lands of an area in need of drainage and contributing to that need, and (2) to provide the authority and the procedure, and to assign the responsibility for the design, financing, construction, and maintenance of drainage systems.

Noteworthy powers of a drainage district are: (1) the power to include in the district all lands to be benefited by the drainage system, and thus assure strength of the district and ability to promote equity in its dealings with landowners; (2) the power to tax lands of the district and enforce tax collections. The second power carries the authority to foreclose on tax-delinquent land and sell it if necessary.

Drainage districts have legal authority to carry out all the functions and activities pertaining to drainage of farm land including financing, design, and installation of drains, and their operation and maintenance.

There must be a need for drainage in the area proposed for the district, and it must be shown that the benefits to the included lands will exceed the costs; the desires of a majority of the landowners to participate must be expressed; the specified organizing procedure must be carried out. Land ownership within a drainage district is the usual requirement for membership and participation in district activities.

After a district has been organized, its taxing procedure set in operation, its capital financing provided, and its drainage system installed, its problems concern largely management, operation, and maintenance, and the discharging of financial obligations.

Frequently, a board of three supervisors constitutes the governing body of the drainage district and carries the responsibility for all the district affairs. Supervisors are appointed by the local elected officials usually upon nomination by the district landowners. The supervisors elect from their number a president, secretary, and treasurer. They must ascertain the needs for drainage in the area and determine the nature, extent, and probable cost of a drainage system, designed to meet those needs. It is also their responsibility to provide capital financing for the installation of the drainage system, usually involving the issue and sale of district bonds, and to set up an equitable taxing system to provide adequate and dependable finances to meet the annual revenue requirements. The ability and willingness of supervisors to serve the district may determine the success or failure of the enterprise.

The large irrigation enterprises, irrigation districts, and mutual water users' associations are developing more and more interest in the drainage of irrigated lands. The trend is toward combining responsibilities for irrigation and drainage systems into one organization, particularly where legal authority is adequate for an irrigation company to assume responsibility for irrigation and drainage.

PROBLEMS AND QUESTIONS

CHAPTER 1

1. Define irrigation.
2. List six purposes for applying irrigation water to the soil.
3. How is irrigation accomplished?
4. What are the four major sources of irrigation water? Define and explain each.
5. What precautions should be kept in mind in determining the amount of water available from precipitation?
6. Is all of the precipitation that falls on cropland available to the crops? Defend your answer.
7. What atmospheric conditions, other than precipitation, will produce significant amounts of supplemental water?
8. Does ground water contribute directly to the water needs of plants? When? When is it harmful?
9. What are some of the advantages of irrigation in humid areas?
10. How does the need for irrigation in humid areas differ from that in arid regions?
11. What are the major factors which should be considered in the development of an irrigation project?
12. Answer the following questions in relation to the agricultural area most familiar to you.
    (a) What is the present status of irrigation?
    (b) How do the climate, water supply, and soils influence irrigation practices?
    (c) In what manner will the agricultural economy be improved by more extensive and improved irrigation?

CHAPTER 2

1. Why is accurate information regarding water supply important to the development of an irrigation system?
2. What is a snow survey?
3. What are the advantages of small ponds and reservoirs for irrigation? What are some of the disadvantages?
4. What factors should be considered when selecting a site for a reservoir?
5. How can reservoir silting be reduced or eliminated? List four ways and describe the operation of each.
6. What can be done to reduce loss of irrigation water by evaporation?
7. What are phreatophytes? Are they beneficial? Why?
8. What are the two principal methods of "rainmaking"? Explain each.

9. Discuss the feasibility of saline water conversion for irrigation purposes.
10. What are some of the sources of water that can be used for recharging ground water reservoirs?
11. What are the methods used in ground water recharge operations?
12. What does "safe-yield" refer to?
13. List three approaches to ground water investigations and describe the techniques used in each.

CHAPTER 3

1. Is it practicable for the irrigation farmer to greatly modify the texture of his soils? Why?
2. What soil structure is best suited to irrigation and crop production? Describe ways in which the farmer can maintain a favorable structure in his soil.
3. Distinguish between the *real* and the *apparent* specific gravity of a soil. Is it possible for the apparent specific gravity to be equal to or larger than the real specific gravity? Explain.
4. What substances occupy the pore space of a soil? Is the percentage pore space of a field soil influenced by its water content?
5. Why is the rate of water-flow into soils of importance in irrigation practice?
6. For a soil of given texture and structure, will a 1.4 meter depth of well-drained root-zone soil hold twice as much irrigation water as one of 0.7 meter depth? Assume that the water table is 10 meters or more below the land surface. Give reasons for your answers.
7. What properties of the soil determine the percentages of these three classes of moisture in the soil: hygroscopic, capillary, and gravitational?
8. Are irrigated soils that are naturally well drained ever completely saturated? Explain.
9. Why are the actual heights to which water will rise by capillary action in a soil usually less than the theoretical heights computed from Equation 3.5?
10. How can the concept of field capacity be determined and used even though there is no point on the moisture drainage curve that uniquely defines field capacity?
11. Why is the moisture content at which a crop permanently wilts a function of consumptive-use rate as well as soil texture?
12. A sharp-edged cylinder 150 millimeters in diameter is carefully driven into the soil so that negligible compaction occurs. A 200 millimeter column of soil is secured. The wet weight is 5525 grams and the dry weight is 4950 grams.
    (a) What is the percent moisture on a dry weight basis?
    (b) What is the apparent specific gravity of the soil?
13. A cylinder was carefully pushed into the soil without compressing or disturbing the soil. The cross-sectional area of the cylinder was 0.025

square meters. The length of the column of soil within the cylinder was 0.30 meters. The weight of the soil within the cylinder was 9.5 kilograms when it was dried. The weight of the soil before drying was 11.4 kilograms. Determine P_w, A_s, P_v.

14. A stream of 115 liters per second is used to apply 130 hectare-millimeters of water per hectare to a 3.5-hectare field. How long will it take?

15. An irrigator uses a stream of 100 liters per second for two days (48 hours) to irrigate 12 hectares of sugar beets. What is the average depth of water applied?

16. A farmer desires to irrigate a border which is 12 meters wide and 150 meters long. He wants to apply an average of 75-millimeter depth of water to the area with a stream of 60 liters per second. How long will it take him to irrigate this border?

17. The soil moisture at field capacity is 27.2 percent and the moisture content at the time of irrigating is 19.0 percent. The apparent specific gravity is 1.3 and the depth of soil to be wetted is 1 meter.
    (a) How many hectare-millimeters per hectare of water must be applied?
    (b) How long will it take to irrigate the 5 hectares with a 115 liter per second with a stream?

18. Soil samples indicate an average moisture and apparent specific gravity in the soil as follows:

Depth	P_w (dry wt)	A_s
0– 300 mm	14.7	1.34
300– 600 mm	15.3	1.39
600– 900 mm	17.6	1.32
900–1200 mm	18.2	1.30

Compute the depth of water held in the first 1.2 meters.

19. A farmer owning "bench" land, in which the sandy loam soil averaged about 1.25 meters in depth and was underlain by gravel and coarse sand to a depth of 10 meters or more, discovered in March by borings with a soil auger that the light winter precipitation had penetrated the soil to a depth of only 150 millimeters. He at once applied a 150 liter per second stream of flood water to his 10-hectare tract and kept the stream well spread out on the land for a period of four 24-hour days in order to give the soil a good soaking. There was no surface runoff. Find approximately what percentage of the water applied was lost by deep percolation.

CHAPTER 4

1. Why is measurement of soil moisture important?
2. What characteristics should be considered when deciding whether to use a spiral-bit auger or a posthole auger to obtain a soil moisture sample?

3. Concerning the Oakfield probe:
   (a) Under what conditions does it work best?
   (b) What conditions make it difficult to use?
4. For what purposes would you consider feel and appearance of the soil sufficiently accurate as an indication of soil moisture content?
5. What are the principal advantages and disadvantages of adsorption blocks for determining soil moisture?
6. What characteristics should be considered when deciding whether resistance blocks or tensionmeters should be used?
7. Describe the principle of operation of the neutron method of measuring soil moisture.
8. List eight methods of estimating soil moisture.

CHAPTER 5

1. Show by computation that kinetic energy can usually be ignored for flow-through soils.
2. From Equation 5.5 and the definition following, it is seen that A is the gross area at right angles to the flow direction. What approximate change would occur in the computed magnitude of permeability k for flow-through loam soil if A were taken as net cross-sectional area?
3. Imagine a soil column of unit cross-sectional area at right angles to the direction of flow of water and state whether or not it is practicable to measure accurately the net cross-sectional area of the channels through which flow occurs: (1) for a saturated soil; (b) for an unsaturated soil. Give reasons for your answer.
4. Consider a vertical soil column of 0.25-square meter cross-sectional area and 1 meter long. If 0.4 cubic meter of water percolate through the column in 36 hours from a supply pipe which permits the water to flow onto the soil just fast enough to keep the soil surface covered, what is the permeability in meters per 24-hour day?
5. Measurement of the permeability of a 15-meter stratum of saturated clay soil overlying a water-bearing gravel shows that $k = 0.5$ mm/hr. If the pressure head in the gravel is 25 meters of water (measured at the lower surface of the clay) and 0 m near the soil surface, water is flowing vertically upward through the clay. Compute the flow in cubic feet per second through a block of clay 15 meters thick and 250 hectares in area.
6. A contour map of water pressures overlying an artesian basin shows an average fall in pressure head of 5 meters per kilometer. Assume a mean thickness of water-bearing gravel of 8 meters and that $k = 0.06$ mm/s. Compute the underground flow in cubic feet per second through a section of gravel 300 meters long at right angles to the direction of flow.
7. Assume that a 16-hectares tract of land is irrigated frequently and given enough water to keep the soil practically saturated below the 2-meter depth but that the water table is 30 meters deep. If the aver-

age $k = 35$ l/s per ha, compute the number of hectare-millimeters of water that flow vertically downward to the water table each month.

8. A soil sample weighing 455 grams when brought from the field in a cylinder of 300 cubic centimeter volume, weighed 375 grams when dried. The soil particles displaced 145 cubic centimeter of water. Compute the following:

(a) Proportion of the total volume filled with water.

(b) Percentage of the pore space filled with water.

9. The following intake data were taken prior to the designing of an irriation system:

Elapsed Time (min)	Accumulated Intake (mm)
2	7.1
4	10.6
6	11.9
10	15.2
30	27.9

Determine the values of a and n in Equation 5.8

10. Intake data obtained using cylinders and buffers will most closely approximate the actual intake resulting from what method of irrigation?

11. Discuss the characteristics of the three basic zones in the phenomena of water movement through soil during irrigation.

CHAPTER 6

1. Explain why humid-region soils do not contain excessive amounts of alkali.

2. How does the amount of water moving through the soil affect the salinity or lack of salinity in the soil?

3. Water having a conductivity of 500 micromhos per centimeter has how many tons of salt per hectare-millimeter? What is the concentration in milliequivalents per liter and in milligrams per liter?

4. Is sodium carbonate a black salt? What is black alkali? What salts give rise to the occurrence of black alkali?

5. Why is exchangeable-sodium percentage important in reclamation of soils?

6. Provided one-half of the 831 milligrams per liter of alkali salts in lower Sevier River irrigation water were deposited in the upper 1 meter of soil each year, how many years would it take to add 0.5 percent total salts to the soil provided 0.67 meter of water is applied to the soil each year? Assume $A_s = 1.40$.

7. A drain tile main outlet from a 400-hectare tract discharges an average of 30 liters per second during each of the 12 months of the year.

If the average salt content of the drainage water is 1200 milligrams per liter, the irrigation water applied to the tract is practically free from salinity, and the mean depth of drains is 2 meters, what is the annual reduction in salt content of the soil in terms of the percentage of the weight of the dry soil?

8. What added precautions are necessary in reclaiming an alkali soil that need not be considered when reclaiming a saline soil?

9. Explain why it may be easier to leach an alkali soil with salty water than with pure water.

10. Explain fully why lowering of the water table is helpful in the prevention of accumulations of soluble salts on the surface of the soil.

11. In addition to the lowering of the water table, describe other means of preventing, or at least decreasing, the accumulations on the surface of the soil.

12. Under what conditions, if any, is it advisable to use, for irrigation purposes, water that contains appreciable percentages of soluble salts? What precautions are necessary to minimize the danger of using saline irrigation water?

13. Are the texture and the structure of soils related to the salinity and alkali problems? If so, explain.

14. What is the primary reason for adding chemical amendments to the soil?

15. What is the most important reclamation procedure for saline soil?

16. Of what value is water-quality data?

17. Define an alkali soil. Define a saline soil.

18. Describe an agricultural area where in your judgment proper management of a high water table would be a better solution economically than drainage.

19. Discuss the practicality of each of the three methods for temporarily controlling salts on irrigated land.

20. In the permanent reclamation of saline and alkali lands, what four conditions should be maintained? Why are all four necessary?

21. What management precautions must be taken when using saline waters for irrigation?

22. Is the following statement true? If the irrigation water is salty, less water should be applied because this will result in less salt being applied to the soil. Discuss.

CHAPTER 7

1. Are studies of the consumptive use of water in irrigation likely to increase in importance as time advances? Why?

2. Are the terms "consumptive use" and "evapotranspiration" the same? Explain.

3. Draw a graph showing how both transpiration and evaporation vary during the growing season for a particular crop. Explain briefly the trends shown.

4. List all of the factors that affect consumptive use.

5. List those factors affecting the evaporation component of consumptive use and state which factors managment can alter, thereby changing evaporation.

6. How is the wilting of a plant related to the amount of available energy in the atmosphere?

7. Will a light rainfall reduce the consumptive use? Explain.

8. What are the principal liabilities and limitations of the following methods of determining consumptive use:
   (a) Tanks and lysimeters?
   (b) Soil-moisture studies?

9. What kind of valley is best suited for determining consumptive use by the inflow-outflow method?

10. What climatic data is required for each of the following formulas: Hargreaves, Jenson-Haise, and Blaney-Criddle?

11. Would you expect a Piche evaporometer and an atmometer to indicate a higher evaporation ratio than an evaporation pan? Why?

12. Explain why consumptive use rates will vary from day to day because of the stage of growth of the plants.

13. Estimate the value of the consumptive use-evaporation ratio for the following stages of crop growth:
    (a) Tasseling of corn.
    (b) Flowering of potatoes.
    (c) Filling of pea pods.
    (d) Ripening of tomatoes.
    (e) Cutting of alfalfa.
    (f) Digging of sugar beets.
    (g) Cutting of sugar cane.
    (h) Harvesting of lettuce.
    (i) Picking of cotton.

14. A field of corn about 1.5 meters tall is growing vigorously, and tasseling has just started. What would you estimate the consumptive use-evaporation ratio to be? If records indicate the evaporation to be 5 millimeters per day, what would the consumptive use be?

15. Using the Penman method, compute the consumptive use in millimeters per day for spring wheat at Logan, Utah.
    Air temperature = mean max. = 27°C, mean min. = 10°C.
    Dewpoint = 7°C.
    Wind speed, U_2 = 175 km/day at 2m.
    Average solar radiation = 585 langleys/day.
    Average clear day solar radiation = 800 langleys/day.

16. Using the Jenson-Haise method, compute the average daily consumptive use in inches of water per day for the same conditions: as Problem 15.

17. Using the Hargreaves method, compute the average daily consumptive use in millimeters per day for the same conditions as Problem 15.

18. Using the Blaney-Criddle method, calculate the annual consumptive use for cotton growing, with the temperatures and percent daytime hours given in Table 7.4. Assume an average k = 0.70.

CHAPTER 8

1. List the three principal factors which influence time of irrigation and amount of water to apply.
2. Outline the reasons for reduced yield when the soil is too wet or too dry.
3. List several crops which show by their appearance that soil moisture is deficient. Describe the appearance of each crop listed.
4. Why is it poor practice to wait until a crop shows signs of drought before irrigating?
5. By early spring snow surveys show that irrigation water supplies will be critically short during the latter part of the growing season. The crops normally grown are alfalfa, sorghum, and cotton. Outline the adjustment in cropping practice that you would recommend to meet the impending drought.
6. What irrigation and fertility practices should be followed when vegetative growth is desired?
7. What are the major purposes of irrigating soils during the nongrowing or dormant season?
8. Under what conditions, if any, is it justifiable to divert water from partly filled storage reservoirs during the nongrowing or dormant season for irrigation purposes?
9. Discuss the feasibility and need for fall and winter irrigation in your area.
10. Why does early spring irrigation often retard growth of plants?
11. Describe the probable difference in root systems of crops which have been irrigated by frequent, light irrigations and widely-spaced, heavy irrigations.
12. A crop of alfalfa is being produced on a net water use of 600 millimeters. Each of four irrigations applies 150 millimeters of water over the surface.
    (a) Assuming the soil moisture extraction pattern shown in Fig. 8.6, calculate the average depth of water stored in each 200 millimeters of the 1.2 meter of root zone.
    (b) If the total water-holding capacity of the soil is 160 millimeters per meter, what percentage is stored in each meter during each irrigation?
13. What fertilizer element is most essential during (a) vegetative growth? (b) fruiting?
14. How should irrigation practice be modified during vegetative, wet fruit, and dry fruit stages of plant growth?
15. Explain how irrigation practice may influence the availablility to plants of nitrogen and of phosphate fertilizer in the soil.
16. A rather limited water supply had been applied to a field by sprinkler irrigation for several years, allowing only 25 millimeters of water to enter the soil per irrigation. Explain why the fertile, deep-loam soil exhibited droughty characteristics and was considered to be shallow and low in fertility under these conditions of management.
17. Calculate the consumptive use rate of a crop that is just entering the

wet fruit stage of growth when the evaporation is 8 millimeters per day. If 9 days have elapsed since the last irrigation, how much water has been depleted from the soil? What depth of water should be applied if the irrigation efficiency is to be 60 percent?

18. Enumerate the conditions in the order of their importance, which you consider most essential to the attainment of community economical use of irrigation water.

19. (a) State three major conditions which tend to satisfy irrigation farmers with a low water-application efficiency.

(b) State three major conditions which tend to stimulate irrigators to attain a high water-application efficiency.

20. The soil of an irrigated farm is a clay loam of comparatively uniform texture to a depth of 1.8 meters, below which there is a coarse gravel to a great depth. Moisture determinations before irrigation and again 48 hours after irrigation showed an average of 115 hectare-millimeters per hectare irrigation water stored in the soil from an irrigation in which the irrigator used a stream of 85 liters per second continuously for 24 hours on a 4-hectare tract of alfalfa. Neglecting consumptive use between completion of irrigation and the taking of samples for moisture determinations, what was the water-application efficiency? What was the efficiency considering that 7.5 millimeters per day was used on each of the two days between irrigation and sampling?

21. The average apparent specific gravity of the soil of the tract considered in Problem 20 is 1.3. Provided the mean increase in moisture content to a depth of 1.8 meters equals 5.35 percent, what is the water-application efficiency?

22. Three cubic meters per second are diverted from a river into a canal. Of this amount 2.25 cubic meters per second are delivered to the farms. The surface runoff from the irrigated area averages 450 liters per second. The contribution to the ground water is 300 liters per second.

(a) What is the water-conveyance efficiency?

(b) What is the water-application efficiency?

23. A farmer irrigates 2 hectares of wheat the first week in July when the average depth of rooting was 1.10 meters. Two days after irrigation he takes a soil auger to the field and by boring holes into the soil determines that the average depth of penetration in each ¼ hectare of the 2 hectares is as follows:

Average Depth of Penetration (m)

.82
.98
1.10
1.22
1.07
.90
.85
1.10

(a) What is the water-distribution efficiency?

(b) What is the water-storage efficiency?

24. Determine the water-application efficiency, the water-storage efficiency, and the water-distribution efficiency for the following conditions:

Stream of 85 liters per second delivered to the field for 5 hours.

Runoff averaged 40 liters per second for 2.5 hours.

Depth of root zone was 1.8 meters

Depth of penetration varied linearly from 1.8 meters at one end to 0.6 meters at the other end of the field.

25. A stream of 175 liters per second was diverted from the river and 105 liters per second were delivered to the field. An area of 2 hectares was irrigated in 8 hours. The root-zone depth was 1.8 meters. The runoff averaged 50 liters per second for 3 hours. The depth of water penetration varied linearly from 1.8 meters at the head of the field to 1.2 meters at the end of the field. Determine the water-conveyance efficiency, the water-application efficiency, the water-storage efficiency, and the water-distribution efficiency.

CHAPTER 9

1. How do excessive intake rates interfere with good surface irrigation?
2. Explain how an irrigation stream too small to irrigate efficiently using surface methods can often be applied very efficiently by sprinkler irrigation.
3. Explain why less skilled labor can be used for sprinkler irrigation than for surface irrigation.
4. Why is it easier to apply small depths of water more efficiently by sprinkler than by surface irrigation methods?
5. For what reasons would frequent light irrigations be desirable?
6. What precautions should be taken when using sprinkler irrigation for frost protection?
7. List the possible advantages in applying fertilizers and soil amendments to the soil through sprinkler systems.
8. Make sketches of the ways in which fertilizers can be injected into a sprinkler system.
9. Why is it advisable and often necessary to operate the sprinklers for a few minutes before and after fertilizer is applied? Consider the system, the crop, and the soil.
10. Which of the three general types of sprinklers is used most extensively? Why?
11. Tabulate the average pressures and rates of application for the three sprinkler types.
12. What are the advantages of medium-pressure sprinklers?
13. What crops would most nearly justify use of semipermanent sprinkler systems?
14. What are the advantages of portable sprinkler systems?

15. What natural conditions would favor use of a gravity sprinkler system?
16. Why are silt and debris in irrigation water objectionable when using sprinkler systems?
17. Why is debris removal from irrigation water difficult?
18. Discuss the relative importance of initial and annual costs of irrigation systems?
19. Larger sprinkler nozzles and higher pressures require more power and less labor for irrigation. How would this fact influence sprinkler design in an underdeveloped country?
20. What information regarding soils is needed for a good sprinkler irrigation design?
21. (a) When is the soil most susceptible to puddling and crusting from droplet impact from sprinklers?
    (b) What can the irrigator do to minimize damage from a given sprinkler system?
22. Explain how the cropping pattern which will develop when the sprinkler system is in full use should influence the design of the system.
23. What are the advantages of trickle irrigation? The disadvantages.

CHAPTER 10

1. Why does the method of irrigation used depend upon the value of the crop being produced, the cost of water, and the general economy of the area?
2. Which of the methods of irrigating is the oldest and the least efficient?
3. What crops are generally irrigated by flooding?
4. What are the essential points of difference between border-strip flooding and check flooding?
5. What are the essential points of difference between corrugations and furrows used for irrigation?
6. Level borders have been used for irrigating relatively flat lands in areas not strictly arid or humid. Explain why heavy rains would create a problem.
7. What natural conditions favor subirrigation?
8. List the reasons why the hydraulics of surface irrigation are much more complicated than the hydraulics of open-channel flow.
9. What ten criteria must be considered when designing an irrigation system?

CHAPTER 11

1. (a) What are the major functions of diversion structures?
   (b) What forces are permanent diversion structures required to resist?

(c) Are the dimensions of farm diversion structures, i.e., lengths, depths, and widths, influenced by the soils in which they are built? Explain.

2. In the selection of a permanent farm conveyance structure, give the conditions which would influence your choice between a flume, ditch, surface pipe, and underground pipe.

3. (a) Why is it necessary to have a larger bottom width and depth in order to convey a given quantity of water through an earth ditch than through a concrete flume of the same slope?

(b) If you were going to build an 200 millimeter pipe on a slope of 0.4 meters per kilometer and you wanted to get the largest possible quantity of water through it, neglecting differences in cost, what kind of pipe would you select? Why?

4. (a) What major objectives should influence the irrigator's selection of irrigation water-distribution structures?

(b) Does the cost of water influence the selection of a distribution structure?

(c) Do the soil properties influence the selection of distribution structures? Explain briefly.

5. In measuring water to a farmer from a variable canal stream will a submerged orifice or a weir be most helpful toward delivery of a flow as nearly uniform as practical? Why?

6. To what extent are shop facilities needed to maintain and repair irrigation implements and structures?

8. Explain why a good plowing job is essential if the land is to be irrigated.

9. (a) What crops might be damaged by deep furrowing as a result of tearing up shallow roots?

(b) Would deep cultivation also damage the same crops?

10. Explain the statement that good water control and therefore adequate water-control structures are essential before irrigation water can be used efficiently.

11. Discuss the relative practicality and economics of using wood, concrete, brick and mortar, metal, and prefabricated components for water-control structures in your locality.

12. What are the advantages of using buried pipe for water conveyance?

13. How would you start water flowing in a 50-millimeter siphon used to convey water from a ditch to a furrow?

CHAPTER 12

1. Discuss the advantages and disadvantages of earth canals.

2. Why should amortized annual cost of lining a canal be considered instead of the initial cost?

3. How might canal lining be related to drainage?

4. Consider a farm irrigation ditch in a loam soil having the following dimensions:

(a) Bottom width, 0.5 meters.

(b) Total depth, 0.60 meters.

(c) Side slopes of 1 horizontal to 1 vertical.

(d) Depth of water, 0.40 meters.

Find the following properties:

(a) Cross-sectional area of the stream.

(b) Wetted perimeter.

(c) Hydraulic radius of the stream.

5. If the bottom of the ditch described in Problem 4 has a uniform slope of 1 meter per kilometer, and if the bottom and sides are kept smooth and free from weeds, what will be the mean velocity of flow and the discharge? Use $n = 0.02$.

6. If the canal described in Problem 4 were permitted to grow weeds on the sides and bottom, what would be the velocity and discharge? Use $n = 0.04$.

7. For a canal of the same dimensions as in Problem 4 built in earth on a slope of 2 meters per kilometer, determine the velocity and discharge.

8. How do you account for the fact that the velocity and discharge are not doubled when the slope is doubled?

9. For a concrete-lined ditch in good condition having the same dimensions and slope as the ditch in Problem 7, determine the velocity and discharge. Use $n = 0.014$.

10. Rework Problem 7 using the Chezy equation and Fig. 5.2. Use an e value of 0.3 millimeter and consider the water temperature to be 14°C. (Alternative—Consider ν to be $1.124 \times 10^{-6}\,\text{m}^2/\text{s}$.)

11. Compute C and f for the conditions of Problem 6.

12. Compute C and f for the conditions of Problem 10.

13. Show that a square, covered box flowing 0.95 full carries more water than the same box flowing full. Explain.

14. Given a canal with 1 to 1 side slopes and cross-section area of 10 square meters, find the dimensions for the best hydraulic cross section.

15. List the principal materials used as canal linings. Discuss the advantages and disadvantages of each.

16. Describe the "inflow-outflow" method for determining seepage losses from a canal.

CHAPTER 13

1. List the three principal methods of drilling wells and explain each. Discuss the advantages and disadvantages of each.

2. What are the purposes of well screens and of gravel packing a well? When would you use each? Should they be used together? If so, when? If not, why?

3. Why is the development of a well so important?

4. List the three principal methods of well development and explain the operation of each.

5. What is the difference between a confined and an unconfined well?

6. Is a confined or artesian well necessarily a flowing well? Explain.
7. What are some of the assumptions made in the development of well equations?
8. What discharge can be expected from an unconfined well 380 millimeters in diameter when the draw-down is 6 meters in an aquifer saturated to a depth of 15 meters? Assume the coefficient of permeability to be 21 meters per day and the radius of influence to be 180 meters.
9. What diameter well is required to deliver a discharge of 55 liters per second from a confined aquifer 15 meters thick with a draw-down of 21 meters? Assume a radius of influence of 120 meters and a permeability of 15 meters per day.
10. Determine the height of the seepage face for an unconfined well 300 millimeters in diameter, a permeability of 6 millimeters per second, a discharge of 50 liters per second, and a depth of water in the well of 4.5 meters.

CHAPTER 14

1. Define horsepower, theoretical horsepower, brake horsepower, water horsepower, and pumping plant efficiency.
2. Define static head and draw-down.
3. What are the characteristic curves of a pump? Of what value are they?
4. Under what conditions would you use a centrifugal pump? What are some of the problems connected with its use?
5. Define total dynamic head.
6. An irrigator desires to lift a stream of 30 liters per second a vertical height of 12 meters. If the loss of head in the casing and pump results in a 62 percent over-all pump efficiency and the electric motor as an efficiency of 91 percent, how many horsepower will his motor need? How many kilowatts will it use while pumping? Assume 1 hp = 0.746 kW.
7. With the same height of lift and the same efficiencies as given in Problem 6, how many kilowatts would a motor require in order to deliver a stream that would supply enough water in 30 hours to cover a 4-hectare tract to a depth of 150 millimeters? Assume 1 hp = 0.746 kW.
8. The gross water requirement for orchard irrigation is assumed to be 900 hectare-millimeters per hectare. If a pumping plant (motor and pump) operates at an efficiency of 57 percent, how many kilowatt-hours of energy will be required to lift water 9 meters for each hectare?
9. Compute the horsepower required to pump:
   (a) A stream of 60 liters per second against a head of 12 meters, assuming 100 percent efficiency.
   (b) Actual pumping plant efficiency is 59 percent. What is the horsepower requirements?

(c) Using the "Electric Service Schedules and Costs," Chapter 14, what would the charge be per month (30 days) if the motor is run continuously? Assume that voltage and term discount are received. Assume 1 hp = 0.746 kW.

10. Would it be advisable to use the centrifugal pump whose characteristics are shown in Fig. 14.2 when it is desired that 100 liters per second be pumped against a head of 13 meters? What is the efficiency? Why is a high efficiency desirable? What horsepower would be used?

11. A farmer pumps 30 liters per second for 24 hours each day of the irrigation season. The static head is 6 meters and the draw-down while pumping 30 liters per second is 1.5 meters. If his pumping plant is 60 percent efficient, what does it cost him per month if he pays the rates as given in the article "Electric Service Schedules and Costs"?

12. Define kilowatt hour, horsepower hour, and brake horsepower hour.

13. If the water horsepower of a pump is 50 hp, the efficiency of the pump is 75 percent, and the efficiency of the motor is 90 percent, how many kilowatthours are consumed in one month's operation? How many brake horsepower hours are consumed?

14. Compute the monthly cost for operating the pump in Problem 13 if:
   (a) An electric motor is used and the power schedule is as given in the text.
   (b) A diesel engine is used and diesel fuel costs 11¢/liter.
   (c) A gasoline engine is used and gasoline costs 15¢/liter.
   (d) A propane engine is used and propane costs 10¢/liter.

CHAPTER 15

1. Discuss the statement that irrigation and drainage are complementary practices in arid regions.

2. What is the relative importance of surface and of subsurface drainage in arid and in humid climates?

3. Explain why the need for surface drainage increases as the annual rainfall increases.

4. List the benefits of drainage.

5. How can knowledge of the source of irrigation water be useful in solving a drainage problem?

6. Under what conditions would you use piezometers instead of auger holes to measure the hydraulic characteristics of the ground water?

7. What practical means can frequently be taken to reduce the severity of drainage problems by controlling the source of the drainage waters?

8. Discuss the statement that shallower water-table depths are economically feasible under excellent management.

9. A field investigation of drainage deals with what four subjects? Why is each one important?

10. The soil excavated from an open drain is all placed upon one side. Make assumptions regarding side slopes and widths and prepare a graph showing the value of land removed from production by a 300 meter drain excavated to 1.5-, 3.0-, and 4.5-meter depths.

11. List in sequence the operations performed by a large trenching machine, including the use of related equipment such as a tractor or bullodozer.

12. What can be done to maintain a tile drainage system in good operating condition?

13. Why is reuse of drainage water important?

14. Prepare a list of advantages and disadvantages of tile drains versus open drains.

15. In Equation 15.2 list the quantities R, k, H, L, and h in the order of ease of measurement, placing first those that are most easily measured.

16. Refer to Fig. 15.6 and Equation 15.2 and explain why the flow of ground water to the drain is proportional (approximately) to the square of the effective depth of drain.

17. A new open drain is not drawing enough water from the soil to lower the water table sufficiently. To increase the drain discharge would you increase the bed slope, make the drain wider, deeper, longer, or use a combination of these remedies? Give reasons for your answer.

18. Determine the flow from the soil into a 3-meter depth open drain 120 meters long when the drains are spaced 45 meters apart. The depth of pervious stratum is 4.5 meters, and the depth of the water table midway between drains is 1 meter below the ground surface. The average permeability of the pervious stratum is 1.5×10^{-4} m/s. The depth of water in the drain is negligible.

19. For a soil of great depth and uniform permeability, with all other conditions as in Problem 18, determine the flow into 150-millimeter diameter tile drain flowing two-thirds full, by means of Equation 15.7.

20. In Problem 18, what will be the change in flow Q, toward the drain, if all other factors remain unchanged, but: (a) the spacing of the drain lines S is doubled? (b) the permeability of soil k is doubled? (c) the depth of the drain is increased from 3 to 3.75 meters?

21. Sandy loam soil to be drained is 15 meters deep and has a permeability of 3×10^{-5} m/s. The water table is 1.2 meters below the ground surface. Two and two-tenths meter-depth tile drains 150 meters long have a slope of 1/2000. Determine the discharge of each drain and the spacing for 150-millimeter-diameter tile flowing half full. Assume the flow toward the drain is through semicylindrical surfaces.

22. In a field drainage experiment by pumping from an artesian aquifer the following data were obtained:
    (a) Flow of water to well or pump discharge　　　　　$Q = 120$ l/s
    (b) Radius at maximum pressure head　　　　　　　　$R = 450$ m

(c) Radius at minimum pressure head $r = 5.5$ m
(d) Pressure head at maximum radius $H = 8$ m
(e) Pressure head at minimum radius $H = 2.1$ m
(f) Depth of water-bearing aquifer $D = 5$ m

CHAPTER 16

1. What are the principal advantages and disadvantages of the following water measuring devices?
   (a) Orifice
   (b) Weir
   (c) Parshall flume
   (d) Trapezoidal flume
   (e) Current meter

2. When using a current meter at 0.6 of the depth, do you measure 0.6 from the water surface or from the bottom of the stream?

3. Explain why more flow will occur over a dull-crested weir than over a sharp-crested weir, even though the depth of water over the crest is the same.

4. (a) Find the theoretical velocity of a jet of water flowing out of a square orifice in a large tank if the center of the orifice is 600 millimeters below the water surface.
   (b) If the orifice opening considered in Problem 4(a) is 150 by 150 millimeters, what is the theoretical discharge in liters per second?
   (c) What are the probable maximum and probable minimum actual discharges in cubic meters per second?

5. (a) In measuring the water that flows through a submerged orifice, is it necessary to know the vertical distance from the upstream water surface to the center of the orifice? Explain.
   (b) Find the discharge in liters per second through a rectangular standard submerged orifice 450 millimeters long (horizontal dimension) by 200 millimeters deep (vertical dimension) if the upstream surface is 175 millimeters vertically above the downstream water surface. First use the appropriate equation and check your result by use of a table.

6. (a) In using weirs to measure water, is it essential to make direct measurement of the velocity of the water as it flows through the weir notch? Explain.
   (b) By means of the appropriate equation, compute the liters per second over a rectangular weir having suppressed end contractions if the weir crest is 600 millimeters long and the water surface at a point 2.5 meters upstream from the weir is 140 millimeters vertically above the weir crest. Check your result with a weir table.
   (c) If the weir described in Problem 6(b) has complete end contractions, would the discharge be more or less than your computed result? How much?

7. For the same length of weir crest and depth of water over the crest as in Problem 6(b), compute the discharge over a trapezoidal weir. Check your results with a table.

8. For a right-angle triangular notch weir, what is the discharge when the depth of water vertically above the apex of the weir notch is 180 millimeters at a point 1.5 meters upstream from the weir?

9. Show that doubling the effective head, causing discharge through a submerged orifice, increases the discharge approximately 41 percent.

10. Show that doubling the head over a rectangular or trapezoidal weir makes the discharge 2.8 times greater.

CHAPTER 17

1. Why are most successful irrigation and drainage enterprises owned and operated by the water users?

2. Why has the commercial irrigation corporation been less influential than the mutual irrigation corporation?

3. Differentiate between private, quasipublic, and public irrigation and drainage enterprises.

4. What are the distinctive features of a mutual irrigation company?

5. How does a mutual irrigation company differ from an irrigation district?

6. Can a mutual irrigation corporation sell the land owned by its delinquent stockholders in order to collect payments of irrigation assessments?

7. Can an irrigation district sell the land owned by its delinquent members in order to collect payments of irrigation assessments?

8. What are the two conflicting water right doctrines? How do they differ?

9. What are the salient features of the (a) doctrine of riparian rights, (b) doctrine of appropriation, (c) common law doctrine, (d) reasonable use concept?

10. What creates priority under (a) riparian right doctrine, (b) appropriation right doctrine?

11. Why is a drainage district normally necessary before land can be adequately drained?

12. What powers are vested in a drainage district?

13. What advantages would result from combining an irrigation district with the drainage district organized for the same land?

14. Explain who settlers ultimately gain ownership of a Bureau of Reclamation project.

APPENDIX A

AMERICAN–ENGLISH–METRIC CONVERSION CONSTANTS

American–English–Metric Conversion Constants

Length and Distance	Inches	Feet	Yards	Statute Miles	Nautical Miles	Centi-meters	Meters	Kilo-meters
1 inch equals	1	0.083	0.027	—	—	2.54	—	—
1 foot	12	1	0.333	—	—	30.48	0.305	—
1 yard	36	3.0	1	—	—	91.44	0.914	—
1 statute mile	—	5280	1760	1	0.865	—	1609	1.61
1 nautical mile	—	6080	2027	1.15	1	—	1853	1.85
1 centimeter	0.394	0.0328	0.0109	—	—	1	0.01	—
1 meter	39.37	3.281	1.094	—	—	100	1	0.001
1 kilometer	—	3281	1094	0.6214	0.5396	—	1000	1

Useful Approximations: 5 cm = 2 in.; 10 cm = 4 in.; 30 cm – 1 ft.; 10 m = 33 ft; 10 km = 6 miles; 16 km = 10 miles; 1 fathom = 6 ft; 120 fathoms = 1 cable length.

Areas	Square Inches	Square Feet	Square Yards	Acres	Square Centimeters	Square Meters	Hectares
1 square inch	1	0.007	—	—	6.45	0.00064	—
1 square foot	144	1	0.1111	—	—	0.0929	—
1 square yard	1296	9	1	—	—	0.8361	—
1 acre	—	43560	4840	1	—	4050	0.405
1 square centimeter	0.155	—	—	—	1	0.0001	—
1 square meter	1550	10.76	1.20	—	10.000	1	0.0001
1 hectare	—	107.600	11.955	2.47	—	10.000	1

Useful Approximations: 1 ha = 2½ acres; 1 in.^{2} = 6½ cm^2; 1 m^2 = 11 ft^2; a hectare is 330 ft on each side.

Table continued on next page.

American–English–Metric Conversion Constants, continued

Volume and Capacity	Cubic Inches	Cubic Feet	American Gallons	Imperial Gallons	Cubic Meters	Liters	Quarts
1 square inch	1	0.00057	0.0043	0.0035	—	0.0163	0.0173
1 square foot	1728	1	7.48	6.23	0.0283	28.3	29.9
1 American gallon	231	0.134	1	0.833	0.0038	3.78	4
1 Imperial gallon	277.4	0.161	1.201	1	0.0045	4.55	4.8
1 cubic meter	61030	35.3	264.2	220	1	1000	1057
1 liter	61.03	0.0353	0.264	0.220	0.001	1	1.057

Useful Approximations: 16.4 cm^3 = 1 in.^{3}; 35 ft^3 in a m^3; 1 register ton = 100 ft^3; 1 U.S. shipping ton = 40 ft^3 or 32.14 U.S. bushels or 31.14 Imp. bushels.

Weights	Grains	Grams	Ounces (avoir.)	Pounds (avoir.)	Kilos	Short Ton	Long Ton	Metric Ton
1 grain	1	0.0648	0.0023	0.00014	—	—	—	—
1 gram	15.4	1	0.035	—	—	—	—	—
1 ounce	437.5	28.35	1	0.0625	0.0283	—	—	—
1 pound	7000.—	454.—	16	1	0.4536	—	—	—
1 kilogram	15432.	1000.	35.27	2.205	1	—	—	—
1 short ton	—	—	—	2000.—	907.	1	0.893	0.907
1 long ton	—	—	—	2240	1016	1.12	1	1.016
1 metric ton	—	—	—	2205	1000	1.102	0.984	1

Air Pressure: In terms of atmospheres—1 atm equals 14.7 psi or 1.033 kilo/cm^2. Normal tire pressure is 2 atm = 29 psi.

EQUATIONS IN ENGLISH UNITS

$$qt = ad \tag{3.9}$$

where q = the size of stream, in ft³/s (or acre-in./hr)
t = the time, in hours, required to irrigate the area
a = the area, in acres
d = the depth, in in., that the volume of water used would cover the land irrigated, if quickly spread uniformly over its surface

$$t = \frac{P_w A_s D a}{100q} \tag{3.11}$$

where t = the time, in hours, required to irrigate the area
P_w = the dry-weight basis moisture percentage of the soil
A_s = the apparent specific gravity of the soil
D = the depth, in in., of the soil
a = the area, in acres
q = the size of the stream in ft³/s (or acre-in./hr)

$$U = KF = K\Sigma \frac{tp}{100} \tag{7.21}$$

where the following quantities must be determined for the same period:

U = consumptive use of crop in in., for a given time period
K = empirical crop consumptive use coefficient (annual, irrigation season, or growing season)
F = sum of the consumptive use factors for the period (sum of the products of mean temperature and percent of annual day-time hours $(t \times p)/100$
t = mean temperature, in °F
p = percentage of day-time hours of the year, occurring during the period. (See Table 7.3).

$$u = kf = k \frac{tp}{100} \tag{7.22}$$

where u = monthly consumptive use, in in.
k = monthly crop coefficient
f = monthly consumptive-use factor, $(t \times p)/100$

$$V = \frac{16}{\sqrt{f}} \sqrt{R \frac{h_L}{L}} = \frac{16}{\sqrt{f}} \sqrt{RS} \qquad (12.6)$$

where V = the average velocity, in ft/s
 f = the coefficient of friction loss
 R = the hydraulic radius, in ft
 h_L = the head loss, in ft
 L = the length between the two points, in ft
 S = the head loss per unit length between points or the slope of the piezometric head line

$$V = C \sqrt{RS} \qquad (12.7)$$

where V = the average velocity, in ft/s
 C = the Chezy roughness coefficient
 R = the hydraulic radius, in ft
 S = the head loss per unit length between points or the slope of the piezometric head line

$$V = \frac{1.49}{n} R^{2/3} S^{1/2} \qquad (12.8)$$

where V = the average velocity, in ft/s
 n = the Manning roughness coefficient
 R = the hydraulic radius, in ft
 S = the head loss per unit length between points or the slope of the piezometric head line

$$C = \frac{16}{\sqrt{f}} = \frac{1.49}{n} R^{1/6} \qquad (12.9)$$

where C = the Chezy roughness coefficient
 f = the coefficient of friction loss
 n = the Manning roughness coefficient
 R = the hydraulic radius, in ft

$$HP = \frac{62.5Qh}{550} = \frac{Qh}{8.8} \qquad (14.1)$$

where HP = horsepower
 Q = discharge, in ft³/s
 h = vertical lift, in ft

$$Q = 0.61A \sqrt{2gh} \qquad (16.5)$$

where Q = the rate of flow, in ft³/s
 A = the cross-sectional area of flow, in ft²
 g = the acceleration due to gravity (or the force of gravity per unit mass of water), which is 32.2 ft/s²
 h = the depth of water, in ft, or pressure head, causing discharge through the orifice

$$Q = 3.33LH^{3/2} \qquad (16.9)$$

where Q = the rate of flow, in ft^3/s
L = the length of the weir crest, in ft
H = the depth, or head, of water flowing over the weir, in ft

$$Q = 3.37LH^{3/2} \qquad (16.11)$$

where Q = the rate of flow, in ft^3/s
L = the length of the weir crest, in ft
H = the depth, or head, of water flowing over the weir, in ft

$$Q = 2.49H^{5/2} \qquad (16.12)$$

where Q = the rate of flow, in ft^3/s
H = the depth, or head, of water flowing over the weir, in ft

$$Q = 8.8D^{2.5}H^{3.5} \qquad (16.15)$$

where Q = the rate of flow, in ft^3/s
D = the inside diameter of the pipe, in ft
H = the average height of rise of the center of the jet, in ft

$$Q = 5.57D^{1.99}H^{0.53} \qquad (16.16)$$

where Q = the rate of flow, in ft^3/s
D = the inside diameter of the pipe, in ft
H = the average height of rise of the center of the jet, in ft

$$Q = C' \sqrt{\frac{g}{2}} \, \frac{AX}{\sqrt{Y}} = CA \, \frac{X}{\sqrt{Y}} \qquad (16.17)$$

where Q = discharge in ft^3/s
C = coefficient of discharge
g = acceleration of gravity, 32.2 ft/s^2
A = cross-sectional area of water at end of pipe, in ft^2
X = coordinate of the point on the surface measured in ft parallel to the pipe
Y = vertical coordinate, measured in ft

FIGURES IN ENGLISH UNITS

Figure 6.3   Concentration of irrigation waters related to conductivity.

DESIGNED FOR HEADS OF	A	B	C	D	E	LUMBER THICKNESS
1 cfs. – 2 cfs	8'-0"	2'-0"	3'-0"	1'-0"	2'-6"	1"
2 cfs – 5 cfs	9'-0"	3'-6"	3'-0"	2'-0"	3'-0"	1½
5 cfs – 8 cfs.	10'-0"	3'-6"	4'-0"	2'-0"	3'-0"	1½

Figure 11.6   Standard single-wing wooden check gate.

DESIGNED FOR HEADS OF	A	B	C	D	E	F	G
3 cfs – 6 cfs.	9'-0"	3'-6"	3'-0"	2'-0"	3'-0"	3'-0"	2'-0"
8 cfs. – 10 cfs	12'-0"	4'-0"	4'-0"	2'-0"	4'-0"	4'-0"	2'-6"
10 cfs – AND UP	14'-0"	4'-0"	5'-0"	2'-0"	4'-6"	4'-6"	2'-6"

Figure 11.7   Standard double-wing wooden check gate.

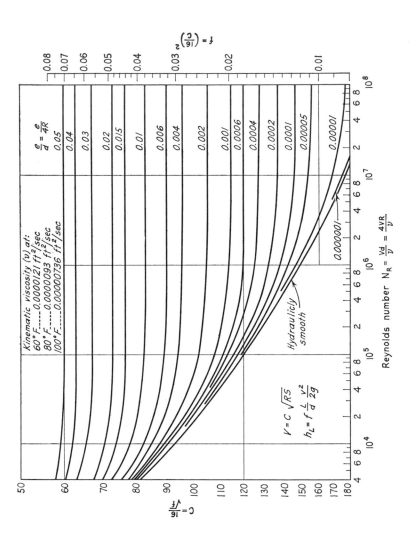

Figure 12.2   Roughness values for turbulent flow in conveyance structures.

For sudden enlargements and sudden contractions the equivalent length is in feet of pipe of the smaller diameter, d.

The dashed line shows the determination of the equivalent length of a 6 - in. standard elbow.

Figure 14.5   Minor friction losses caused by elbows, valves, and other fittings. (From *Oregon Agr. Exp. Sta. Bul.* 181, by Wolfe, 1950).

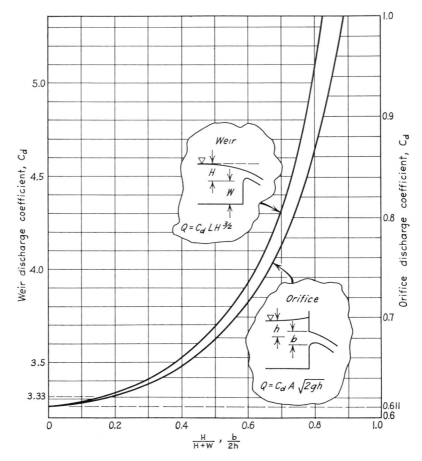

Figure 16.7   Variation of discharge coefficients with boundary form for a rectangular weir and for an orifice.

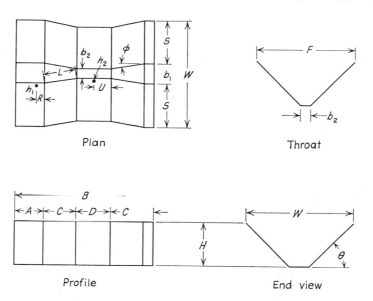

Plan   Throat

Profile   End view

Flume no.	Description	b_1	b_2	A	L	C	D	E	F	H	B	R	S	U	W	θ	ϕ
1	Large 60°-v	2	0	7	7	$6\frac{15}{16}$	7	3	8	$6\frac{3}{4}$	$30\frac{7}{8}$	$1\frac{1}{2}$	4	$3\frac{1}{2}$	10	60°	8.25°
2	Small 60°-v	2	0	5	$4\frac{1}{8}$	$4\frac{3}{64}$	5	2	$4\frac{3}{4}$	4	$20\frac{3}{32}$	1	$2\frac{3}{8}$	$2\frac{1}{2}$	$6\frac{3}{4}$	60°	11.20°
3	2"-60° wsc	$4\frac{7}{8}$	2	8	$8\frac{1}{2}$	$8\frac{13}{32}$	$8\frac{1}{2}$	3	14	$13\frac{1}{2}$	$36\frac{5}{16}$	$1\frac{1}{2}$	6	$4\frac{1}{4}$	$16\frac{7}{8}$	60°	9.22°
4	2"-45° wsc	$4\frac{7}{8}$	2	8	$8\frac{1}{2}$	$8\frac{3}{8}$	$8\frac{1}{2}$	3	$22\frac{3}{16}$	$10\frac{19}{32}$	$36\frac{1}{4}$	$1\frac{1}{2}$	$10\frac{19}{32}$	$4\frac{1}{4}$	$26\frac{1}{16}$	45°	9.91°
5	2"-30° wsc	$4\frac{7}{8}$	2	8	$8\frac{1}{2}$	$8\frac{3}{8}$	$8\frac{1}{2}$	3	$36\frac{41}{64}$	10	$36\frac{1}{4}$	$1\frac{1}{2}$	$17\frac{5}{16}$	$4\frac{1}{4}$	$39\frac{33}{64}$	30°	9.80°
6	4"-60° wsc	8	4	9	10	$9\frac{13}{16}$	10	3	20	$13\frac{7}{8}$	$41\frac{5}{8}$	$1\frac{1}{2}$	8	5	24	60°	11.20°
7	2"-30° csu	10	2	10	$10\frac{3}{4}$	10	10	3	$35\frac{1}{2}$	$9\frac{21}{32}$	43	$1\frac{1}{2}$	$16\frac{3}{4}$	5	$43\frac{1}{2}$	30°	21.80°

Note: Dimensions in inches except as shown. See Fig. 6-16 for the relationship of Q to depth of each of the seven flumes.

Figure 16.15   Details of designs of trapezoidal flumes. (After Robison and Chamberlain.)

Figure 16.16   Calibration curves for trapezoidal flumes under free-flow conditions.

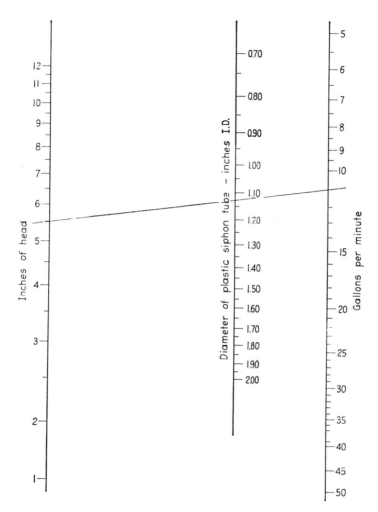

Figure 16.23   Discharge versus head for plastic siphon tubes.

APPENDIX
D
TABLES IN ENGLISH UNITS

TABLE 2.1 Increase of Energy Consumption with Increase of Salinity for a Commercial Membrane Demineralizer, Based on a Product Purity of 500 ppm. (Courtesy of Ionics, Inc.)

Feedwater Salinity (ppm)	Approximate Production Rate (gal/day)	Approximate Energy Consumption (kWh/1000 gal)
900	28,000	3
1,500	14,000	7
2,000	10,000	10
3,000	7,000	15
4,000	5,500	20
5,000	4,500	25
6,000	4,000	30
8,000	3,000	40
10,000	2,500	50
15,000	1,800	80
20,000	1,400	100
35,000	900	180

TABLE 3.1 Representative Physical Properties of Soils

Soil Texture	Infiltration[a] and Permeability (in./hr) I_f	Total Pore Space (%) N	Apparent Specific Gravity A_s	Field Capacity (%) FC	Permanent Wilting (%) PW	Total Available Moisture[b]		
						Dry Weight (%) $P_w = FC - PW$	Volume (%) $P_v = P_w A_s$	in./ft $d = \dfrac{P_w}{100} A_s D$
Sandy	2 (1–10)	38 (32–42)	1.65 (1.55–1.80)	9 (6–12)	4 (2–6)	5 (4–6)	8 (6–10)	1.0 (0.8–1.2)
Sandy loam	1 (0.5–3)	43 (40–47)	1.50 (1.40–1.60)	14 (10–18)	6 (4–8)	8 (6–10)	12 (9–15)	1.4 (1.1–1.8)
Loam	0.5 (0.3–0.8)	47 (43–49)	1.40 (1.35–1.50)	22 (18–26)	10 (8–12)	12 (10–14)	17 (14–20)	2.0 (1.7–2.3)
Clay loam	0.3 (0.1–0.6)	49 (47–51)	1.35 (1.30–1.40)	27 (23–31)	13 (11–15)	14 (12–16)	19 (16–22)	2.3 (2.0–2.6)
Silty clay	0.1 (0.01–0.2)	51 (49–53)	1.30 (1.25–1.35)	31 (27–35)	15 (13–17)	16 (14–18)	21 (18–23)	2.5 (2.2–2.8)
Clay	0.2 (0.05–0.4)	53 (51–55)	1.25 (1.20–1.30)	35 (31–39)	17 (15–19)	18 (16–20)	23 (20–25)	2.7 (2.4–3.0)

Note: Normal ranges are shown in parentheses.

[a] Intake rates vary greatly with soil structure and structural stability, even beyond the normal ranges shown above.

[b] Readily available moisture is approximately 75% of the total available moisture.

TABLE 4.1 Guide for Judging How Much of the Available Moisture Has Been Removed from the Soil

Soil Moisture Deficiency (%)	Feel or Appearance of Soil and Moisture Deficiency in in. of Water per ft of Soil			
	Coarse Texture	Moderately Coarse Texture	Medium Texture	Fine and Very Fine Texture
0 (Field capacity)	Upon squeezing, no free water appears on soil but wet outline of ball is left on hand. 0.0	Upon squeezing, no free water appears on soil but wet outline of ball is left on hand. 0.0	Upon squeezing, no free water appears on soil but wet outline of ball is left on hand. 0.0	Upon squeezing, no free water appears on soil but wet outline of ball is left on hand. 0.0
0–25	Tends to stick together slightly, sometimes forms a very weak ball under pressure. 0.0 to 0.2	Forms weak ball, breaks easily, will not slick. 0.0 to 0.4	Forms a ball, is very pliable, slicks readily if relatively high in clay. 0.0 to 0.5	Easily ribbons out between fingers, has slick feeling. 0.0 to 0.6
25–50	Appears to be dry, will not form a ball with pressure. 0.2 to 0.5	Tends to ball under pressure but seldom holds together. 0.4 to 0.8	Forms a ball somewhat plastic, will sometimes slick slightly with pressure. 0.5 to 1.0	Forms a ball, ribbons out between thumb and forefinger. 0.6 to 1.2

Soil Moisture Deficiency (%)	Feel or Appearance of Soil and Moisture Deficiency in in. of Water per ft of Soil			
	Coarse Texture	Moderately Coarse Texture	Medium Texture	Fine and Very Fine Texture
50–75	Appears to be dry, will not form a ball with pressure.[a] 0.5 to 0.8	Appears to be dry, will not form a ball.[a] 0.8 to 1.2	Somewhat crumbly but holds together from pressure. 1.0 to 1.5	Somewhat pliable, will ball under pressure.[a] 1.2 to 1.9
75–100 (100% is permanent wilting)	Dry, loose, single-grained, flows through fingers. 0.8 to 1.0	Dry, loose, flows through fingers. 1.2 to 1.5	Powdery, dry, sometimes slightly crusted but easily broken down into powdery condition. 1.5 to 2.0	Hard, baked, cracked, sometimes has loose crumbs on surface. 1.9 to 2.5

[a] Ball is formed by squeezing a handful of soil very firmly.

TABLE 6.2 Leaching Requirement Related to Quality of Drainage and Irrigation Water for an Equilibrium Condition, Ignoring Influence of Rainfall

| Quality of Irrigation Water | | | | Percentage Leaching Requirement for Given Qualities of Drainage Water | | | |
Micromhos/cm $EC \times 10^6$	ppm	MEQ/L	Tons/Acre-ft	5 millimhos/cm $EC \times 10^3$	10 millimhos/cm $EC \times 10^3$	15 millimhos/cm $EC \times 10^3$	20 millimhos/cm $EC \times 10^3$
100	63	1	.09	2	1	0.7	0.5
200	125	2	.17	4	2	1	1
400	250	4	.34	8	4	3	2
800	500	8	.82	16	8	5	4
1600	1000	16	1.36	32	16	11	8
3200	2000	32	2.72	64	32	21	16
6400	4000	64	5.44	128	64	43	32

TABLE 7.8 Total Consumptive Use and Peak Daily Use, Western United States (After Woodward)

| | Southern Coastal[a] (1) | | | | South Pacific Coastal Interior and North Coastal (2) | | | | | | | |
| | 300 Days Plus | | 250–300 Days | | 250–300 Days | | 210–250 Days | | 180–210 Days | | 150–180 Days | |
Crops	Season Use (in.)	Daily Use (in./day)	Season Use (in.)	Daily Use (in./day)	Season Use (in.)	Daily Use (in./day)	Season Use (in.)	Daily Use (in./day)	Season Use (in.)	Daily Use (in./day)	Season Use (in.)	Daily Use (in./day)
Alfalfa	36.0	0.20	30.0	0.17	37.0	0.27	32.0	0.22	26.0	0.20	22.0	0.18
Pasture	33.5	0.20	28.0	0.17	33.0	0.27	30.0	0.22	24.0	0.20	20.0	0.18
Grain—small	16.0	0.18	14.0	0.16	17.0	0.22	14.5	0.20	12.0	0.20	10.0	0.18
Beets—sugar	29.0	0.20	25.0	0.18	30.0	0.27	26.0	0.22	—	—	—	—
Beans—field	12.0	0.18	10.0	0.16	13.0	0.22	11.0	0.18	—	—	—	—
Corn—field	—	—	—	—	—	—	19.0	0.25	16.0	0.22	14.0	0.20
Potatoes	—	—	—	—	24.0	0.20	20.0	0.18	—	—	—	—
Peas—green	10.0	0.18	8.0	0.16	11.0	0.20	9.0	0.18	8.0	0.16	7.0	0.16
Legume seed	25.0	0.20	22.0	0.18	26.0	0.25	22.0	0.22	20.0	0.20	18.0	0.18
Tomatoes	18.0	0.16	15.0	0.16	19.0	0.20	16.0	0.18	13.0	0.16	11.0	0.16
Vegetable seed	16.0	0.16	14.0	0.16	18.0	0.18	16.0	0.18	14.0	0.16	12.0	0.16
Beans—pole	—	—	—	—	18.0	0.20	16.0	0.20	14.0	0.18	12.0	0.16
Corn—sweet	16.0	0.18	14.0	0.16	16.0	0.20	14.0	0.18	12.0	0.18	11.0	0.16
Apples	—	—	—	—	24.0	0.20	22.0	0.20	20.0	0.18	18.0	0.18
Cherries	—	—	—	—	24.0	0.20	22.0	0.20	—	—	—	—
Peaches	—	—	—	—	24.0	0.20	22.0	0.20	—	—	—	—
Prunes	—	—	—	—	22.0	0.20	20.0	0.20	—	—	—	—
Apricots	—	—	—	—	22.0	0.20	20.0	0.20	—	—	—	—
Oranges	20.0	0.16	18.0	0.16	22.0	0.18	—	—	—	—	—	—
Avocados	18.0	0.16	—	—	—	—	—	—	—	—	—	—
Walnuts	22.0	0.20	18.0	0.18	24.0	0.25	22.0	0.22	20.0	0.20	—	—
Strawberries	22.0	0.20	18.0	0.18	23.0	0.25	18.0	0.22	16.0	0.20	14.0	0.18
Lettuce	4.0	0.16	4.0	0.16	6.0	0.18	5.0	0.18	—	—	—	—
Mint	—	—	—	—	23.0	0.24	21.0	0.22	19.0	0.20	17.0	0.18
Hops	—	—	—	—	20.0	0.22	20.0	0.20	—	—	—	—

[a] Fog belt.

Table continued on next page.

TABLE 7.8 Total Consumptive Use and Peak Daily Use, Western United States (Continued)

| | Central Valley—California and Valleys East Side of Cascade Mountains (3) | | | | | | | | | | | |
| | 250–300 Days | | 210–250 Days | | 180–210 Days | | 150–180 Days | | 120–150 Days | | 90–120 Days | |
Crops	Season Use (in.)	Daily Use (in./day)	Season Use (in.)	Daily Use (in./day)	Season Use (in.)	Daily Use (in./day)	Season Use (in.)	Daily Use (in./day)	Season Use (in.)	Daily Use (in./day)	Season Use (in.)	Daily Use (in./day)
Alfalfa	40.0	0.30	34.0	0.28	30.0	0.25	26.0	0.22	20.0	0.20	14.0	0.18
Pasture	36.0	0.30	30.0	0.28	28.0	0.25	24.0	0.22	18.0	0.20	13.0	0.18
Grain—small	18.0	0.22	16.0	0.22	15.0	0.20	14.0	0.18	13.0	0.18	12.0	0.16
Beets—sugar	33.0	0.30	28.0	0.25	24.0	0.22	20.0	0.20	18.0	0.18	—	—
Beans—field	17.0	0.22	13.0	0.20	13.0	0.20	12.0	0.18	12.0	0.18	—	—
Corn—field	26.0	0.35	22.0	0.32	22.0	0.30	20.0	0.25	18.0	0.22	17.0	0.20
Potatoes—summer	12.0	0.16	—	—	—	—	—	—	—	—	—	—
Potatoes—fall	—	—	19.0	0.25	18.0	0.22	18.0	0.20	17.0	0.18	16.0	0.16
Peas—green	—	—	8.0	0.18	7.0	0.18	7.0	0.18	7.0	0.16	6.0	0.15
Peas—field	—	—	10.0	0.18	9.0	0.18	9.0	0.18	8.0	0.16	8.0	0.15
Tomatoes	20.0	0.20	18.0	0.18	18.0	0.18	17.0	0.17	16.0	0.16	—	—
Cotton	26.0	0.30	22.0	0.28	—	—	—	—	—	—	—	—
Grain—sorghum	15.0	0.20	13.0	0.18	12.0	0.17	10.0	0.16	9.0	0.15	8.0	0.15
Apples	—	—	26.0	0.20	23.0	0.20	21.0	0.18	—	—	—	—
Cherries	—	—	24.0	0.20	21.0	0.20	19.0	0.18	—	—	—	—
Peaches	22.0	0.22	22.0	0.20	20.0	0.20	18.0	0.18	—	—	—	—
Apricots	20.0	0.22	17.0	0.20	15.0	0.20	—	—	—	—	—	—
Oranges	28.0	0.18	—	—	—	—	—	—	—	—	—	—
Strawberries	24.0	0.20	20.0	0.20	18.0	0.18	—	—	—	—	—	—
Lettuce—winter	4.0	0.20	—	—	—	—	—	—	—	—	—	—
Mint	—	—	20.0	0.22	18.0	0.20	—	—	—	—	—	—
Hops	—	—	18.0	0.20	16.0	0.18	—	—	—	—	—	—
Grapes	30.0	0.25	25.0	0.22	22.0	0.20	—	—	—	—	—	—
Walnuts	24.0	0.22	20.0	0.20	—	—	—	—	—	—	—	—
Almonds	22.0	0.25	20.0	0.22	—	—	—	—	—	—	—	—

Intermountain, Desert, and Western High Plains (4)

Crops	250–300 Days Season Use (in.)	Daily Use (in./day)	210–250 Days Season Use (in.)	Daily Use (in./day)	180–210 Days Season Use (in.)	Daily Use (in./day)	150–180 Days Season Use (in.)	Daily Use (in./day)	120–150 Days Season Use (in.)	Daily Use (in./day)	90–120 Days Season Use (in.)	Daily Use (in./day)
Alfalfa	52.0	0.40	44.0	0.32	36.0	0.29	30.0	0.26	24.0	0.22	19.0	0.20
Pasture	48.0	0.40	40.0	0.30	33.0	0.28	28.0	0.25	22.0	0.22	17.0	0.20
Grain—small	21.0	0.25	18.0	0.22	16.0	0.20	16.0	0.20	16.0	0.20	14.0	0.18
Beets—sugar	37.0	0.30	32.0	0.30	30.0	0.28	26.0	0.25	24.0	0.22	18.0	0.20
Beans—field	22.0	0.25	17.0	0.20	14.0	0.20	14.0	0.18	14.0	0.17	12.0	0.15
Corn—field	—	—	30.0	0.35	26.0	0.30	24.0	0.28	22.0	0.24	—	—
Potatoes—fall	—	—	23.0	0.30	21.0	0.28	20.0	0.25	19.0	0.22	17.0	0.20
Peas—field	—	—	—	—	10.0	0.19	10.0	0.18	10.0	0.17	9.0	0.15
Tomatoes	—	—	20.0	0.22	18.0	0.20	17.0	0.18	16.0	0.17	—	—
Cotton	32.0	0.30	30.0	0.28	—	—	—	—	—	—	—	—
Grain—sorghum	19.0	0.25	18.0	0.20	16.0	0.20	14.0	0.18	12.0	0.17	—	—
Apples	—	—	—	—	28.0	0.22	24.0	0.20	20.0	0.18	—	—
Cherries	—	—	—	—	26.0	0.22	—	—	—	—	—	—
Peaches	—	—	29.0	0.25	27.0	0.22	—	—	—	—	—	—
Apricots	26.0	0.25	24.0	0.25	25.0	0.20	—	—	—	—	—	—
Almonds	22.0	0.25	20.0	0.25	—	—	—	—	—	—	—	—
Vineyards	40.0	0.27	32.0	0.25	26.0	0.22	—	—	—	—	—	—
Legume seed	—	—	—	—	—	—	—	—	16.0	0.18	14.0	0.16
Grass seed	—	—	—	—	—	—	—	—	14.0	0.14	12.0	0.14
Potatoes—seed	—	—	—	—	—	—	—	—	16.0	0.16	14.0	0.15
Grapefruit	45.0	0.20	—	—	—	—	—	—	—	—	—	—
Oranges	36.0	0.18	—	—	—	—	—	—	—	—	—	—
Lettuce—winter	6.0	0.18	—	—	—	—	—	—	—	—	—	—
Melons	22.0	0.25	20.0	0.22	18.0	0.20	16.0	0.18	—	—	—	—
Palm dates	60.0	0.30	—	—	—	—	—	—	—	—	—	—
Truck crops	20.0	0.25	18.0	0.22	14.0	0.20	12.0	0.18	12.0	0.16	10.0	0.15

TABLE 9.1 Irrigation Application Rate for Frost Protection at Different Temperatures and Wind Velocities

Plant Temperatures (°F)	Wind Velocity (miles/hr)		
	0–0.03	0.5–1.5	2–3
	Irrigation Rate (in./hr)		
27	0.10	0.10	0.10
25	0.10	0.14	0.22
23	0.12	0.20	
21	0.14	0.26	
18	0.20		

TABLE 9.2 Factors Affecting the Selection of Sprinkler Irrigation Systems

Type of System	Maximum Slope (%)	Water Application Rate (in./hr) Min.	Max.	Shape of Field	Field Surface Conditions	Maximum Height of Crop (m)	Labor Required (hr/acre per irrigation)	Size of Single System (acres)	Approx. Relative Cost[a] (dollars/ha)	Cooling and Frost Protection	Pesticide Application	Fertilizer Application	Liquid Animal Waste Distribution
Multi-Sprinkler													
Hand-moved:													
Portable set	20	0.10	2.0	Rectangular	No limit	No limit	0.50–1.50	1–40	50–200	No			Not Recommended
Solid set	No limit	0.05	2.0	Any shape			0.20–0.50	1 or more	400–900	Yes			
Tractor-moved:													
Skid-mounted	5–10	0.10	2.0		Smooth enough for safe tractor operation		0.20–0.40	20–40	100–300				
Wheel-mounted	5–10	0.10	2.0				0.20–0.40	20–40	100–300				
Self-moved:													
Side-wheel-roll	5–10	0.10	2.0	Rectangular	Reasonably smooth	4	0.10–0.30	20–80	100–300				
Side move	5–10	0.10	2.0			4–6	0.10–0.30	20–80	125–300				
Self-propelled:													
Center-pivot	5–15	0.20	1.0	Circular Square or Rectangular	Clear of obstructions, path for towers	8–10	0.05–0.15	40–160	125–250	No	Yes	Yes	Yes
Side-move	5–15	0.20	1.0			8–10	0.05–0.15	80–160	125–250				
Single-Sprinkler													
Hand-moved	20	0.25	2.0	Any shape	No limit	No limit	0.50–1.50	20–40	50–200				Not Recommended
Tractor-moved:													
Skid-mounted	5–15	0.25	2.0		Safe operation of tractor		0.20–0.40	20–40	100–300				
Wheel-mounted	5–15	0.25	2.0				0.20–0.40	20–40	100–300				
Self-propelled	No limit	0.25	1.0	Rectangular	Lane for winch and hose		0.10–0.30	40–100	120–250				
Boom-Sprinkler													
Tractor-moved	5	0.25	1.0	Any shape	Safe operation of tractor	8–10	0.20–0.50	20–40	200–300				Yes
Self-propelled	5	0.25	1.0	Rectangular	Lane for boom and hose	8–10	0.10–0.50	40–100	200–300				
Permanent													
Manual or Automatic	No limit	0.05	2.0	Any shape	No limit	No limit	0.05–0.10	1 or more	400–1000	Yes			

[a]Does not include the cost of water supply, pump, power unit, and mainline.
Source: From "Planning for an Irrigation System," published by American Association for Vocational Instructional Materials in cooperation with the Soil Conservation Service, U.S.D.A., June 1971.

401

TABLE 10-1 Factors Affecting the Selection of Surface Irrigation Systems

Type of System	Maximum Slope (%) Humid Areas Nonsod Crops	Humid Areas Sod Crops	Arid Areas Nonsod Crops	Arid Areas Sod Crops	Water Application Rate of Intake Family[a] (in./hr) Min.	Max.	Shape of Field	Row Crops (Row or Bedded)	Adaptable to Sown, Drilled, or Sodded Crops	Weed Control in Rice	Orchards and Vineyards	Labor Required (hr/acre per irrigation)	Approximate Cost[b] (dollars/acre)
Level:													
Level border	Nearly Level				0.1	2.0	Any shape			Yes		0.10–0.5	100–250
Contour levee	0.1				0.1	0.5		Yes	Yes		Yes	0.05–0.5	50–150
Level furrow	Nearly level				0.1	2.0	Rows should be of equal length			No		0.20–0.7	100–250
Graded:													
Graded border	0.5	NA	2.0	4.0	0.3	2.0	Rectangular	No				0.20–1.0	100–300
Contour ditch	NA	4.0	4.0	15.0	0.1	3.0	Any shape		Yes		Yes	1.00–2.0	25–50
Graded furrow	0.5	NA	3.0	NA	0.1	3.0	Rows should be of equal length	Yes		No		0.40–1.2	100–300
Corrugation	NA	NA	4.0	8.0	0.1	1.5	Rectangular	No				0.40–1.2	50–100
Contour furrow	Cross slope	3.0	Cross slope	6.0	0.1	2.0	Rows should be of equal length	Yes	No			0.50–1.5	75–150

[a] Intake family is a grouping of soil by the Soil Conservation Service. It is based on the ability of the soil to take in the required amount of water during the time it takes to irrigate.

[b] Does not include the cost of water supply, pump, power unit, and mainline.

Source: From "Planning for an Irrigation System," published by American Association for Vocational Instructional Materials in cooperation with the Soil Conservation Service, U.S.D.A., June 1971.

TABLE 12.3 Limiting Velocities for Essentially Straight Canals After Aging

Material	Value of Manning's n	Velocity in (ft/s)	
		Clear Water	Water Transporting Colloidal Silts
Fine sand, colloidal	0.020	1.50	2.50
Sandy loam, noncolloidal	0.020	1.75	2.50
Silt loam, noncolloidal	0.020	2.00	3.00
Alluvial sites, noncolloidal	0.020	2.00	3.50
Ordinary firm loam	0.020	2.50	3.50
Volcanic ash	0.020	2.50	3.50
Stiff clay, very colloidal	0.025	3.75	5.00
Alluvial site, colloidal	0.025	3.75	5.00
Shales and hardpans	0.025	6.00	6.00
Fine gravel	0.020	2.50	5.00
Graded loams to cobbles when noncolloidal	0.030	3.75	5.00
Graded silts to cobbles when colloidal	0.030	4.00	5.50
Coarse gravel, noncolloidal	0.025	4.00	6.00
Cobbles and shingles	0.035	5.00	5.50

TABLE 14.1 Horsepower Required to Lift Different Quantities of Water to Elevations of 10 to 80 ft

(Efficiency of pumping plant 50% of theoretical.
Use for estimating only.)

Discharge		Horsepower Required for Elevations of							
gpm	ft³/s	10 ft	20 ft	30 ft	40 ft	50 ft	60 ft	70 ft	80 ft
100	0.22	0.5	1.0	1.5	2.0	2.5	3.0	3.5	4.0
150	0.33	0.8	1.5	2.3	3.0	3.8	4.6	5.3	6.1
200	0.45	1.0	2.0	3.0	4.0	5.0	6.1	7.1	8.1
250	0.56	1.3	2.5	3.8	5.0	6.3	7.6	8.8	10.1
300	0.67	1.5	3.0	4.6	6.1	7.6	9.1	10.6	12.1
350	0.78	1.8	3.5	5.3	7.1	8.8	10.6	12.4	14.1
400	0.89	2.0	4.0	6.1	8.1	10.1	12.1	14.1	16.2
450	1.00	2.3	4.6	6.8	9.1	11.4	13.6	15.9	18.2
500	1.11	2.5	5.0	7.6	10.1	12.6	15.2	17.7	20.2
600	1.34	3.0	6.1	9.1	12.1	15.2	18.2	21.2	24.2
700	1.56	3.5	7.1	10.6	14.1	17.7	21.2	24.8	28.3
800	1.78	4.0	8.1	12.1	16.2	20.2	24.2	28.3	32.3
900	2.01	4.6	9.1	13.6	18.2	22.7	27.3	31.8	36.4
1000	2.23	5.0	10.1	15.2	20.2	25.2	30.3	35.4	40.4
1250	2.78	6.3	12.6	18.9	25.2	31.6	37.9	44.2	50.5
1500	3.34	7.6	15.2	22.7	30.3	37.9	45.4	53.0	60.6

Source: N. Mex. Agr. Exp. Sta. Bul. 237.

TABLE 14.2   Loss of Head, in ft, Due to Friction per 100 ft of 15-Year-Old Ordinary Iron Pipe[a]

Rate of Flow (gpm)	Nominal Diameter of Pipe									
	¾ in.	1 in.	2 in.	2½ in.	3 in.	4 in.	5 in.	6 in.	8 in.	10 in.
1										
2	1.9									
3	4.1	1.26								
4	7.0	2.14								
5	10.5	3.25								
10	38.0	11.7	0.50	0.17	0.07					
15	80.0	25.0	1.08	0.36	0.15					
20	136.0	42.0	1.82	0.61	0.25					
25		64.0	2.73	0.92	0.38					
30		89.0	3.84	1.29	0.54					
35		119.0	5.1	1.72	0.75					
40		152.0	6.6	2.20	0.91	0.22				
45			8.2	2.80	1.15	0.28				
50			9.9	3.32	1.38	0.34				
70			18.4	6.20	2.57	0.63	0.21			
75			20.9	7.1	3.00	0.73	0.24			
100			35.8	12.0	4.96	1.22	0.41	0.17		
120			50.0	16.8	7.0	1.71	0.58	0.24		
125			54.0	18.2	7.6	1.86	0.64	0.27		
150			76.0	25.5	10.5	2.62	0.88	0.36		
175			102	33.8	14.0	3.44	1.18	0.48		
200			129	43.1	17.8	4.40	1.48	0.62		
225				54.3	22.3	5.45	1.86	0.77		
250				66.0	27.2	6.72	2.24	0.95	0.23	
270					31.3	7.70	2.60	1.08	0.26	
275					32.5	8.05	2.70	1.12	0.27	
300					38.0	9.30	3.14	1.32	0.32	
350						12.40	4.19	1.75	0.42	
400						16.00	5.40	2.21	0.55	
450						19.80	6.70	2.37	0.69	0.23
470						21.60	7.22	2.99	0.74	0.25
475						22.00	7.42	3.02	0.76	0.25
500						24.00	8.10	3.34	0.83	0.28
550							9.60	4.02	0.98	0.33
600							11.30	4.91	1.16	0.40
650							13.20	5.40	1.34	0.45
700							15.10	6.20	1.54	0.52
750							17.20	7.12	1.77	0.60
800								8.00	1.90	0.67
850								8.95	2.20	0.75
900								10.11	2.46	0.83
950								11.20	2.71	0.92
1000								12.04	2.98	1.01
1050								13.25	3.22	1.09
1100								14.55	3.55	1.21
1150								15.60	3.83	1.31
1200								17.10	4.17	1.42
1250								18.50	4.43	1.51
1500									6.29	2.14
2000									10.80	3.60

Table continued on next page.

TABLE 14.2 Loss of Head, in ft, Due to Friction per 100 ft of 15-Year-Old Ordinary Iron Pipe[a] (continued)

Rate of Flow (gpm)	Nominal Diameter of Pipe									
	¾ in.	1 in.	2 in.	2½ in.	3 in.	4 in.	5 in.	6 in.	8 in.	10 in.
2500										5.40
3000										7.72
3500										10.00
4000										

[a] Pipe coefficients: The values of friction given in this table are for commercial wrought iron or cast iron pipe of 15 years' service handling soft clear water. In order to be able to use that table for other classes of pipe, the values taken from the table should be multiplied by a coefficient, selected from the list below, to correspond to the required condition:

New smooth brass and steel pipe (Durand) coefficient	=0.6
New smooth iron pipe (Williams & Hazen C = 120)	=0.7
15-year-old ordinary pipe (Williams & Hazen C = 100)	=1.0
25-year-old ordinary pipe (Williams & Hazen C = 90)	=1.2
Portable aluminum with couplers (Williams & Hazen C = 120)	=0.7

Source: Based on original data by Williams and Hazen. Courtesy Goulds Pumps, Inc. and published in Bulletin 481, Oregon Agricultural Experiment Station.

TABLE 14.4 Performance Standards for New Deep-Well Pumping Plants

Energy Source	Rated Load for Representative Power Units (hp-hr/gal)	Performance Standard (WHP-hr/gal)[a]
Diesel	14.75[b]	11.06
Gasoline	11.30[b]	8.48
Tractor fuel	11.08[b]	8.31
Propane	8.92[b]	6.69
Natural gas	81.9[c]/1000 ft³	61.4/1000 ft³
Electric	88% efficient	0.885[d]/kWh

[a] Based on pump efficiency of 75%.
[b] Taken from Test D of Nebraska Tractor Test Reports. Drive loss for flat belt is included.
[c] Manufacturer's data corrected for 5% drive loss.
[d] Not corrected for drive loss. Assume a direct connection.
Source: Paul E. Schleusener, J. J. Sulck, and J. F. Schrunk, "How Efficient is Your Pumping Plant?" *Exp. Sta. Qtrly.*, Lincoln, Nebraska, Spring 1955. Courtesy Nebraska Experiment Station.

TABLE 14.5   Maximum Fuel Requirements for a Good Pumping Plant[a]

Flow of Water (gpm)	Lift (ft)[b]	Water Horse-power	Fuel Required (gal/hr)				
			Propane	Diesel	Gasoline or Tractor Fuel	Natural Gas[c]	Elec-tricity[d]
500	100	13	2	1¼	1½	210	14
	150	19	2¾	1¾	2¼	310	21
	200	25	3¾	2¼	3	410	29
700	100	18	2¾	1¾	2	290	20
	150	27	4	2½	3¼	430	30
	200	35	5¼	3¼	4¼	580	40
800	100	20	3	1¾	2½	330	23
	150	30	4½	2¾	3½	490	34
	200	40	6	3¾	4¾	660	46
1000	100	25	3¾	2¼	3	410	29
	150	38	5¾	3½	4½	620	43
	200	50	7½	4½	6	820	57

[a] Based on standards in Table 14.4.
[b] If you have a pressure at the pump discharge, add the number of feet shown in Table 4.6 to the distance the water must be lifted while pumping.
[c] Cubic feet per hour.
[d] Kilowatt-hours per hour.
Source: Paul E. Schleusener, J. J. Sulck, and J. F. Schrunk, "How Efficient is Your Pumping Plant?" *Exp. Sta. Qtrly.*, Lincoln, Nebraska, Spring, 1955. Courtesy Nebraska Experiment Station.

TABLE 14.6   Relationship Between Discharge Pressure and Lift

Discharge Pressure (psi)	Equivalent "Lift" (ft)
30	69
50	116
70	162
110	254

Source: Paul E. Schleusener, J. J. Sulck, and J. F. Schrunk, "How Efficient is Your Pumping Plant?" *Exp. Sta. Qtrly.*, Lincoln, Nebraska, Spring, 1955. Courtesy Nebraska Experiment Station.

TABLE 16.1 Summary of Orifice and Weir Formulas

Measuring device (all sharp crested)	Views	Formula	Equation Number
Orifice	Front view Side view	$Q = 0.61A\sqrt{2gh}$	(6.5)
Rectangular weir (without contraction)	Top view	$Q = 3.33LH^{3/2}$	(6.9)
Rectangular weir (with contraction)	$(L-0.2H)$ Top view Side view	$Q = 3.33(L - 0.2H)H^{3/2}$	(6.9) and (6.10)
Trapezoidal weir	End view	$Q = 3.37LH^{3/2}$	(6.11)
90° Triangular weir	90° End view	$Q = 2.49H^{5/2}$	(6.12)

TABLE 16.3 Weir-Box Dimensions for Rectangular, Cipolletti, and 90° Triangular Notch Weirs[a]

Rectangular and Trapezoidal Weirs with End Contractions

	H	L	A	K	B	E	C	D	F	G
Flow (s·ft)	Maximum Head	Length of Weir Crest	Length of Box Above Weir Notch	Length of Box Below Weir Notch	Total Width of Box	Total Depth of Box	End of Crest to Side	Crest to Bottom	Hook Gage Distance Upstream	Hook Gage Distance Across Stream
½ to 3	1.0	1	6	2	5½	3½	2¼	2	4	2
2 to 5	1.1	1½	7	3	7	4	2¾	2½	4½	2
4 to 8	1.2	2	8	4	8½	4½	3¼	2¾	5	2½
6 to 14	1.3	3	9	5	12	5	4½	3¼	5½	3
10 to 22	1.5	4	10	6	14	5½	5	3½	6	3

90° Triangular Notch Weir

	H	L	A	K	B	E	C	D	F	G
½ to 2½	1.00	—	6	2	5	3	2½	1½	4	2
2 to 4⅓	1.25	—	6½	8½	6½	3¼	3¼	1½	5	2½

[a] All dimensions are in feet. The letters at the heads of the columns in this table refer to Fig. 16.10.

TABLE 16.4 Free Flow Through Parshall Measuring Flume

Upper Head H_a		Throat Widths												
Feet	Inches (Approx.)	1 in.	2 in.	3 in.	6 in.	9 in.	1 ft	2 ft	3 ft	4 ft	5 ft	6 ft	8 ft	10 ft
							Flow (ft³/s)							
0.10	1³/₁₆	0.009	0.019	0.03	0.05	0.09								
0.12	1⁵/₁₆	0.013	0.025	0.04	0.07	0.12								
0.14	1¹/₁₆	0.016	0.032	0.05	0.09	0.15								
0.16	1⁵/₁₆	0.020	0.039	0.06	0.11	0.19								
0.18	2³/₁₆	0.024	0.047	0.07	0.14	0.22								
0.20	⅜	0.028	0.055	0.08	0.16	0.26	0.35	0.66	0.97	1.26				
0.22	⅝	0.032	0.065	0.10	0.19	0.30	0.40	0.77	1.12	1.47				
0.24	⅞	0.037	0.074	0.11	0.22	0.35	0.46	0.88	1.28	1.69				
0.26	3⅛	0.042	0.084	0.12	0.25	0.39	0.51	0.99	1.46	1.91	2.36	2.80		
0.28	⅜	0.047	0.094	0.14	0.28	0.44	0.58	1.11	1.64	2.15	2.65	3.15		
0.30	⅝	0.052	0.105	0.15	0.31	0.49	0.64	1.24	1.82	2.39	2.96	3.52	4.62	
0.32	1³/₁₆	0.058	0.116	0.17	0.34	0.54	0.71	1.37	2.02	2.65	3.28	3.90	5.13	
0.34	4¹/₁₆	0.064	0.127	0.19	0.38	0.59	0.77	1.50	2.22	2.92	3.61	4.30	5.66	
0.36	⁵/₁₆	0.069	0.139	0.21	0.41	0.64	0.84	1.64	2.42	3.19	3.95	4.71	6.20	
0.38	⁹/₁₆	0.075	0.151	0.22	0.45	0.70	0.92	1.79	2.64	3.48	4.31	5.13	6.76	
0.40	1³/₁₆	0.082	0.163	0.24	0.48	0.76	0.99	1.93	2.86	3.77	4.68	5.57	7.34	9.1
0.42	5¹/₁₆	0.088	0.176	0.26	0.52	0.81	1.07	2.09	3.08	4.07	5.05	6.02	7.94	9.8
0.44	¼	0.095	0.189	0.28	0.56	0.87	1.15	2.24	3.32	4.38	5.43	6.48	8.55	10.6

Table continued on next page.

Table 16.4 (continued)

									Throat Widths					
Upper Head, H_a														
Feet	Inches (Approx.)	1 in.	2 in.	3 in.	6 in.	9 in.	1 ft	2 ft	3 ft	4 ft	5 ft	6 ft	8 ft	10 ft
							Flow (ft³/s)							
0.46	½	0.101	0.203	0.30	0.61	0.94	1.23	2.40	3.56	4.70	5.83	6.96	9.19	11.4
0.48	¾	0.108	0.217	0.32	0.65	1.00	1.31	2.57	3.80	5.03	6.24	7.44	9.8	12.2
0.50	6	0.115	0.230	0.34	0.69	1.06	1.39	2.73	4.05	5.36	6.66	7.94	10.5	13.1
0.52	¼	0.123	0.245	0.36	0.73	1.13	1.48	2.90	4.31	5.70	7.09	8.46	11.2	13.9
0.54	½	0.130	0.260	0.38	0.78	1.20	1.57	3.08	4.57	6.05	7.52	8.98	11.9	14.8
0.56	¾	0.138	0.275	0.40	0.82	1.26	1.66	3.26	4.84	6.41	7.97	9.52	12.6	15.7
0.58	¹⁵⁄₁₆	0.145	0.290	0.43	0.87	1.33	1.75	3.44	5.11	6.77	8.43	10.1	13.3	16.6
0.60	7³⁄₁₆	0.153	0.306	0.45	0.92	1.40	1.84	3.62	5.39	7.15	8.89	10.6	14.1	17.5
0.62	⁷⁄₁₆	0.161	0.322	0.47	0.97	1.48	1.93	3.81	5.68	7.53	9.37	11.2	14.8	18.5
0.64	¹¹⁄₁₆	0.169	0.338	0.50	1.02	1.55	2.03	4.01	5.97	7.91	9.85	11.8	15.6	19.5
0.66	¹⁵⁄₁₆	0.177	0.355	0.52	1.07	1.63	2.13	4.20	6.26	8.31	10.3	12.4	16.4	20.4
0.68	8³⁄₁₆	0.186	0.372	0.55	1.12	1.70	2.23	4.40	6.56	8.71	10.9	13.0	17.2	21.5
0.70	⅜	—	0.389	0.57	1.17	1.78	2.33	4.60	6.86	9.11	11.4	13.6	18.0	22.5
0.72	⅝	—	0.406	0.60	1.23	1.86	2.43	4.81	7.17	9.53	11.9	14.2	18.9	23.5
0.74	⅞	—	0.424	0.62	1.28	1.94	2.53	5.02	7.49	9.95	12.4	14.9	19.7	24.6
0.76	9⅛	—	0.442	0.65	1.34	2.02	2.63	5.23	7.81	10.1	12.9	15.5	20.6	25.7
0.78	⅜	—	0.459	0.68	1.39	2.10	2.74	5.44	8.13	10.8	13.5	16.2	21.5	26.8
0.80	⅝	—	—	0.70	1.45	2.18	2.85	5.66	8.46	11.3	14.0	16.8	22.4	27.9
0.82	¹³⁄₁₆	—	—	0.73	1.50	2.27	2.96	5.88	8.79	11.7	14.6	17.5	23.3	29.0
0.84	10¹⁄₁₆	—	—	0.76	1.56	2.35	3.07	6.11	9.13	12.2	15.2	18.2	24.2	30.2
0.86	⁵⁄₁₆	—	—	0.79	1.62	2.44	3.18	6.33	9.48	12.6	15.8	18.9	25.1	31.4
0.88	⁹⁄₁₆	—	—	0.81	1.68	2.52	3.29	6.56	9.82	13.1	16.3	19.6	26.1	32.5
0.90	¹³⁄₁₆	—	—	0.84	1.74	2.61	3.41	6.80	10.2	13.6	16.9	20.3	27.0	33.7
0.92	11¹⁄₁₆	—	—	0.87	1.81	2.70	3.52	7.03	10.5	14.0	17.5	21.0	28.0	35.0
0.94	¼	—	—	0.90	1.87	2.79	3.64	7.27	10.9	14.5	18.1	21.8	29.0	36.2
0.96	½	—	—	0.93	1.93	2.88	3.76	7.51	11.3	15.0	18.8	22.5	30.0	37.5
0.98	¾	—	—	0.96	2.00	2.98	3.88	7.75	11.6	15.5	19.4	23.2	31.0	38.7
1.00	12	—	—	0.99	2.06	3.07	4.00	8.00	12.0	16.0	20.0	24.0	32.0	40.0
1.02	¼	—	—	1.02	2.12	3.17	4.12	8.25	12.4	16.5	20.6	24.8	33.0	41.3
1.04	½	—	—	1.05	2.19	3.26	4.25	8.50	12.8	17.0	21.3	25.6	34.1	42.6
1.06	¾	—	—	1.09	2.26	3.36	4.37	8.76	13.2	17.5	21.9	26.3	35.1	44.0
1.08	¹⁵⁄₁₆	—	—	1.12	2.32	3.45	4.50	9.01	13.5	18.1	22.6	27.1	36.2	45.3
1.10	13³⁄₁₆	—	—	—	2.40	3.55	4.62	9.27	13.9	18.6	23.3	27.9	37.3	46.7
1.12	⁷⁄₁₆	—	—	—	2.46	3.65	4.75	9.54	14.3	19.1	23.9	28.8	38.4	48.0
1.14	¹¹⁄₁₆	—	—	—	2.53	3.75	4.88	9.80	14.7	19.7	24.6	29.6	39.5	49.4
1.16	13¹⁵⁄₁₆	—	—	—	2.60	3.85	5.01	10.1	15.1	20.2	25.3	30.4	40.6	50.8
1.18	13³⁄₁₆	—	—	—	2.68	3.95	5.15	10.3	15.6	20.8	26.0	31.3	41.8	52.3
1.20	⅜	—	—	—	2.75	4.06	5.28	10.6	16.0	21.3	26.7	32.1	42.9	53.7
1.22	⅝	—	—	—	2.82	4.16	5.41	10.9	16.4	21.9	27.4	33.0	44.1	55.2
1.24	⅞	—	—	—	2.89	4.27	5.55	11.2	16.8	22.5	28.1	33.8	45.2	56.6
1.26	15⅛	—	—	—	—	4.37	5.69	11.5	17.2	23.0	28.9	34.7	46.4	58.1
1.28	⅜	—	—	—	—	4.48	5.82	11.7	17.7	23.6	29.6	35.6	47.6	59.6
1.30	⅝	—	—	—	—	4.59	5.96	12.0	18.1	24.2	30.3	36.5	48.8	61.1
1.32	¹³⁄₁₆	—	—	—	—	4.69	6.10	12.3	18.5	24.8	31.1	37.4	50.0	62.7
1.34	16¹⁄₁₆	—	—	—	—	4.80	6.25	12.6	19.0	25.4	31.8	38.3	51.2	64.2
1.36	⁵⁄₁₆	—	—	—	—	4.92	6.39	12.9	19.4	26.0	32.6	39.2	52.5	65.7
1.38	⁹⁄₁₆	—	—	—	—	5.03	6.53	13.2	19.9	26.6	33.3	40.1	53.7	67.3
1.40	¹³⁄₁₆	—	—	—	—	—	6.68	13.5	20.3	27.2	34.1	41.1	55.0	68.9
1.42	17¹⁄₁₆	—	—	—	—	—	6.82	13.8	20.8	27.8	34.9	42.0	56.2	70.5
1.44	¼	—	—	—	—	—	6.97	14.1	21.2	28.5	35.7	42.9	57.5	72.1
1.46	½	—	—	—	—	—	7.12	14.4	21.7	29.1	36.5	43.9	58.8	73.7
1.48	¾	—	—	—	—	—	7.26	14.7	22.2	29.7	37.3	44.9	60.1	75.4
1.50	18	—	—	—	—	—	7.41	15.0	22.6	30.3	38.1	45.8	61.4	77.0
1.52	¼	—	—	—	—	—	7.57	15.3	23.1	31.0	38.9	46.8	62.7	78.7
1.54	½	—	—	—	—	—	7.72	15.6	23.6	31.6	39.7	47.8	64.0	80.4
1.56	¾	—	—	—	—	—	7.87	15.9	24.1	32.3	40.5	48.8	65.4	82.1
1.58	¹⁵⁄₁₆	—	—	—	—	—	8.02	16.3	24.6	32.9	41.4	49.8	66.7	83.8

Table continued on next page.

TABLE 16.4 (Concluded)

| Upper Head H_n | | | | | | | Throat Widths | | | | | | | |
| Feet | Inches (Approx.) | 1 in. | 2 in. | 3 in. | 6 in. | 9 in. | 1 ft | 2 ft | 3 ft | 4 ft | 5 ft | 6 ft | 8 ft | 10 ft |
							Flow (ft³/s)							
1.60	19³/₁₆	—	—	—	—	—	8.18	16.6	25.1	33.6	42.2	50.8	68.1	85.5
1.62	⁷/₁₆	—	—	—	—	—	8.34	16.9	25.5	34.3	43.0	51.8	69.5	87.2
1.64	¹¹/₁₆	—	—	—	—	—	8.49	17.2	26.0	34.9	43.9	52.8	70.9	89.0
1.66	¹⁵/₁₆	—	—	—	—	—	8.65	17.6	26.5	35.6	44.7	53.9	72.3	90.7
1.68	20³/₁₆	—	—	—	—	—	8.81	17.9	27.0	36.3	45.6	54.9	73.7	92.5
1.70	⅜	—	—	—	—	—	8.97	18.2	27.6	37.0	46.4	56.0	75.1	94.3
1.72	⅝	—	—	—	—	—	9.13	18.5	28.1	37.7	47.3	57.0	76.5	96.1
1.74	⅞	—	—	—	—	—	9.29	18.9	28.6	38.3	48.2	58.1	77.9	97.9
1.76	21⅛	—	—	—	—	—	9.46	19.2	29.1	39.0	49.1	59.1	79.4	99.7
1.78	⅜	—	—	—	—	—	9.62	19.6	29.6	39.7	49.9	60.2	80.8	101.4
1.80	⅝	—	—	—	—	—	9.79	19.9	30.1	40.5	50.8	61.3	82.3	103.4
1.82	¹³/₁₆	—	—	—	—	—	9.95	20.2	30.7	41.2	51.7	62.4	83.8	105.3
1.84	22¹/₁₆	—	—	—	—	—	10.1	20.6	31.2	41.9	52.6	63.5	85.3	107.1
1.86	⁵/₁₆	—	—	—	—	—	10.3	20.9	31.7	42.6	53.6	64.6	86.8	109.0
1.88	⁹/₁₆	—	—	—	—	—	10.5	21.3	32.3	43.3	54.5	65.7	88.3	110.9
1.90	¹³/₁₆	—	—	—	—	—	10.6	21.6	32.8	44.1	55.4	66.8	89.8	112.9
1.92	23¹/₁₆	—	—	—	—	—	10.8	22.0	33.3	44.8	56.3	67.9	91.3	114.8
1.94	¼	—	—	—	—	—	11.0	22.4	33.9	45.5	57.3	69.1	92.8	116.7
1.96	½	—	—	—	—	—	11.1	22.7	34.4	46.3	58.2	70.2	94.4	118.7
1.98	¾	—	—	—	—	—	11.3	23.1	35.0	47.0	59.1	71.4	95.9	120.6
2.00	24	—	—	—	—	—	11.5	23.4	35.5	47.8	60.1	72.5	97.5	122.6
2.05	⅝	—	—	—	—	—	11.9	24.3	36.9	49.7	62.5	75.4	101.4	127.6
2.10	25³/₁₆	—	—	—	—	—	12.4	25.3	38.4	51.6	64.9	78.4	105.4	132.7
2.15	¹³/₁₆	—	—	—	—	—	12.8	26.2	39.8	53.5	67.4	81.4	109.5	137.8
2.20	26⅜	—	—	—	—	—	13.3	27.2	41.3	55.5	69.9	84.4	113.6	143.0
2.25	27	—	—	—	—	—	13.7	28.1	42.7	57.5	72.4	87.5	117.8	148.3
2.30	⅝	—	—	—	—	—	14.2	29.1	44.2	59.6	75.0	90.6	122.0	153.7
2.35	28³/₁₆	—	—	—	—	—	14.7	30.1	45.7	61.6	77.6	93.8	126.0	159.1
2.40	¹³/₁₆	—	—	—	—	—	15.2	31.1	47.3	63.7	80.3	97.0	130.7	164.6
2.45	29⅜	—	—	—	—	—	15.6	32.1	48.8	65.8	82.9	100.2	135.1	170.2
2.50	30	—	—	—	—	—	16.1	33.1	50.4	67.9	85.6	103.5	139.5	175.8

TABLE 16.5   Standard Dimensions of Parshall Measuring Flumes from 3- to 9-in. Throat Width

Dimension Letter[a]	Dimensions in ft and in. for Throat Widths (W) of				
	1 in.	2 in.	3 in.	6 in.	9 in.
A	1 ft 2⁹/₃₂ in.	1 ft 4⁵/₁₆ in.	1 ft 6⅜ in.	2 ft ⁷/₁₆ in.	2 ft 10⅝ in.
2/3A	0 ft 9¹⁷/₃₂ in.	0 ft 10⅞ in.	1 ft ¼ in.	1 ft 4⁵/₁₆ in.	1 ft 11⅛ in.
B	1 ft 2 in.	1 ft 4 in.	1 ft 6 in.	2 ft 0 in.	2 ft 10 in.
2/3B	0 ft 9⁵/₁₆ in.	0 ft 10⅝ in.	1 ft 0 in.	1 ft 4 in.	1 ft 10⅝ in.
C	0 ft 3²¹/₃₂ in.	0 ft 5³/₁₆ in.	0 ft 7 in.	1 ft 3½ in.	1 ft 3 in.
D	0 ft 6¹⁹/₃₂ in.	0 ft 8⅝ in.	0 ft 10³/₁₆ in.	1 ft 3½ in.	1 ft 10⅝ in.
E	0 ft 9 in.	1 ft 0 in.	1 ft 3 in.	1 ft 6 in.	2 ft 0 in.
F	0 ft 3 in.	0 ft 4½ in.	0 ft 6 in.	1 ft 0 in.	1 ft 0 in.
G	0 ft 8 in.	0 ft 10 in.	1 ft 0 in.	2 ft 0 in.	1 ft 6 in.
K	0 ft ¾ in.	0 ft ⅞ in.	0 ft 1 in.	0 ft 3 in.	0 ft 3 in.
N	0 ft ⅛ in.	0 ft 1³/₁₆ in.	0 ft 2¼ in.	0 ft 4½ in.	0 ft 4½ in.
X	0 ft ⁵/₁₆ in.	0 ft ⅝ in.	0 ft 1 in.	0 ft 2 in.	0 ft 2 in.
Y	0 ft ½ in.	0 ft 1 in.	0 ft 1½ in.	0 ft 3 in.	0 ft 3 in.

[a] Letters refer to Fig. 16.14.

TABLE 16.6   Standard Dimensions of Parshall Measuring Flumes from 1 to 10 ft Throat Width

Throat Width, W (ft)	Dimensions in ft and in.[a]					
	A	2/3 A	B	2/3 B	C	D
1.0	4 ft 6 in.	3 ft 0 in.	4 ft 4⅞ in.	2 ft 11¼ in.	2 ft 0 in.	2 ft 9¼ in.
2.0	5 ft 0 in.	3 ft 4 in.	4 ft 10⅞ in.	3 ft 3¼ in.	3 ft 0 in.	3 ft 11½ in.
3.0	5 ft 6 in.	3 ft 8 in.	5 ft 4¾ in.	3 ft. 7⅛ in.	4 ft 0 in.	5 ft 1⅞ in.
4.0	6 ft 0 in.	4 ft 0 in.	5 ft 10⅝ in.	3 ft 11⅛ in.	5 ft 0 in.	6 ft 4¼ in.
5.0	6 ft 6 in.	4 ft 4 in.	6 ft 4½ in.	4 ft 3 in.	6 ft 0 in.	7 ft 6⅝ in.
6.0	7 ft 0 in.	4 ft 8 in.	6 ft 10⅜ in.	4 ft 6⅞ in.	7 ft 0 in.	8 ft 9 in.
7.0	7 ft 6 in.	5 ft 0 in.	7 ft 4¼ in.	4 ft 10⅞ in.	8 ft 0 in.	9 ft 11⅜ in.
8.0	8 ft 0 in.	5 ft 4 in.	7 ft 10⅛ in.	5 ft 2¾ in.	9 ft 0 in.	11 ft 1¾ in.
10.0	9 ft 0 in.	6 ft 0 in.	8 ft 9⅞ in.	5 ft 10⅝ in.	11 ft 0 in.	13 ft 6⅜ in.

[a] Letters refer to Fig. 16.14 in which other dimensions for these flumes are shown.

TABLE 16.7 Approximate Coefficients of Discharge for Use in the Coordinate Method for Measuring the Rate of Water Flow from Horizontal Pipes

d/D	A. Flowing Partially Full at the End							
	X/D							
	1.00	1.50	2.00	2.50	3.00	4.00	5.00	8.00
0.2	1.02	1.01	1.00					
0.3	1.11	1.06	1.03	1.02	1.01	1.00		
0.4	1.17	1.10	1.06	1.03	1.02	1.01	1.00	
0.5	1.22	1.13	1.07	1.04	1.03	1.01	1.01	1.00
0.6	1.26	1.15	1.09	1.05	1.03	1.02	1.01	1.00
0.7	1.30	1.17	1.10	1.06	1.04	1.02	1.01	1.00
0.8	1.33	1.18	1.11	1.07	1.04	1.02	1.01	1.00

Y/D	B. Flowing Completely Full at the End							
	X/D							
	1.00	1.50	2.00	2.50	3.00	4.00	5.00	8.00
0.5	1.44	1.28	1.18	1.13	1.10	1.06	1.03	1.00
1.0	1.37	1.24	1.17	1.12	1.09	1.06	1.03	1.00
2.0		1.11	1.09	1.08	1.07	1.05	1.03	1.00
3.0			1.04	1.04	1.04	1.04	1.03	1.00
4.0			1.01	1.01	1.02	1.03	1.02	1.00
5.0			0.97	0.99	1.00	1.01	1.01	1.00

INDEX